Richard,

For a criti[illegible]

of planted fores[illegible]

for a great friend.

Tom

Planted Forests: Contributions to the Quest for Sustainable Societies

Edited by:

James Reid Boyle

Jack K. Winjum

Kathleen Kavanagh

and

Edward C. Jensen

Forest Resources Department, College of Forestry, Oregon State University, Corvallis, OR, USA

KLUWER ACADEMIC PUBLISHERS

DORDRECHT / BOSTON / LONDON

Library of Congress Cataloging-in-Publication Data

Planted forests : contributions to the quest for sustainable societies
/ edited by James Reid Boyle ... [et al.].
p. cm. -- (Forestry sciences ; v. 56)
Papers presented at a symposium, held June 1995 in Portland,
Oregon.
Includes index.
ISBN 0-7923-5468-0 (alk. paper)
1. Tree farms--Congresses. I. Boyle, James Reid. II. Series.
SD391.P7 1999
634.9'5--dc21 98-31689

ISBN 0-7923-5468-0

Published by Kluwer Academic Publishers,
P.O. Box 17, 3300 AA Dordrecht, The Netherlands

Sold and distributed in the U.S.A. and Canada
by Kluwer Academic Publishers,
101 Philip Drive, Norwell, MA 02061, U.S.A.

In all other countries, sold and distributed
by Kluwer Academic Publishers Group,
P.O. Box 322, 3300 AH Dordrecht, The Netherlands

Printed on acid-free paper

Printed in the Netherlands

PLANTED FORESTS

FORESTRY SCIENCES

Volume 56

The titles published in this series are listed at the end of this volume.

CONTENTS

New Forests **17:** 1–2, 1999.

Preface

Planted forests in many forms are common and essential parts of our forested landscapes and of our human societies. Planted forests occur in many forms: plantations of fast-growing eucalypts and hybrid poplars in rows; extensively-planted Douglas-fir trees in the mountainous Pacific Northwest, and loblolly pine on old agricultural lands in the southeast of the U.S.A.; Scots pines in Finland and in Scotland; *Pinus patula* in Swaziland; Radiata pine in New Zealand and Chile; and many more. The myriad, diverse forms and contributions of these forests, plantations, and tree farms have received relatively little positive discussion in recent times as concerns about forestry have focused on naturalness and biodiversity. The papers of this volume explore the values and challenges of planted forests.

The symposium from which these papers are derived, held in June 1995 in Portland, Oregon, gathered more than 200 people from 13 countries to explore the diversity and utility of planted forests. Sponsors were: College of Forestry, Oregon State University; The Ward K. Richardson Family Fund; Boise Cascade Corp.; BLM, USDI; Oak Ridge National Lab Biofuels program; Pacific Northwest Station, USFS; Weyerhaeuser Co. Foundation; Blount, Inc.; Crown Pacific Management; Oregon Dept. of Forestry; Georgia-Pacific Corp.; Willamette Industries, Inc.; Starker Forests, Inc.; Stimson Lumber Co.; Columbia Forest Products, Inc.; NCASI; Sun Studs, Inc.; Davidson Industries, Inc.; WTD Industries.

At the symposium we spent three days in presentations and discussions, and a day dispersed in the planted forest landscapes of Oregon and Washington. The papers in this issue of New Forests represent the topics of the symposium: the variety, types and significance of planted forests; technical and social considerations for planted forests; and case studies of planted forests on five continents. These papers, and the ongoing work of their authors, reflect a wide cross-section of the diversity, contributions and issues of forests and trees planted to fulfill social and economic goals of sustaining human communities and cultures.

The members of the Planted Forests Symposium organizing committee included Bill Atkinson, Bob Buckman, Nick Chappell, Mike Cloughsey, Ray Craig, Dean DeBell, Rick Fletcher, Dave Hibbs, Norman E. Johnson, Joseph Lint, Cindi Miner, Jerry Otto, George Stankey, Barte Starker, John Tappeiner and Jack Walstad. The organizing committee contributed to the success of bringing together the authors of these papers. We also acknowledge the work of editor Rosanna Mattingly and of our staff at the College of Forestry who contributed to the assembly, preparation and editing of these papers: Becca Berkey, Kristi Souther, Amy Nordlander, graphic artist Sandie Arbogast, and media assistant Betsy Littlefield.

Illustrations of planted forests, supplied by many of the authors, are available at our Planted Forests web site, created by Tad Buford with assistance of Betsy Littlefield and Kathy Howell at: http://www.cof.orst.edu/pubs/cof/plntdfor/index.htm

James R. Boyle
Jack Winjum
Kathleen Kavanagh
Edward Jensen
Guest Editors

Introduction to Planted Forests

New Forests **17:** 5–9, 1999.

Planted forests: views and viewpoints

JAMES R. BOYLE
Forest Resources Department, College of Forestry, Oregon State University, Corvallis, USA

Key words: plantations, sustainability

Abstract. Planted forests in many forms are parts of landscapes in many regions of the world. These forests contribute to the economic welfare and potential sustainability of human communities. They provide wood, forage, wildlife habitats, watershed protection, recreational settings, aesthetic vistas, and ecological conditions for many other forest values. Forests and other versions of tree plantations have been planted to replace harvested Douglas-fir forests in Oregon, to provide fuelwood and forage in Senegal, for pulpwood in Brazil, for lumber in New Zealand, for watershed protection in Nepal, and for aesthetic enhancement in Scotland and Denmark. As human pressures on native forests continue to increase – for reasons ranging from fuelwood needs to desires to preserve old-growth ecosystems – it is clear to many of us that planted forests, intensively and extensively managed, are essential to our present and future societies and cultures, whatever forms they take.

The title and topic of this volume, *Planted Forests: Contributions to the Quest for Sustainable Societies*, are as much a challenge as they are a statement. Can we really plant a "forest"? Can we develop truly "sustainable" societies? Beyond the semantics of "forest" and "sustainable" we must ask, what are the crucial issues to be addressed in order to have planted forests continue to be so very important in our world today. And, how can planted forests continue to contribute to our mutual quests for sustainable societies? Sustainable societies for all peoples of the world, not just for Americans or Danes or New Zealanders. As I consider forests, forestry, planted trees and potentially sustainable societies, three "facts" are very clear to me:

1. planted forests are essential;
2. wild, native forests are vital to our cultures and as references in our work toward sustainable forestry; and
3. sustainability is a quest, not a current reality.

Planted forests, of many sorts, are essential in our world today to provide solid wood and wood fibers and numerous other products and values. The FAO (FAO 1994) estimated that there were about 100 million hectares of forest plantations in the world, and new ones were being established in the tropics alone at a rate of at least 2.6 million hectares per year. We would not have current versions of our industrialized or agrarian cultures without planted forests. Wood for our structures, fibers for our papers, protection for watershed soils, and fuelwood for many home and industrial uses come today from planted forests. Just as we and most peoples of the world cannot sustain ourselves without intensive agriculture, we cannot maintain any versions of our current lifestyles without planted forests. Planted forests currently occupy significant portions of the land of many nations, from 26% of Japan and 11% of Sweden to 8.4% of the United Kingdom, 1.3% of the U.S.A. to 0.1% of Zambia (South 1998). Experts have estimated that an additional 50 to 100 million hectares of appropriately located planted forests would be needed to meet the projected industrial wood requirements of the developing countries by the year 2010 (FAO 1994). Roger Sedjo (this volume) documents recent expansions of plantations. In our "developed" nations, with more and more native forest land being reserved from wood production, and with urbanization daily covering former forest lands, high-yielding planted forests are also needed. We simply can not meet our demands for solid wood, wood pulp, fuelwood and other forest values from naturally-growing native forests alone. We're dependent on planted forests.

Native, "natural", forests will continue to be essential for practical values and for our cultural heritage, as refugia for numerous organisms and associated processes, and to provide "encyclopedias" from which scientists can continue to learn about forest properties and functions. These features, in addition to their recreational, aesthetic and spiritual resources, carbon storage, organism habitats and watershed protection constitute important "ecological services". Native forests will provide baselines against which we can measure potential sustainability of managed and planted forests. We are challenged to apply the ecology that is known, to expand that knowledge, and to adapt our forest and landscape management to preserve, enhance and recapture essential and desirable forest ecosystem attributes (e.g. Kohm and Franklin 1997).

The idea of "sustainability" and the concept of "sustainable development" are currently a quest, a process, a set of ideas and actions in gestation, development and evolution. They are not a certainty. I use "sustainability" in the sense of the Brundtland report (World Commission on Environment and Development 1987) – meeting the needs of people today without jeopardizing the abilities of future generations to meet their needs. This is the idea of sustainable development. The potentials for truly sustainable societies in most places on earth today – in "overdeveloped" industrialized nations, and in overpopulated, resource-poor regions – are not at all clear. To meet our needs and provide viable options for future generations, we must continually assess our interactions with and impacts on our ecosystems and environment. We must assess our "ecological footprints" (Rees 1996) in the world. And, in the sense of Garrett Hardin (Hardin 1986) we must evaluate the "cultural carrying capacities" of our landscapes and nations. In forestry we must expand our understandings of what's required to sustain the growth of trees – and of other forest ecosystem values – while sustaining human communities.

Planted forests and wildland native forests are complementary and compatible in many landscape and regional contexts. Consider the varieties of planted forests in my home region, the Pacific Northwest, USA, where forests have traditionally been planted in mountainous wildlands as native forests have been cut for wood products. Planted forests are intermingled in complex patterns and large expanses among native forests and other land uses. These planted forests have been tended with various levels of intensity, depending on ownership and management objectives. On most federal lands, in the past, forests were managed for the multiple-use values of wood, wildlife, water, recreation and grazing, with strong emphasis on production of wood for societal uses in building homes and communities. In the past decade, emphasis on federal lands in the Northwest and nationwide has shifted from wood production to other "ecosystem" values, often with reduction in management intensity. On forest industry lands, in many state-owned forests, and in many smaller private forests, management intensity has increased in the past twenty years. With emphasis on very efficient production of wood, land owners use intensive site preparation, careful planting with genetically-improved seedlings, control of competing vegetation, thinnings to control stand density and to yield intermediate economic harvests, and shortened times to final harvest – to as little as 40 years versus 60 to 100 or more years in times past. Large areas of such managed, younger forests have changed the texture and fabric of Northwestern landscapes, while continuing to support communities of people directly and indirectly involved with management and uses of forests.

On other lands, in Oregon, Washington, California and elsewhere, agricultural-like systems of hybrid poplar and eucalypts provide wood fiber in five to ten years. Cultural techniques for these tree crops mimic those of high-technology agriculture, including automatic drip irrigation with water and nutrients in as least one system in central Oregon. Whether such "fiber farms" are "planted forests" is a matter of definition. They do represent one end of the spectrum of planted trees grown for societal values, in this case paper, and are included in our considerations of planted forests. One of the most extensive and successful examples of such systems, in Brazil, is described by Edgard Campinhos (in this volume).

Elsewhere, planted forests of the exotic radiata pine in New Zealand have become a mainstay of a national economy and a major factor in preservation of native forests (Sutton this volume). Many European forests have been planted, often with exotics such as Douglas-fir, Sitka spruce and lodgepole pine, and managed at various levels of intensity (Hermann and Lavender this volume). In numerous developing countries community forests and roadside and farmland plantings contribute myriad values to families and cultures as depicted by Julian Evans, Alan Long and P. K. R. Nair, and Guy Cellier (all in this volume).

Concerns about planted forests addressed in the following papers range from broad philosophical and social issues discussed by Niels Elers Koch and J. P. Skovsgaard to details of basic physiological and ecosystem properties and processes discussed by Melvin Cannell. And, in many cases, issues encompass combinations of these and other facets, as Julian Evans discusses for regions of the wet and dry tropics. Authors discuss attributes and values of these forests in many of their forms – from extensively planted forests in mountainous wild lands to intensively cultured fiber plantations in agricultural settings. Authors discuss the variety, nature and significance of planted, managed forests in our societies. They discuss opportunities and challenges associated with establishing and managing these trees and forests – forests as diverse as the cultures in which they occur.

Aside from quibbling over whether a given area of planted trees is a forest, those of us interested in and dependent on trees and forests should be able to agree to advance the science, management and feasibility of planted forests and tree farms of all forms. The authors of the papers that follow, and many others working to advance forestry and wood fiber production, provide solid groundwork in the quest for sustainable forests, sustainable forestry and sustaining the cultural richness of human communities on the earth. Planted forests are essential; wild, native forests are vital to the sustainability of our cultures; and, sustainability is an ongoing quest. These collected papers are a checkpoint for reflection and action about planted forests and sustainable societies. Planted forests are currently contributing to attempts for sustainability, as well as sustaining current societies that are in evolutionary states whose fates will only be known by those of our grandchildren's and great-grandchildren's generations. As necessarily optimistic foresters, we continue to plant trees and sustain the quest.

References

Dixon, John A. and Fallon, Louise A. 1989. The concept of sustainability: origins, extensions, and usefulness for policy. Society and Natural Resources 2: 73–84.

Food and Agriculture Organization of the United Nations, SARD 1994. The Road From Rio: Moving Forward in Forestry. Food and Agriculture Organization of the United Nations, Rome. 30 pages. ISBN 92-5-103587-3.

Hardin, Garrett 1986. Cultural carrying capacity: a biological approach to human problems. BioScience 36: 599–606.

Kohm, Kathryn A. and Franklin, Jerry F. 1997. Creating a Forestry for the 21st Century, The Science of Ecosystem Management. Island Press, Washington D.C. & Covelo, CA. 475 p.

Rees, William E. 1996. Revisiting carrying capacity: Area-based indicators of sustainability. Population and Environment 17(3): 195.

South, D.B. 1998. The tree nursery – mother of plantations. In: Zabkiewicz, J.A. (Ed) Proc. Weedworks '93. FRI Bulletin (In press). (Proceeding of a Symposium held at Rotorua, New Zealand, March 1993).

World Commission on Environment and Development 1987. Our Common Future. Oxford University Press, Oxford, New York, Toronto. 400 pages. ISBN 0-19-282080-X.

New Forests **17:** 11–22, 1999.

Sustainable management of planted forests: some comparisons between Central Europe and the United States

NIELS ELERS KOCH[1] and J.P. SKOVSGAARD[2]
[1]*Danish Forest and Landscape Research Institute DK-2970, Hørsholm, Denmark;*
[2]*Department of Forestry, Danish Forest and Landscape Research Institute DK-2970, Hørsholm, Denmark*

Key words: forestry objectives, multiple-use forestry, silviculture, afforestation, reforestation

Abstract. Following the introduction of planned forestry and the regular high forest system more than 250 years ago, forests in Central Europe became increasingly shaped by plantation silviculture. Many natural woodlands were replaced by planted forests, and forest plantations are still being established through afforestation of extensive land areas. Nowadays, forests are managed for many different purposes, including wood production, recreation, ecological, cultural, and amenity values, biodiversity, and soil and groundwater protection. This brings new challenges to forest management and silviculture. To reach a sustainable forestry, in the sense of Brundtland (WCED 1987), we are now reshaping European forestry toward a more nature oriented silviculture. Maybe forestry in the United States and elsewhere can benefit from the mistakes made in Europe during the last two centuries and take a shortcut toward sustainable forestry.

Introduction

Some 250 years ago, heavy exploitation of Central European forest resources led to the genesis of planned and well-regulated forestry. In Denmark, for example, the unregulated exploitation was carried very far before efficient central control reversed the development (Figure 1). Similar trends may be found in many different countries, only the time and area scales differ. In the United States the curve moves in time in accordance with the colonization and industrialization of the West.

In this paper we highlight some key issues that we believe to be common to the development of forestry and forestry paradigms in industrialized countries, namely the temporal path from exploitation of natural forests through regulation-oriented forest management over an ecologically based paradigm to so-called integrated multiple-use forestry (Koch 1991; Skovsgaard

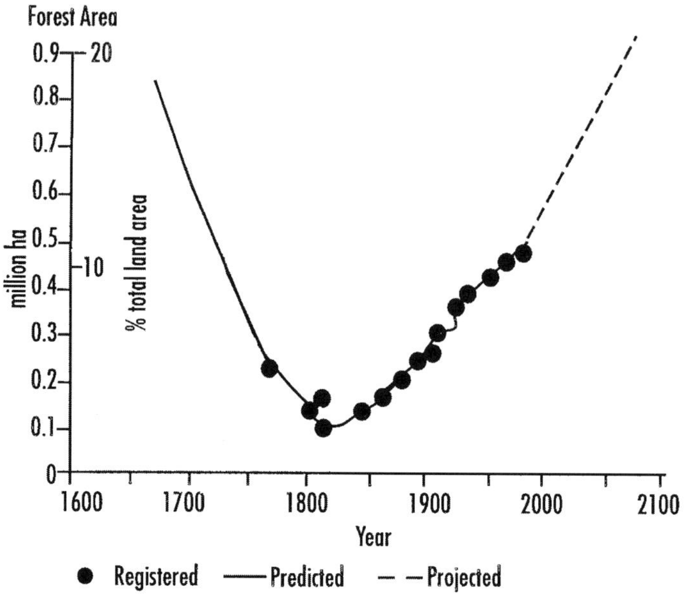

Figure 1. The forest area (at least 60 percent stocked with trees) of Denmark between 1600 and 2100 – an example of exploitation and recovery of forest resources.

1995). During this development, natural forests are often replaced by planted stands, and past deforestations are compensated for through afforestations. Consequently, forest management becomes characterized by plantation-like approaches. The traditional regulation-oriented approach may succeed in sustaining some forest values, especially wood production, but leaves many others degraded. As a consequence, a site specific and ecologically based forest management evolves. This, however, may not satisfy society's requirements for sustainability, sensu Brundtland (WCED 1987), regarding such issues as biodiversity, aesthetics, and cultural and spiritual values.

Nowadays, the majority of the adult population in industrialized countries, born and raised in cities, is employed in the service and information sectors, and has different value concepts and relationships with nature than did their ancestors. Unlike their agrarian ancestors, very few are directly utilizing nature for a livelihood. Instead, urbanites enjoy the luxury of focusing on romantic, idealistic, and symbolic forest values. This has led to major changes in forest policy in several countries (Koch and Kennedy 1991), and we believe that it will lead to more changes in the future. Based on current social and silvicultural challenges in Europe, we suggest possible shortcuts toward a sustainable management of planted forests that may also apply to the United States.

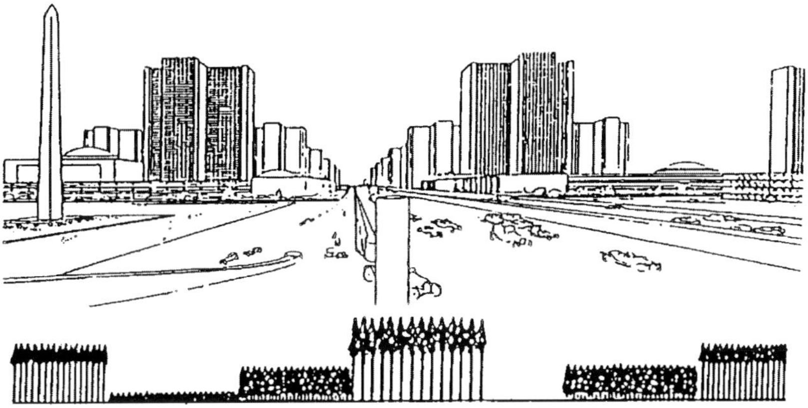

Figure 2. The functionalistic nature of plantation forestry, compared with Les Corbusiers's plan for a part of Paris (*Source*: Frivold 1991).

Planned forestry in Central Europe

The industrial revolution of the nineteenth century led most European countries to focus on wood production on a sustained-yield basis to fulfill the increasing and changing demands for wood products. A highly regulated plantation forestry soon proved to be an efficient means of achieving and securing this narrowly focused single-use objective. Agrarian forest grazing practices were forbidden, remaining forest areas were preserved and intensively managed, forest lakes and bogs were drained and afforested, and other afforestations took place. Often exotic tree species or provenances were used.

The aim was to increase wood production and to change forest compositions to favor the industrial use of wood. Often this was achieved by turning the forests into easily manageable, highly productive, pure and even-aged stands. In many cases, biological diversity and multiple-use forest values were reduced in the chase for wood production efficiency.

Plantation forestry is functionalistic, and has been compared to the functionalism in architecture illustrated by the architect Les Corbusiers's plan for a part of Paris (Figure 2). An approach of this kind, irrespective of its merits, implies uniformity and thus drawbacks, such as risk of ecological instability. In response, an ecologically based paradigm, notably for silviculture, has evolved during the twentieth century. According to this paradigm, forest management should be more site specific and stand treatments should be based on a close correspondence with natural forest ecosystems. However, being largely focused on wood production, this initial nature-oriented approach has not kept pace with changing societal value concepts.

Integrated, sustainable, multiple-use forestry

To adapt to changing social values of forests, Central European forestry now aims for a so-called integrated, sustainable, multiple-use forest management concept. This concept implies an integration of forest functions such as wood production, biodiversity, recreation, amenity values, groundwater protection, and carbon fixation.

In his 1989 Starker Lecture, Plochmann (1989) summarized the basic principles: the land base of Central Europe is too small, and the population too dense, to segregate different land uses in different areas. The product mix for each management unit, some of them quite small, must be planned according to natural, political, economic, and social parameters. This does not mean that all goods and services should be produced at every single unit of the forest, as some interests are conflicting. A superior target and priority setting that also allows for primary (or dominant) use is necessary (e.g., Daniels 1987). The concept implies, however, that the decision-maker, aiming at the long-term optimal combination for society, at each decision takes into account all the present and future goods and services that forest land could provide (Figure 3).

This may lead to an economic loss to the private forest owner because many non-timber values cannot be sold. However, the losses are seldom large and are often compensated for by the government, or looked upon as an investment made by the private forest owner to avoid further restrictions in timber production.

A good Danish example of integrated multiple-use forestry is the intensively managed national forest Jægersborg Dyrehave, approximately 15 km north of Copenhagen. This forest is about 1,200 ha large, has 3 million recreational visits/yr, a population of 2,000 fallow deer and red deer, and an annual timber harvest of 8,500 m^3.

Tables 1 and 2 give some statistical comparisons of forestry in Denmark and Oregon, and of the national forests in Denmark and the United States. From this it can be seen that the outputs per unit area are much larger in Denmark, even though the growth conditions in general are not as favorable as in Oregon. The corresponding inputs in labor and capital are also much larger. However, Denmark is only producing one-fourth of its own wood consumption, and Central Europe, two-thirds of its total consumption. So, Denmark and Central Europe as a whole export some of their forestry-related environmental problems, and draw heavily on the forest resources of other countries and regions. Thus, nationally or regionally, sustainable forestry may result in adverse effects on sustainability in other parts of the world. This is one of the reasons for intensive forest management in Europe, and for plans to convert more agricultural land to forest.

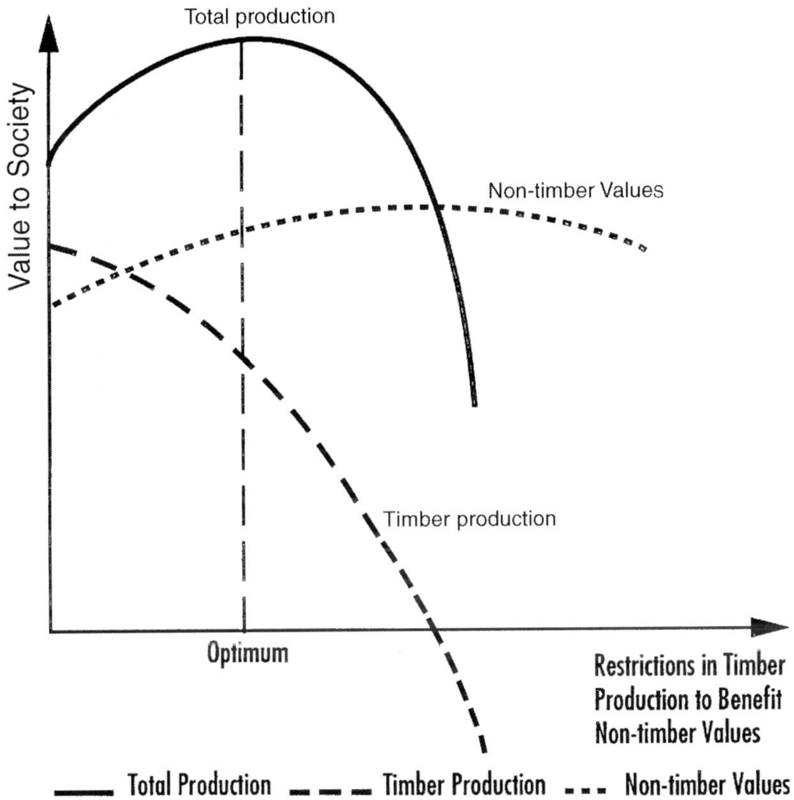

Figure 3. Model describing the relationship among total production, timber production, and non-timber values.

Segregation or integration?

It seems to us, as Europeans, that most large conflicts over forest land in the United States have been handled by "slicing up the pie into pieces" (e.g., national parks, wilderness areas, wild and scenic rivers, spotted owl habitat conservation areas, and tree farms). This is probably a logical reaction in a large country with an abundance of natural resources and land areas. The size of the pie, however, is not infinite. Surprisingly, the United States, in spite of the country's vast forest resources, for several years has only produced about 85 percent of its own wood consumption (U.S.D.A. Forest Service 1989). The United States, like Central Europe, is to a large extent relying on the forest resources of other parts of the world. These two regions are so wealthy that they may import wood, and thereby export some of their environmental problems related to forestry. In doing so, are we truly following the motto of the Brundtland Report (WCED 1987), "Think globally, act locally"?

Table 1. Comparison of forestry in Denmark and Oregon (*Source*: Koch 1991; Danmarks Statistik and Skov- og Naturstyrelsen 1994; Statistics Denmark 1996).

	Denmark	Oregon
Population (millions)	5.3	2.8
Land area (km^2)[a]	43,094	248,640
Islands (number)	406	10
Highest point (m)	171	3,416
Lowest point (m)	−8	0
Climate	Temperate Insular	Temperate Maritime & Continental
Precipitation (mm)	660	203–3,302
Frost-free days (number)	165–215	60–200
Forest (% land area)	11	46
Forested area (million ha)	0.45	11.34
Forested area per capita (ha)	0.1	4.1
Commercial timberland (million ha)	0.42	8.91
Owner distribution of forest land		
Public, national, and local (%)	31	63
Private estates/forest industry (%)	47	21
Farmers (<50 ha)/nonindustrial private	22	16
Area distribution by timber type		
Hardwoods (%)	34	7
Conifer (%)	61	93
Christmas trees (%)	5	0.2
Timber harvest		
Total (million m^3/yr)	2.0	44.85
Per unit area (million m^3/yr/ha)	4.5	3.9
Forest area/graduate forester (ha)	735	8,100
Forest area/forest worker (ha)	102	810
Visitors to forests		
Total (million visits/yr)	50	60
Visitor days/yr (million)	8	40
Visitor days/yr/ha	17.3	2.5

[a] Land and inland water area, with Greenland and Faroes 2,220,000 km^2.

As a consequence of the "slicing up the pie into pieces" policy, multiple-use forestry in the United States often is realized by segregation rather than integration of forest functions (Figures 4A and 4B). In Europe, we have approached similar forest-interest conflicts from the opposite side, i.e., not from an abundance of forest resources, but, on the contrary, from a lack of forest resources (Figure 1). This has taught Europeans the hard way to use

Table 2. Comparison of the National Forest and Nature Agency of Denmark and the U.S.D.A. Forest Service (*Source*: Koch 1991; Danmarks Statistik and Skov- og Naturstyrelsen 1994).

	National Forest and Nature Agency	U.S.D.A. Forest Service	Factor of difference
Area			
Total area (million ha)	0.11	77.35	703
Commercial timberland (%)	62	45	0.7
Net land acquisition (ha)	0.81	37.66	47
Area per inhabitant (ha)	0.022	0.312	14
Timber			
Harvested (million m^3/yr)	0.59	62.54	106
Commercial harvest (m^3/yr/ha)	12.37	4.47	0.4
Reforested (1,000 ha/yr)	0.07	12.79	183
Commercial reforested (%/yr)	1	0.5	0.5
Outdoor recreation			
Annual visits (million)	20	135	7
Visitor days (million)	3	242	80
Intensity of use (visitor days/ha)	7.5	1.3	0.2
Employees			
Full-time equivalents	514	36,143	70
Full-time equivalents/ha	3	0.5	0.15
Forest workers (million hr)	2	–[a]	–
Expenses			
Expenditures ($billion)	0.1	2.7	27
Receipts ($billion)	0.05	1.6	32
Total ($billion)	−0.05	−1.1	22
Net cost of national forest system			
Total ($billion)	−0.03	−0.6	20
Per unit area ($billion)	−185.2	−7.4	0.04

[a] Unknown number of entrepreneurs.

forest resources in an integrated manner for multiple purposes (illustrated in Figure 4C).

The silvicultural challenge in Europe

The current silvicultural challenge in Europe is to create and maintain forest types that favor both production of wood and the non-wood benefits of the

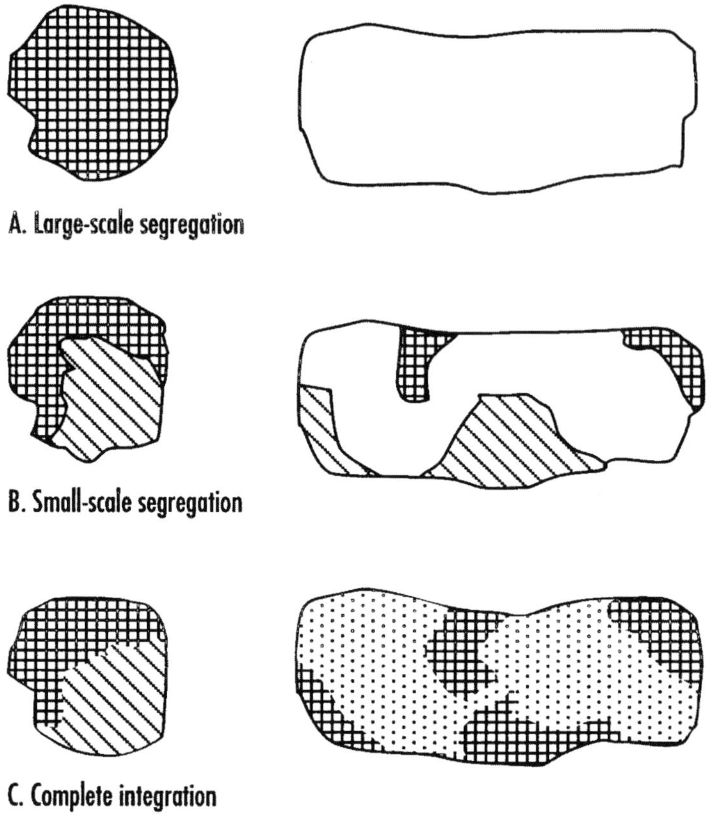

Figure 4. Segregation or integration of different uses of forest resources in the landscape. ▦ Non-timber uses only, ▧ Non-timber uses dominant, ▒ Timber dominant, □ Timber only.

forests. This challenge is by no means new, but for two centuries the main thrust in Central European forest management has favored a silviculture for pure, even-aged stands.

Now it is realized that a nature-oriented silviculture may provide a viable basis for an integrated and sustainable forest management, sensu Brundtland (WCED 1987). With afforestation, an expeditious and well-targeted effort toward integration is particularly important to minimize uniformity of plantations and related drawbacks, as well as potentially negative environmental effects. In terms of silvicultural systems, the introduction of a nature-oriented approach may be illustrated as in Figure 5. In the temperate broadleaved zone, silviculture should move away from the clearcutting system toward a single tree selection system. In this process the size of regeneration units decreases,

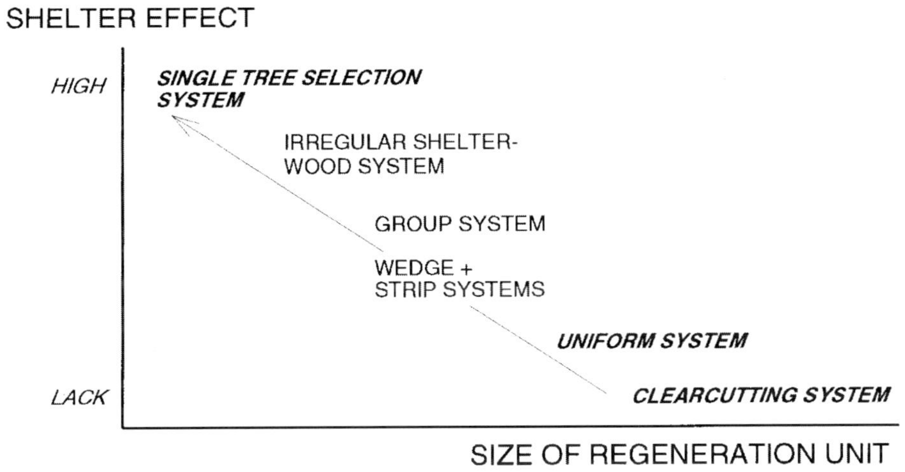

Figure 5. The gradient of silvicultural systems, illustrated by two main characteristics: the degree of shelter and the size of regeneration units (*Source*: Skovsgaard 1995). Note: The silvicultural challenge to create and maintain forest types that favor both production and environmental targets may for temperate forests in Europe, in many cases, be met by moving along the path toward a single tree selection system.

the shelter effect increases, and regeneration becomes diffuse in space and time.

Care is required, however, not to regard the single tree selection system as the standard or the goal for a nature-oriented silviculture. Depending on site and other circumstances, other silvicultural systems, and even those at the other extreme of the scale, may provide for close-to-nature management. This holds, for example, in regions where forest ecosystems are regenerated by "nature" through burning. Likewise, the time scale needed to alter silvicultural practices will depend on site conditions and should be considered carefully. Irrespective of the local target level for a nature-oriented silviculture, forest management will be less simple and a more intensive administration will be required than in the situation where timber is the main issue.

The past European influence

The single-use forestry paradigm was brought to North America, Australia, New Zealand, Japan, and India by European (mostly German) foresters (Behan 1975). Because of the wood scarcity experienced in many European countries in the eighteenth and nineteenth centuries, sustained flows of wood products were the primary social-value contribution of early "modern" forestry. The professional value system that developed included a belief that the answers to forest management questions, because they were seen as biological and engineering issues, were to be found by foresters in the forest itself. It also included a tendency to disregard people and their "eccentric" forest values, such as beauty and quiet (Duerr 1979).

Bernhard Fernow, a German-born and -educated pioneer of American forestry, wrote in 1902 (Fernow 1902): "The first and foremost purpose of a forest growth is to supply us with wood material; it is the substance of the trees itself, not their fruit, their beauty, their shade, their shelter, that constitutes the primary object ..."

In the United States, however, there was early concern with multiple-use social values, and a recognition that conflicting interests over forest management could arise. As dean of America's first forestry school (Cornell University, 1897–1903), Fernow learned this in 1903 when his Forestry Department was denied funding and terminated by the New York legislature as a result of clearcutting on its school's experimental forest (Rodgers 1968).

The current European influence

In forestry, positive influence often originates from negative experience. If there is a lesson to be learned from the European history of plantation or plantation-like forestry, it might be summarized as follows:

1. Preserve natural forests.
2. Do not convert seminatural forests to even-aged, pure stands, but try to move directly to a more holistic, nature-oriented silviculture.
3. Create and maintain a forest plantation that as fast as possible develops into a "real" forest. This is the most important issue relating to a "successful" afforestation (successful in terms of sustainability, sensu Brundtland (WCED 1987)). The basic idea behind this is that you should not only grow and manage trees, but also manage the whole forest. This differs from the "pure" plantation philosophy with its uniformity. The challenge is to manage for the single tree and the whole forest ecosystem at the same time, i.e., "To Follow and Support Nature in Its Effects," as stated by the Danish forester D. N. von Warnstedt in 1781.
4. Carefully consider how the new plantation is adjusted to the ecology of the whole landscape when planning an afforestation. Inside the plantation, the design at forest level as well as at stand level should, from the start, mimic and accommodate what is considered to be the "final" outcome.
5. Prefer local tree species and local provenances that already have proven their ecological stability.

6. Be patient. It will take several decades to create a forest ecosystem from a plantation.

For more detailed suggestions we recommend *Creating new native woodlands* by Rodwell and Patterson (1994) and *The design of forest landscapes* by Lucas (1991).

Closing comments

Based on more than 200 years of experience with planned and well-regulated forestry, it is evident that developing more holistic approaches in silvicultural practices and in forest management is the biggest current challenge for forest science as well as for practice. This will most certainly have an important bearing on future research activities, and requires enhanced cooperation across scientific disciplines, across professions, and across countries. Changes in silvicultural practices already influence research needs and will require additional funding that may not be proportionate to changes in economics, whether they be considered negative, as on the corporate level, or possibly positive, as on the socioeconomic level. Politicians, policy-makers, forest managers, and forest scientists all have to acknowledge that these challenges may be as far-reaching as those of our predecessors at the introduction of planned forestry in the mid-1700s.

References

Behan, R.W. 1975. Forestry and the end of innocence. Am. For. 81: 16–19.

Daniels, S.E. 1987. Rethinking dominant use management in the forest-planning era. Envirn. Law. 17: 483–505.

Danmarks Statistik and Skov- og Naturstyrelsen (Eds) 1994. Skove og Plantager 1990.

Duerr, W.A. 1979. American forest resource management, pp. 9–20. In: Duerr, W.A., Teeguearden, D.E., Christiansen, N.B. and Guttenberg, S. (Eds) Forest Resource Management. W. B. Saunders Company, Philadelphia.

Fernow, B.E. 1902. Economics of Forestry. Crowell, New York.

Frivold, L.H. 1991. Synen på blandskog genom tiderna. Skog & Forskming, Sveriges Skogsvårdsförbund 1991(2): 6–10.

Koch, N.E. 1991. Sustainable forestry: Some comparisons of Europe and the United States. In: The 1990 Starker Lectures, 41-53. Corvallis, College of Forestry, Oregon State University.

Koch, N.E. and Kennedy, J.J. 1991. Multiple-use forestry for social values. Ambio 20: 330–333.

Lucas, O.W.R. 1991. The Design of Forest Landscapes. Oxford University Press, Oxford.

Plochmann, R. 1989. The forests of Central Europe: A changing view. In: The 1989 Starker Lectures, 1-9. Corvallis, College of Forestry, Oregon State University.

Rodgers, A.D., III 1968. Bernhard Eduard Fernow – a Story of North American Forestry. Hafner, New York.

Rodwell, J. and Patterson, G. 1994. Creating New Native Woodlands. H.M.S.O, London. For. Comm. Bull. No. 112.

Skovsgaard, J.P. 1995. Challenges for the modeling of growth and yield in mixed stands with a naturally-oriented silviculture, based on experiments in pure even-aged stands and on evidence from nonintervention forest sites: The present situation in Denmark, pp. 166–172. In: Deutscher Verband Forstlicher Forschungsanstalten, Sektion Ertragskunde (Ed) Proceedings of 1995 Meeting.

Statistics Denmark (Ed) 1996. Statistical Yearbook 1996.

USDA (United States Department of Agriculture) Forest Service 1989. An Analysis of the Timber Situation in the United States. USDA Forest Service, Washington, DC.

WCED (World Commission on Environment and Development) 1987. Our Common Future. Oxford University Press, Oxford.

The Nature and Uses of Planted Forests

New Forests **17:** 25–36, 1999.

Planted forests of the wet and dry tropics: their variety, nature, and significance

JULIAN EVANS
Tropical Forestry, Imperial College, London, UK

Key words: sustainability, plantation forestry, rural development forestry, environmental forestry, silviculture

Abstract. Planting trees in tropical countries is an increasingly important forestry activity. Although much attention has been paid, and rightly so, to rapid deforestation in the tropics, the contribution and role that planted forests can play are now at the center of policy and planning in most national forestry strategies. The changing emphasis from exploitative management of natural forests to managed natural forests and plantation forests, seen in temperate regions over the last 100 years, has been taking place in tropical countries largely over the last 20 years. In neither temperate nor tropical regions has this evolution finished.

Planted forests will increasingly furnish wood resources that meet domestic wood needs and provide many environmental services. In this paper I seek to place in perspective the contribution that planted forests will make in the tropics. I emphasize (1) the wide variety of purposes that planted forests serve; (2) the sustainability of plantation silviculture; and (3) the overriding importance of due regard to people's participation. I am optimistic about the great value that planted forests, rightly used, can provide for a great many hard-pressed peoples across the tropics and beyond.

Extent and definition of planted forests

Although it is the act of planting a tree that defines the subject, the spectrum of types of planted forests is broad, and ranges from industrial monocultures to enrichment planting of natural woodland formations or growing fruit trees in an African compound. I confine remarks here to locations where planting has led to the establishment of at least a small group of trees that would not otherwise be there. Sawyer (1993), in her analysis of the environmental concerns of plantations in the tropics, has a helpful chapter on this question of definitions.

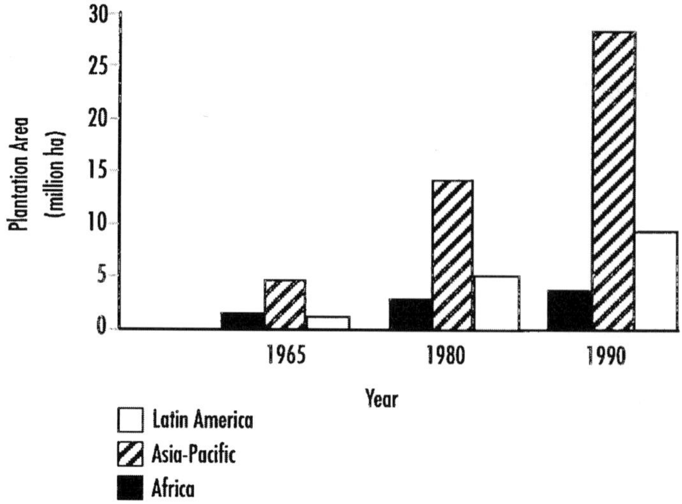

Figure 1. Areas covered by tree plantations in the tropics and hotter subtropics.

Area of planted forests

Data assembled for my own book (Evans 1992) and by the Food and Agriculture Organization of the United Nations (FAO 1993) indicate that there are between 40 and 50 million ha of forest plantations of one kind or another in the tropics and warmer subtropics. Figure 1 summarizes data by continent and shows the large increase since 1965. Such area data are based on reports and country returns collated from many different sources; thus, they should be treated critically. The overall magnitude of planting is correct and the evidence of increased planting is correct. However, individual figures, particularly figures published by countries, are notoriously variable because of different reporting systems, different methods of data collection, and the all-too-frequent practice of claiming land planted based on seedlings raised in the nursery rather than on surveys of successful tree planting.

This last problem bedevils many tropical forestry statistics, particularly those regarding the success of extension programs where optimistic reports simply do not match reality. Far too great an emphasis is placed on numbers and counting, far too little on follow-up and assessing what has really been achieved. Although I know of no survey to prove the point, my guess would be that no more than one seedling out of every three, carefully raised and nurtured in a nursery, survives 1 year in village and extension forestry programs, and, at best, no more than two seedlings out of three in managed industrial or commercial forestry programs.

Despite the caveats above, a lot more tree planting is taking place in the tropics than in the past. A figure of around 2 million ha/yr, compared with only 1 million ha/yr a decade ago, is a realistic estimate.

Geography of plantations

At this point it is helpful to remark on two dichotomies. One is in the title of this paper, the distinction between the wet and the dry tropics, which can crudely be delineated as above or below about 800 mm of rainfall per year; the other is the distinction between highland and lowland, which reflects temperature differences. In summary, planted forests are most successful in the cool, moist, highland tropics and subtropics; they are, perhaps, most needed in parts of the dry tropics.

Variety of tree planting

Planted forests may be classified in a number of ways, such as by their species' composition, by their scale, by their complexity, or by the purpose for which they are established. Different classifications serve different purposes, but the last of these is the most helpful distinction to emphasize the contrast with temperate plantations. Planting trees in tropical countries is undertaken for three principal reasons: industrial plantations, rural development, and environmental benefit.

Industrial plantation forestry

Industrial plantations have been successful, particularly in the cooler highland tropics and on former savanna, grassland, and similar land types. Almost all such planting has been for industrial wood purposes, such as low-grade lumber, pulpwood, and some panel products. Cultivation of quality hardwood timbers, with the exception of teak (*Tectona grandis* L. f.), has proven to be more intractable, and this is elaborated on later.

Rural development forestry

Almost all tropical countries actively promote tree planting as part of rural development. This arises from the important role that trees play in meeting several crucial domestic needs, such as those for fuelwood, fodder, sticks and poles for building materials and fences, and a range of other products used at the village level as part of everyday life. This dependence on trees for day-to-day needs highlights a key difference between tropical and many temperate countries. Additionally, such trees in rural development provide shade, shelter, and a means of soil improvement through such processes as nitrogen fixation.

Rural development forestry embraces the terms "social forestry" and "community forestry," which have fallen somewhat into disfavor because of the politics that have surrounded such programs, e.g., in India, Thailand, and Kenya. For many tropical countries today, the bulk of tree planting activity is for these purposes.

One important strategy in rural development forestry is planting trees in association with food crops, or as a source of food or fodder themselves. Although the term used for these plantings, "agroforestry," is relatively new, the practices are frequently ancient and much profitable research is focused on investigating the success and science of these long-practiced husbandries.

Environmental planting

A particular feature of many tropical plantations is their use to provide a specific environmental service or to harness trees with the principal aim of a non-timber benefit. Tree planting to assist soil erosion control, whether in support of bunds on steep hillsides or as part of shelterbelts to reduce wind erosion, is widely practised. Reclamation of wasteland and degraded areas and the less interventionist rehabilitation (according to the classification of Lamb 1994) are far more widespread in the tropics than, for example, in my own country of Great Britain.

Parallel to this are three further purposes for which the main objective is to bring about environmental benefit: (1) as carbon offsets, though obviously forest products may be obtained; (2) as compensatory plantations to divert pressure away from natural forest systems (again forest products are available, but these plantations provide a means of saving forests elsewhere); and (3) as buffer zones around areas of natural forest, i.e., a specific locational example of compensatory plantations.

In several instances, more than one of these purposes is well served by planted forests. The distinctions are drawn because the three broad roles planted forests can play are much more explicit in many tropical countries. All of the benefits that trees and woodlands can provide are much more carefully thought through and used.

Plantation silviculture

The principles that underpin successful plantation forestry in temperate countries apply equally to the tropics. Planted forests in the tropics, indeed any type of tree planting, do not differ materially in their technological requirements from those in temperate regions. Plantation forestry requirements involve six key features: species, planting stock, stocking levels, establishment, protection, and thinning and pruning.

Species choice

For both industrial plantations and tree planting in rural development, exotic species predominate. Indeed, eucalypts (*Eucalyptus* spp.) and pines (*Pinus* spp.) alone account for well over half of all planting that has taken place in the tropics; pines are used predominantly for industrial purposes, eucalypts for both industrial purposes, e.g., pulpwood and pit props, and village use in supplying sticks, poles, and firewood. Legume species, particularly of the genera *Acacia* and *Prosopis*, are important in social and village forestry. Some individual species have also found a widespread niche, e.g., *Grevillea robusta* A. Cunn. (silky oak) as a pole and timber tree in hedgerows and field boundaries; *Leucaena leucocephala* (Lam.) de Wit (Ipil ipil) as a fodder and fuelwood species; *Casuarina equisetifolia* L. (she-oak) in soil erosion control and as a nitrogen-fixing species, and *Azadiracta indica* A. Juss. (neem) for shade, fuel, and wood. The predominant genera and species from a compilation made in 1980 remain the important ones today (see Table 1).

There are relatively few plantations of high quality tropical hardwoods for a variety of reasons, notably because of such problems as *Hypsipyla* shoot borer damage to the *Meliacaea* – mahoganies, khayas, and tropical cedars, and problems resulting from inadequate understanding of the ecology of many tropical hardwoods found in rain forest ecosystems. The review by Appanah and Weinland (1993) of experience in peninsula Malaysia well illustrates the difficulties. The major exception is teak, which has been successfully planted in many tropical countries and occupies more than 2 million ha. There is no doubt that improved understanding of how to grow quality hardwoods will be one way of helping to alleviate pressures resulting in tropical deforestation.

As in temperate plantation silviculture, the matching of species with site is crucial. Successful use of planted forests in the tropics has depended on selecting the appropriate provenance for a site, developing land races, and developing some astonishingly successful tree breeding strategies. Among tropical pines, eucalypts, and, for example, the phyllode acacias, remarkable improvements have been made as a result of attention to genetics. This remains a priority.

Complementary to selection of genetic material is the proper characterization of sites to match species' requirements. Many examples exist, such as the successful Australian Centre for International Agricultural Research

Table 1. Tree genera and species[a] used in tropical forestry plantations in 1980 (*Source*: Evans 1992).

Genus/group	Species[b]	Relative use (%)
Pinus	*P. patula, P. caribaea, P. elliottii, P. merkusii, P. kesiya, P. oocarpa,* and others	33.7
Araucaria, Cupressus, and other conifers	*A. cunninghamii, A. angustifolia, C. lusitanica,* and others	3.0
Eucalyptus	*E. grandis, E. camaldulensis, E. globulus,*[c] *E. saligna, E. tereticornis, E. robusta, E. citriodora, E. urophylla, E. deglupta,* and others	37.5
Tectona	*T. grandis*	14.2
Acacia, Leucaena, Gmelina, Grevillea, Albizzia, Meliacaea, Terminalia, Prosopis, Casuarina, Cordia, Triplochiton, and other hardwoods		11.6

[a] This analysis does not include data from southern China, where the main planted species are *Cunninghamia lanceolata, Eucalyptus exserta, E. citriodora, Pinus massoniana,* and *P. yunnanensis*.
[b] Order indicates relative importance.
[c] Only successful at high elevations such as in Colombia, Ecuador, Ethiopia, India, and Peru.

programs of Plantgro and Bioclim, which have refined considerably our understanding of where Australian species will grow well (Booth et al. 1989). There have also been major species' failures. For example, trials of southern pines in the humid tropics failed from being off-site climatically, and radiata pine (*Pinus radiata* D. Don) and *Leucaena leucocephala* have become more restricted as a result of *Dothistroma* sp. infection and psyllid defoliation, respectively.

Planting stock

Planted forests in the tropics are almost wholly established with seedlings that have been grown in containers in nurseries. Very little forest establishment in either wet or dry tropics is achieved through direct sowing or aerial seeding techniques. All of the container systems familiar to temperate foresters are available and used in the tropics, but the black polyethylene bag or tube, or local variants such as those made from banana leaves, veneer curls, and old tin cans, are the stock-in-trade for extension nurseries.

One other nursery practice, the use of hardwood cuttings, has proven to be particularly successful in the tropics. Teak is the most widely used example of a hardwood cutting that can be kept for many weeks in a shady place and then successfully inserted into the ground to root and grow. Indeed fence posts of teak will often root and develop into trees! Such "pole size" cuttings are promoted in parts of India as a quick means of tree planting, but more generally the use of small hardwood cuttings can be a cheap and easy-to-apply technology for villagers and farmers. The high technology equivalent of micropropagation and related tools is used in a number of major commercial plantations.

Tropical tree nurseries vary enormously in size from large production units, covering several hectares and producing several million plants each year, to home or compound nurseries, where the owners grow just a handful of trees. In Britain we buy our gardening materials, trees, shrubs, and plants from garden centers; in the tropics villagers bring seed and cutting material to their homes and cultivate them there.

Stocking

Compared to temperate plantations, there is often a wider range of planting density in the tropics, and this reflects the wider range of purpose for which the trees are being grown. In many compounds and near villages throughout the tropics, trees will be planted at spacings closer than 1 m or stockings of more than 10,000/ha. These high densities are used to grow sticks for building huts and fencing materials, and also to provide easily handled firewood. Most industrial plantations are established at around 2.5–3.0 m spacing, i.e., about 1,000–2,000 trees/ha. In agroforestry and enrichment planting silvicultures, much wider spacings are used. In both situations rows of trees may be 10 m or more apart, and sometimes trees in each row will be equally widely separated. In the tropics, therefore, tree planting densities can vary from a few tens to many thousands/ha, this depending entirely on the purpose for which trees are being planted.

Establishment

The nature of the site being planted largely determines the requirements for site preparation and planting practices. This is not the place to cover such a broad subject, but the following two points are important:

1. On moist, lowland, equatorial sites, vegetation growth, including lianas, can overwhelm even the most vigorous of plantation species. This must be addressed in the planning stage. (As an aside, clearing forest simply to provide land for tree planting is deprecated. It is neither necessary nor desirable when so much abandoned land suitable for planting already exists.)

2. Much tree planting in the tropics serves for both wood production and the provision of environmental services, especially control of soil erosion. Care must be exercised that site preparation and planting activities do not themselves worsen the very erosion problem that the project is seeking to overcome. Sometimes minimum cultivation and weed control, along with concomitant slower establishment of the planted tree, will be preferable to applying traditional silvicultural practices.

Protection

The problems that confront planted trees in tropical countries do not differ materially from those elsewhere. Young trees are at great risk of damage from browsing, trampling, uprooting, and an array of insect pests and fungal diseases. Without due regard to these threats, all the careful work of matching species with site and of preparing the site well will be wasted. Again I would highlight just two issues of special relevance to the tropics, as follows:

1. Villagers' livestock and the common practice of grazing extensively through the bush and taking animals to water represent an enormous threat to tree plantations, particularly those that are part of community or social forestry initiatives. Cattle, camels, sheep, and goats can all easily wander into plantations and do immense harm – though, of course, splendidly nourishing the beast itself! Prevention of such damage may depend on fencing, but more usually involves the use of guards or shepherds whose duty it is to keep out such livestock. In the Bilate project in Ethiopia the local villagers employed 55 shepherds to guard a 30-km strip of woodland they had helped to establish (Evans 1992).
2. Two insect pests are serious threats at the time of establishment. Throughout much of the tropics, but especially in Africa, termites can cause up to 100 per cent mortality in a new plantation. There are no easy cures, though wise choice of species, such as avoidance of certain eucalypts, can help. Leaf-cutting ants (*Atta* spp.) are important pests in tropical America, especially in more humid areas.

Management of stands

Compared to temperate plantations one sees conflicting trends in stand management in the tropics. On the one hand, the lower labor costs make operations such as thinning and high pruning in industrial plantations more attractive. Many fine stands producing high-quality sawlogs and even veneer timber are being grown in tropical and subtropical plantations. On the other hand, where trees are planted to meet domestic needs within the home or village, informal management of cutting the trees when their size is right leads to a "people's selection system"! The woodland is a bank of material cut and used as if it were a natural stand.

One other practice, i.e., pollarding, is widespread, but rarely receives sufficient attention. This is the decapitation of trees 3 or 4 m above ground, and the harvesting of branch material that sprouts from just below the cut surface. All across the tropics pollarding is used as a means of obtaining fodder, sticks, and firewood from trees in a way that does not kill the tree and that ensures that the regrowth is sufficiently elevated to be above browse height, especially from such livestock as cattle and sheep. Where there are camels and goats, the height has to be 5 m or more.

Where tree planting is used for enrichment of natural woodland, great attention must be paid during the first few years to ensure that the planted tree is not swamped, and expensive cleaning operations are essential for this kind of silviculture. The principles of Dawkins (1958) still apply.

Social and economic roles

In the tropics there are some remarkably successful and remarkably unsuccessful plantation projects. This applies to both industrial plantations and tree plantings for rural development. Three issues, more than any other, have led to failures; these are as follows

1. Inadequate attention to questions of land tenure, tree ownership, and the culture of property in the country concerned. If this is not clarified at the outset, untold problems can arise over people's rights and dues. This can threaten the success of a project, both small and large.

2. People's participation. This expression may be cliché, but it is no less important for that. All successful plantation projects depend, to a greater or lesser extent, on the willing cooperation and interest of local people, whether directly or through their locally elected representatives. In forestry projects for rural development, involvement of local villagers, from the idea stage through identifying the land and its boundaries to the subsequent establishment and management of the woodland, is crucial. Many projects failed in the 1970s and 1980s through patronizing attitudes of "give 'em eucalypts" without ever asking whether or not they were wanted.

3. Planning, budgeting, and market evaluation. Because plantations are in the tropics and their encouragement is used to serve aid and development objectives of developing countries, proper planning, budgeting, and market evaluation must not be allowed to take backstage. This applies particularly to large-scale plantation development, which has sometimes been pursued more for prestige and political reasons than for market economics. In several

places in the tropics, plantations have been created and well established only to be abandoned within a few years because the market did not materialize, or because prices changed.

The parallel on a local scale is seen in the choice between planting trees and better management of existing natural woodland. The first reaction, usually by outsiders, to wood shortage is to suggest more planting, whereas often more careful management, such as controlled use of woodland or modifying the cycle of coppicing or pollarding, will be much more beneficial. Even the introduction of fuel-efficient, wood-burning stoves or similar improvements may be a better intervention than the expensive and long-delayed solution of planting trees.

Plantation forestry in the tropics, in all its forms, has much to offer. Failure to address the preceding three issues will override all considerations of species choice, silviculture, and management.

Sustainability

In almost every case, tree planting in the tropics represents a more intensive land use than that practised previously. High rates of growth and short rotations suggest the possibility that sites will be overstressed, which leads to diminishing fertility and poorer yields. Practices such as high stocking and coppicing may add to this risk. Few studies have been carried out to look at sustainability of such plantation practices (Evans 1990), but what evidence does exist suggests that the problem may not be as serious as it might appear. In Swaziland three rotations of *Pinus patula* Schlect. & Cham. have been grown on the same site, with each rotation appearing more productive than its predecessor, with only limited genetic improvement and without adding fertilizer (Evans 1994). Elsewhere in the tropics and subtropics, examples of yield decline often reflect weed mismanagement or mismatching of species to site, rather than an inherent shortcoming of plantation forestry practices. Even the most intensively worked tree plantations generally fall short in the demands for soil nutrients compared to food crops such as maize and yams.

In the narrow sense the outlook for sustainability appears reasonably optimistic. In the broad sense the question of sustainability is more open. There is an expectation that reliance on natural forest formations to provide goods and services will switch to greater dependence on plantations. As attention is switched from the complex and diverse to the simple and more uniform, there will, inevitably, be a reduction in the variety of products obtained (see Kanowski's 1995 analysis). This, in itself, may threaten the existence and livelihoods of some of the tribal peoples inhabiting the remoter parts of tropical countries.

Using plantations for conservation and rehabilitation

Tree planting is seen as increasingly important for protecting natural ecosystems through developing buffer zones, diverting pressure away from the natural resource, acting as a barrier around islands of refugia, and so on. Thus, planting trees becomes a tool to modify people's use of the natural ecosystem to be conserved. Harnessing such practices as agroforestry, making small-scale plantings in the home compound or village, and planting trees beside roads, along canals, and around fields all help to fulfill this role. The contrast with temperate forestry is that these policies seek to change where people go for their forest products, rather than, as in temperate forestry, to use plantations as more efficient means of providing them.

Similar to this role of diverting attention away from natural forests, plantations can be used to begin the process of reclamation and rehabilitation. Planting with useful species to enrich poor quality woodland has been mentioned already, but there is now much interest in using tree plantations to reclaim degraded grasslands, industrial waste sites, and other ground unfit for agriculture. Such plantings can be astonishingly successful. As a by-product they are often much more rich in carbon than was the land before, and therefore are a net carbon sink while they are growing.

Some conclusions

Planted forests in the wet and dry tropics are set to increase. In the next two or three decades they may well begin to supersede natural forests as the main sources of forest products in most tropical countries. Successful development of tree plantations fundamentally depends on people's participation. If local people are committed, the planted forest will succeed. Technical requirements for success in the tropics are as firmly based on sound silviculture as they are for temperate plantation forestry. Planted forests are a sustainable and flexible tool for development in the tropics. They are not a replacement for natural forest formations, but a complement to them.

References

Appanah, S. and Weinland, G. 1993. Planting Quality Timber Trees in Peninsular Malaysia. Institute of Malaysia, Kuala Lumpur. Malayan Forest Record No. 38.

Booth, T.H., Stein, J.A., Hutchinson, M.F. and Nix, H.A. 1989. Mapping regions climatically suitable for a particular species: An example using Africa. For. Ecol. Manage. 28: 19–31.

Dawkins, H.C. 1958. The Management of Natural Tropical High Forest with Special Reference to Uganda. Commonwealth Forestry Institute, Oxford. Commonwealth Forestry Institute Paper No. 34.

Evans, J. 1990. Long-term productivity of forest plantations – status in 1990. Proceedings of the Nineteenth World International Union of Forest Research Organisations Congress 1: 165–180.

Evans, J. 1992. Plantation Forestry in the Tropics (second edition). Oxford University Press, Oxford.

Evans, J. 1994. Long-term experimentation in forestry and site change, pp. 83–94. In: Leigh, R.A. and Johnson, A.E. (Eds) Long-Term Experiments in Agricultural and Ecological Sciences. CAB International, Wallingford, United Kingdom.

FAO (Food and Agriculture Organization of the United Nations).1993. Forest Resources Assessment 1990: Tropical Countries. FAO, Rome. FAO Forestry Paper No. 112.

Kanowski, P. 1995. The complex future of plantation forestry, pp. 483–87. In: Potts, B.M., Borralho, N.M.G., Reid, J.B., Cromer, R.N., Tibbitts, W.N. and Raymond, C.A. (Eds) Eucalypt Plantations: Improving Fibre Yield and Quality. Proceedings of the Cooperative Centre for Temperate Hardwood Forestry-International Union of Forestry Research Organizations Conference. The Cooperative Centre for Temperate Hardwood Forestry, Hobart, Tasmania, Australia.

Lamb, D. 1994. Reforestation of degraded tropical forest lands in the Asia-Pacific region. J. Trop. For. Sci. 7: 1–7.

Sawyer, J. 1993. Plantations in the Tropics: Environmental Concerns. International Union for Conservation of Nature, Gland, Switzerland and Cambridge, United Kingdom.

New Forests **17:** 37–52, 1999.

Eucalypt plantations

JOHN W. TURNBULL
Center for International Forestry Research, Bogor, Indonesia

Key words: planted forest, sustainability, industrial forestry, social forestry, nonindustrial forestry

Abstract. This paper reviews the historical development of the use of the eucalypt over 200 years, from its curiosity status in the botanical gardens of Europe to its extensive use as a fuelwood for the wood-burning locomotives of the national railway systems, and then to its more recent use as a major source of biomass for paper pulp, fiberboard, industrial charcoal, and fuelwood. Ecological and biological aspects of the genus *Eucalyptus* have made it successful as an exotic in industrial monocultures and as a multipurpose tree of benefit to small landholders. Social, policy, and economic aspects of growing *Eucalyptus* are examined, as are prospects for using the eucalypt in the twenty-first century as an industrial plantation tree and as a component of farming systems in the rural landscape.

Introduction

In the late eighteenth century the French botanist, Charles Louis L'Heritier De Brutelle, applied the name *Eucalyptus* to an Australian tree. At that time there was little appreciation of the potential of eucalypts to become a major source of forest products. The wood was difficult to saw and season, and it was commonly considered to have value only as firewood. Remarkably, now at the end of the twentieth century, eucalypts have become the most widely planted hardwood species in the world.

This paper aims to describe the role of eucalypts in world forestry and to emphasize the social and economic aspects of eucalypt planting. It points to changes in management of eucalypt plantations to ensure sustainability, productivity, and social acceptance as we move into the twenty-first century.

Eucalypts as exotics

Eucalypts were initially regarded as botanical curiosities and were introduced to botanical gardens and private arboreta in Europe. Once in cultivation, their potential was recognized and eucalypts were taken to many parts of the world.

Some were planted for their ornamental value, and others, for windbreaks, land reclamation, and leaf-oil production.

In Brazil and South Africa, eucalypts were planted along railways to provide fuel for wood-burning locomotives. In the early 1900s, eucalypts were seen as the answer to the shortage of hardwood timber in the United States and were heavily promoted and extensively planted in California. Planting was curtailed when the difficulties in sawing eucalypt wood became evident.

The reasons for planting eucalypts have changed significantly over time, and the end uses to which the species have been applied are diverse. Today, eucalypts provide sawntimber, mine props, paper pulp, fiberboard, poles, firewood, charcoal, essential oils, honey, tannin, shade, and shelter.

Reliable global estimates of areas of planted eucalypts are difficult to obtain, but published reports suggest that there are at least 12 million ha. Over 90 percent of these forests have been established since 1955, and about 50 percent in the past decade (Turnbull 1991). An assessment of plantation areas in the tropics indicates over 10 million ha of eucalypts at the end of 1990 (FAO 1993). There are large plantation areas in tropical America (4 million ha) and in tropical Asia (5 million ha). The American statistic is dominated by Brazil, where there are an estimated 3 million ha of eucalypt plantations (Campinhos 1998); some 2.3 million ha of these plantations survive from an area of 2.9 million ha approved for plantations with government incentives between 1967 and 1984 (Iuseum et al. 1988; Bazett 1993). In tropical Asia there are over 1 million ha in India (Kumar 1991). In addition, there are substantial plantations in countries with more temperate climates; they include Chile with 170,000 ha (Jayawickrama et al. 1993), China with 600,000 ha (Wang 1991), Morocco with 200,000 ha (Marien 1991), Portugal with 550,000 ha (Borralho 1990), South Africa with 477,000 ha (Schonau and Gardener 1991), and Spain with 396,000 ha (San Miguel 1988).

Increased plantation areas are projected in several countries. Plantings will continue in Brazil, but not at the very high rates of the recent past because there will be more effort to increase the productivity and quality of existing areas. Both China and India have active reforestation programs and, although there has been some resistance to eucalypt plantations in the latter, the great demand for wood will undoubtedly ensure that planting continues. Over 100,000 ha of *Eucalyptus camaldulensis* Dehnh. have been established in Thailand, and there are major planting programs in the Congo, Indonesia, and Vietnam. Extensive planting of eucalypts in the lowland humid tropics has been inhibited by the incidence of pathogens and insect pests which reduce productivity. Only a few species, such as *E. urophylla* S. T. Blake, *E. deglupta* Blume, and *E. pellita* F. Muell., appear to be adapted to hot, humid conditions. There are also active plantation programs in the more temperate areas. In Chile, the country's target is 300,000 ha of eucalypts, mainly *E. globulus* Labill., by the year 2000 (Anonymous 1990). There is also an active eucalypt plantation program with *E. nitens* (Deane and Maiden) Maiden and *E. globulus* in Australia. The area planted to 1994 was about 125,000 ha, and the current annual rate of planting is about 15,000 ha (Cromer et al. 1995). These trends suggest that eucalypt plantation establishment will continue through the remainder of the 1990s, and that the total area could increase to 15–20 million ha by the year 2000.

Ecological features

There are more than 500 species of *Eucalyptus* in Australia and in the neighboring islands of Indonesia, Papua New Guinea, and the Philippines. They occur mainly on acidic soils and extend from the hot, humid, tropical lowlands to the cool, temperate highlands. Some species have high growth rates, wide adaptability to soils and climate, ease of management through coppicing, and valuable wood and nonwood products. The seeds are orthodox, easily stored and transported, and readily propagated. However, there are very few examples of exotic eucalypts exhibiting "weediness." They are relatively free of pests and diseases, and are not palatable to grazing animals. The fast growth rate is mainly the result of indeterminate growth, and of being able to grow whenever there is adequate moisture and a suitable temperature. These characteristics have made eucalypts prime candidates for planting for many purposes in many countries.

Extensive planting has not always been welcomed. In an article on eucalypt planting in Thailand, Lohmann (1990) refers to the use of eucalypts for plantations as follows: "*Eucalyptus* is not only a soil degrader but also a crop destroyer, a depriver of fodder, flattener of natural forests, destroyer of security, creator of poverty and dependence, and killer of knowledge of future generations about how to live on one hand and livelihood and environment on the other."

Controversial aspects of the environmental effects of eucalypts include impacts on soil water, soil nutrients, soil erosion, and biodiversity. These issues have been reviewed by Poore and Fries (1985) and debated by others (e.g., Davidson 1985; ACIAR 1992). These authors and a Food and Agriculture Organization of the United Nations (1993) conference on the effects of eucalypts conclude that, where adverse effects are present, they can be minimized through proper plantation design and management (Ball 1995). It is important to distinguish the ecological effects of large block plantings of eucalypts from the effects of single trees or single rows of trees.

Social, economic, and policy dimensions

Enhancement of people's livelihoods and national economies by successful tree planting can be achieved through industrial plantations, social forestry plantings, agroforestry, and land rehabilitation. Eucalypts can play a role in all of these strategies, but they have been used most in industrial and social forestry.

Are eucalypts "industrial," single-purpose trees, or are they multipurpose trees with a role in social forestry and agroforestry? According to Kanowski and Savill (1990), eucalypts are industrial species unsuitable for meeting local multiple needs, whereas Evans (1985) regards eucalypts as having an "almost universal role as the social forestry tree." The following examples from several countries suggest that some eucalypts, such as *E. grandis* W. Hill ex Maiden, are excellent industrial species; others, such as *E. camaldulensis, E. citriodora* Hook, and *E. globulus*, can be either industrial or multipurpose according to the management applied to them. The majority of the 500 species have no industrial use and have had little or no role as planted trees.

Industrial plantations

Large-scale industrial plantations of eucalypts are a phenomenon of the last three decades. Major industrial plantings of eucalypts have taken place in Brazil, South Africa, Spain, and Portugal. Brazil has become synonymous with large-scale, fast-growing eucalypts, and South Africa is one of the best examples of the economic benefits that can flow from such plantations when they are well managed. The following brief description of the use of eucalypts in Brazil indicates how government policy can influence plantation development and subsequent economic, social, and environmental outcomes.

There is no sector of world forestry that is expanding as rapidly as the industrial use of eucalypts. Most plantations are being established to provide medium- to low-density, short-fibered pulp for paper. Brazil is the largest producer of eucalypt pulp; its plantations, mainly *E. grandis*, annually yield over 15 million m^3 of pulpwood at low cost (Rodrigues et al. 1989). The plantation-grown wood is usually harvested after 5–10 years and provides a uniform material with high brightness, and good opacity and bulk, which make the pulp suitable for the production of copying, printing, writing, and tissue papers. Demand for these products is rising, and with it the demand for eucalypt pulp. In 1990 almost half of the internationally traded bleached hardwood pulp came from eucalypts.

Brazil's success was catalyzed by a fiscal incentive policy implemented in 1965 and continued through the 1970s. This policy resulted in the establishment by private enterprises of 6 million ha of plantations, of which about 4 million ha survive, the majority of them eucalypts. The planting program helped Brazil to avoid what would otherwise have been a major wood supply crisis, when most accessible natural forests had been exploited by the 1980s (Bazett 1993).

Aracruz Celulose S. A. is an example of the very high levels of investment required for the eucalypt pulp and paper industry. Since 1973 the company has invested heavily in research, plantation management, and a processing plant. Tree improvement with an emphasis on clonal forestry has increased the productivity of plantations from 5.9 to 10.9 air dried t/ha/yr (Bertolucci et al. 1995). The company invested U.S.$1.15 billion in 1990 in an expansion of its pulpmill to lift production to over 1 million t/yr (Kyotöla 1989; Aracruz Celulose S. A. 1991). In 1995 Aracruz had sales of over U.S.$700 million, mainly from kraft paper. Aracruz exports about 90 percent of its production and currently accounts for 52 percent of Brazilian eucalypt pulp exports and 22 percent of the world's marketed eucalypt pulp (Campinhos 1998). Aracruz Celulose is a major contributor to the domestic economy through its export earnings. The company also has contract arrangements with small farmers for 15,000–20,000 ha of eucalypt plantations.

Although Brazil is the world's largest producer of eucalypt pulpwood, the greater part of its eucalypt plantations is used to provide high-quality industrial charcoal for iron and steel production. Minas Gerais State, the center of Brazil's iron and steel industry, uses 70 percent of the wood from its eucalypt forests for industrial charcoal. Federal legislation has been passed to require that almost all charcoal be produced from plantations, to prevent further loss of natural forests. There are several private companies in the state, each of which manages 150,000–200,000 ha of eucalypt plantations. There will be more demand for charcoal from plantations if potential industrial development is implemented in northern Brazil (Fearnside 1989). The main species used in charcoal plantations is *E. grandis*, but *E. camaldulensis*, *E. citriodora*, *E. cloeziana* F. Muell., and *E. paniculata* Smith are also significant. Many plantations have been set up in areas of low population density and rural development. The charcoal producers have been assisting small farmers in planting eucalypts on land that is unsuitable for agriculture (Calle et al. 1992).

A number of Brazilian companies now use selected clones for plantations. Aracruz Celulose currently has 85 percent of its plantings from cuttings (Bertolucci et al. 1995). Selected clones are also being used routinely in eucalypt plantations in the Congo, Morocco, and South Africa. Better land preparation, optimum use of fertilizer, improved nutrient uptake with selected mycorrhizae, and good weed control offer additional opportunities for productivity gains. Greater economic benefits will accrue from clonal plantations and more intensive management practices. However, these benefits need to be balanced with the increased risks of loss by cold, fire, nutrient depletion, and pests and diseases in plantations with a narrower genetic base and a higher investment per unit area.

Eucalypts have proven attractive to farmers in both developed and developing countries in many parts of the world. In most instances the trees are grown to provide products and services to the rural population, but there are also plantations established by farmers primarily to supply wood to industrial enterprises. In Spain, there is an example of a marriage between industrial and social forestry, with farmers cooperating with a private pulp and paper company, Celulosas de Asturias S. A.; the farmers grow eucalypts on their land to sell to the mill while also making money from the production of honey and essential oils (CEASA 1994). A similar scheme operates at the Paper Industry Corporation of the Philippines, where several thousand farmers grow *E. deglupta* and other species, and supply about one-third of the company's wood needs.

Nonindustrial plantations

The rural landscapes in Ethiopia, China, India, Malawi, Peru, and elsewhere are dominated by eucalypts established to a very large extent by farmers primarily for local consumption. Eucalypts are a prominent part of the rural landscape in southern China. There are large-scale plantations, as well as dispersed plantings on farms and on the borders of roads, railways, and waterways. In addition to the estimated 600,000 ha of eucalypt plantations, some 1 billion individual trees are planted beside farm houses and along roads and waterways (Wang 1991). The change in government policy which introduced the "responsibility" system in the early 1980s and allocated land to farmers to be managed for personal profit encouraged much private tree planting in China. The main eucalypts used are *E. exserta* F. Muell., *E. citriodora*, *E. globulus*, and a hybrid of unknown origin referred to as "*E. leizhou* No. 1." They are used because of their relatively fast growth, even on the degraded lateritic soils of southern China. The wood is used for posts and poles, furniture, farm tools, and fuel; essential oils and tannins are extracted from the leaves; and the trees support honey production. They have played an important role in improving the quality of life and income of farmers (Zheng 1988). More recently, plantations have been established on the larger forest farms for pulpwood production, and such species as *E. urophylla* and *E. grandis* are favored.

Population growth has led to a greater demand for fuelwood, poles, and other wood products. In countries like India, where fuelwood is the main source of energy for the rural population, the annual shortfall in supply is over 200 million m^3 (Puri 1990). The scale of wood shortages in many countries is such that governments alone cannot meet the needs. Small farmers and other landholders are becoming involved in providing wood from new on-farm plantings of eucalypts. Numerous surveys have found that fuelwood scarcity rarely provides sufficient incentives for people to plant trees. Although fuelwood may be valued as a secondary product, farmers are more interested in eucalypts as a source of such products as construction wood, poles, posts, essential oils, and honey, especially when these can be sold for cash. Planting for shade, shelter, and the marking of farm boundaries is also important for some farmers.

Farmers in Ethiopia plant large numbers of *E. globulus* seedlings on small areas of land and manage them to yield a variety of products, including leaves and small branches for fuelwood, and poles and posts for house building and other farm uses. Farmers who have insufficient land to have woodlots nevertheless often grow a few larger trees which can be harvested and sold when cash is required. The use of trees, especially eucalypts, as a living bank account, to be harvested when there is a need for cash, is widespread. Many people in Ethiopia are absolutely dependent on eucalypts as a source of fuel and house building material. In the West African countries of Burkina Faso, Cameroon, Ivory Coast, and Senegal, *E. camaldulensis*, originally grown to provide industrial wood, has been found to be suitable for small landholdings. For the first 3 years after planting, the trees can be grown with such crops as groundnut, maize, and cotton, thus providing growers with a source of income while reducing plantation maintenance costs. Every 4 years, 1,000 construction poles can be harvested from each hectare, and yield significant income.

Eucalypts have been grown on an increasing scale by farmers in India since 1982 and have had a major impact. The large Indian population has an immense requirement for wood for both industrial and domestic use. Annual demand for firewood is 306 million m^3 and for industrial wood, 32 million m^3 compared to the production of only 40 million m^3 and 12 million m^3, respectively (Kumar 1991). *Eucalyptus tereticornis* Smith, known locally as *Eucalyptus* hybrid or Mysore gum, is the preferred species for planting. It is used extensively for village woodlots, on farms, and along the boundaries of roads, canals, and railways. It has also been widely planted on what are broadly termed "wastelands" or degraded lands. Its popularity is based on its capacity for rapid growth, lack of palatability to cattle, great adaptability to varying environmental conditions, ready market at profitable prices, and suitability of the wood for fuel, rural housing, fences, and other farm uses (Rajan 1987; Shah 1988). Farmers have established block plantings and windbreaks on the margins of the fields, and sometimes grow eucalypts in combination with agricultural crops. As in some other countries, eucalypts are often planted to meet future contingencies and have functioned as savings banks of the rural poor (Kumar 1991). Economic benefits have generally come from eucalypt plantations, although excessive planting by farmers in Haryana and the Punjab resulted in a local glut of fuelwood and small-sized poles, and this led to falling prices. Farmers who planted early made large profits, whereas those planting later did not get similar rewards.

In India and some other countries, there is continuing controversy over the use of eucalypts for social and industrial forestry. In each country the debate takes a slightly different form, but the basic issues are very similar. There are questions about the diversion of cropland from food production, the reduction in employment, and the utilization of common lands by the more powerful members of the community. Although eucalypt growing makes economic sense from the farmer's or entrepreneur's viewpoint, the landless poor may endure increased hardship and reduced welfare when farmers grow eucalypts (Gregersen et al. 1989). In India and elsewhere the cut-off point between beneficiaries and non-beneficiaries of eucalypt planting was not between large farmers and small farmers, or between rich and poor, but between those who owned land and the landless, i.e., between the poor and the very poor (Raintree and Lantican 1993). Although the debate is frequently framed in terms of the beneficial and negative ecological impacts of eucalypts, the major problem in fact relates to land availability, tenure, and management.

An international consultation, sponsored by the Food and Agriculture Organization of the United Nations in 1993 in Bangkok to review recent findings on the social, economic, and environmental effects of eucalypts, concluded that the problems and conflicts blamed on *Eucalyptus* arise mostly from insensitive government policies on afforestation and from social injustice. Legislation and policies on planted forest development need to be considered within the framework of national forest laws and national economic and social development policies (Ball 1995).

Clearly, if eucalypts are to realize their full potential to provide benefits to a community, there will need to be much more careful analysis of ecological and social factors before extensive planting is undertaken. There is, however, great potential for innovative schemes to involve small landholders in commercial wood growing on their land. It is unlikely that the disadvantaged landless, marginal farmers, and other minority groups, will benefit from eucalypt planting unless there is the development of processing and manufacturing to provide opportunities for employment.

The future

As we enter the next century, population growth, climate change, and land degradation will be major issues. With eucalypts there is an enormous genetic resource with which to work to help address some of these issues and to meet the growing demands of developing nations for wood products. By the year 2010 the consumption of fuelwood and poles will have increased substantially in rural areas, especially in India, Kenya, Nigeria, and other densely populated countries of Asia and Africa, where much of the natural forest has already disappeared or is severely degraded (FAO 1991). To meet growing demands in developing nations there will have to be intensification of management of industrial plantations and a greater reliance on wood produced by rural populations for their own use and for industry.

There is increasing concern in developed countries that timber imports should come from forests which are managed "sustainably" and/or "equitably" and conform to certain environmental and social indicators. Most attention has been focused on natural forest management, but plantations will also face similar certification. We can anticipate in the coming years that eucalypt plantations will be increasingly subjected to this scrutiny, and that research to define appropriate sustainability indicators will be necessary.

Some of the challenges and issues facing eucalypt plantation managers and researchers are the biophysical problems that relate to nutrient management, tree improvement through new biotechnologies, and management of pests and diseases. Of at least equal importance are socioeconomic questions and the development of policies and markets which will induce greater participation of rural populations in tree growing.

Nutrient management

The main factors controlling the productivity of eucalypt plantations are soil water, nutrient availability, organic matter, effective rooting depth, and friability, as well as the quality of seed and seedlings, site preparation, and weed control (Schonau and Herbert 1989). Where rainfall is sufficient to grow eucalypts economically, it is likely that nutrient deficiency limits growth. Currently very many eucalypt plantations have suboptimal nutrition. In industrial plantations fertilization at establishment is cost effective, but research, based on previous yields and site index, to determine nutrient budgets will result in the development of computerized nutrient budget programs for various species in the coming years (Noble 1992). There is particular concern about the removal of phosphorus, potassium, calcium, and magnesium from sites during logging and in post-logging burns.

If fast-growing eucalypts are to be managed sustainably, several questions related to nutrition need to be answered. Many questions are related to current understanding of the dynamic interaction of trees and soil, particularly at the root-soil interface. What is the significance of fine root turnover in relation to nutrient dynamics, and to what extent do different tree genotypes have different efficiencies in tapping soil resources? Hendricks et al. (1993) indicate that there is a lower standing biomass of roots on more fertile sites, but much remains to be done to better define the allocation of carbon and nutrients between the belowground and aboveground parts of eucalypts. Studies on other species suggest that the management of soil organic matter through post-logging slash management, including reduced burning, will be important. Intercropping with herbaceous plants and polycultures of trees can also have effects on soil organic matter, and management prescriptions will need to be developed with sustainability as a goal. In the twenty-first century there will have to be much more active nutrition management through site conservation and amelioration practices if sustainability is to be achieved in fast-growing, short-rotation eucalypt plantations.

Tree improvement

There has been rapid development in plant biotechnology in recent years, and powerful tools which can enhance plantation production and utilization are becoming available. Relevant methods include genetic engineering to introduce exotic genes conferring resistance to bacterial or microbial pathogens; marker-assisted breeding and selection; and mass propagation through embryogenesis and other *in vitro* methods (Teasdale 1995). Genetic engineering is already a reality in agricultural products, and will have a major impact on wood production, especially from industrial plantations, in the not-too-distant future. By the turn of the century it is likely that there will be control of sterility in eucalypts, thus paving the way for the introduction of transgenes to provide resistance to attacks by insects, resistance to pathogenic diseases, and resistance to herbicides (Peacock 1995). These transgenic techniques can be applied to traits controlled by single genes, but the use of DNA markers has great potential to cope with multigene controls. The completion of a genetic linkage map for *E. nitens* in Australia has opened up the prospects for selecting trees for specified wood properties and other characteristics at a few months of age by using markers linked to these traits (Byrne and Moran 1994).

Although these new technologies offer exciting opportunities for eucalypts, they are likely to complement rather than replace conventional breeding. It will still be necessary to maintain and renew planted genetic base populations, and to improve populations by recurrent selection in the breeding population (Eldridge 1995). Interspecific hybrids of various combinations of *E. camaldulensis, E. grandis, E. tereticornis*, and *E. urophylla* have been used for a number of years in subtropical areas, but in temperate areas where frosts are a problem there is great scope for new frost-resistant combinations such as *E. nitens* × *E. globulus* (Volker 1995). There seem to be increasing prospects to further the use of controlled hybrids coupled with vegetative propagation to extend eucalypt plantings to new environments. There will also be new demands on tree breeders to produce eucalypt wood with very closely defined characteristics. For example, new pulpmills have to meet strict environmental standards, and wood with chemical characteristics that minimize effluent will be needed (Foelkel and Assis 1995).

Pests and diseases

Forest managers often express concern about the potential impact of pests and diseases on the economic viability of eucalypt plantations. Pryor (1976) suggests that one of the reasons that eucalypts are so successful as exotics is the lack of insect damage, which is so prevalent in natural forests. Nevertheless, there have been instances of serious problems, such as *Gonipterus scutellatus* in southern Africa, *Paropsis* spp. in New Zealand, and *Phoracantha semipunctata* in North Africa and southern Europe. Termites (*Termes* spp.) are a problem in many countries, and in northern Brazil leaf-cutting ants (*Atta* spp.) can completely destroy plantations unless sites are treated before planting (FAO 1979).

In the same way that there have been problems with insect pests, there have also been outbreaks of fungal diseases, although these have rarely been so severe that eucalypt planting has had to be abandoned. In India, "pink" disease, *Corticum salmonicolor*, has been the scourge of *E. globulus, E. grandis, E. citriodora*, and *E. tereticornis* (Toky and Bisht 1988). In Brazil, the stem canker, *Cryphonectria cubensis*, is an example of fungi that have caused problems (Zobel et al. 1987). And in Vietnam and Thailand, there has been major defoliation of *E. camaldulensis* by leaf pathogens in the more humid areas.

As the extent of eucalypt plantations increases, special vigilance will be needed to ensure that new pests are not introduced; if they are, there should be timely application of integrated pest management techniques. These include use of biological control agents, cultural control, resistant tree genotypes, behavioral chemicals, and application of insecticides. The application of biotechnology to produce transgenic trees with insect and pathogen resistance is likely only to be of relevance for a few industrial species. Small-scale plantings with a diversity of species will require integrated pest management techniques.

The potential for severe damage by pests and diseases is magnified when species are planted "off site". Insects have greater opportunities to cause problems when trees are under stress from drought, salinity, or other factors. Few eucalypts grow well in very humid conditions because they are susceptible to leaf diseases. Choosing species well adapted to the site's environmental conditions can be of equal importance to selecting species to give the desired wood products. It will also be sensible to retain sufficient genetic diversity in plantations to counter severe outbreaks of pests and diseases. This may be a particular problem in clonal forestry if only a restricted number of clones are used. However, used properly, the advantages of clonal forestry far outweigh the risks due to disease (Wingfield et al. 1991).

Complex plantation forestry

Industrial forestry has made great progress through simple single-species plantations and intensive management with high inputs to provide high productivity and uniform products. These plantations are usually confined to relatively small areas and, although they are economically viable, they offer limited social and environmental benefits. Such an approach has been described for the Aracruz Celulose company in Brazil. Prospects for a more complex model of plantation forestry appear good under some circumstances (Kanowski 1995). In some parts of the world, trees are grown in smaller blocks and are more integrated with other land uses. Reference has already been made to the growing of eucalypts in Brazil, the Philippines, and Spain by small landholders to provide charcoal and pulpwood for industries, as well as to produce nonindustrial products such as honey. There are great opportunities for adapting such arrangements to provide for more direct involvement of local people and to yield a more diverse range of products and services. These smaller units lend themselves to more complex management, and so may have better prospects for maximizing yields and improving the ecological environment by the use of mixtures of different species. Improved growth of eucalypts mixed with the nitrogen-fixing *Paraserianthes falcataria* has been reported (DeBell et al. 1989), and it is probable that *Acacia* and other genera could also grow satisfactorily with eucalypts and produce a beneficial outcome. There are also opportunities for management of understory plants, either as crops or to assist nutrient cycling and increase biodiversity of plantations. Complex plantation forestry will require interdisciplinary approaches to define the economic, social, and environmental contexts in which it is appropriate (Kanowski 1995). However, as large areas of land not subject to tenurial problems become increasingly difficult to locate, the prospects for greater development of more complex systems of wood production involving small landholders will become more feasible and attractive.

References

ACIAR (Australian Centre for International Agricultural Research) 1992. Eucalypts: Curse or Cure. ACIAR, Canberra.

Anonymous.1990. Around the world: Chile. Commonw. For. Rev. 69: 212.

Aracruz Celulose, S. A. 1991. Annual Report for 1990. Aracruz, Rio de Janeiro.

Ball, J.B. 1995. Proceedings of the Regional Expert Consultation on Eucalyptus. Rapa Publication 1995/6. Food and Agriculture Organization of the United Nations Regional Office for Asia Pacific, Bangkok.

Bazett, M.D. 1993. Industrial Wood. Study No. 3. Shell/World Wide Fund for Nature Tree Plantation Review. Shell International Petroleum Company Limited and World Wide Fund for Nature Tree Plantation, London.

Bertolucci, F.L.G., Demuner, B.J., Garcia, S.L.R. and Ikemori, Y.K. 1995. Increasing fiber yield and quality at Aracruz, pp. 31–34. In: Potts, B.M., Borralho, N.M.G., Reid, J.B., Cromer, R.N., Tibbitts, W.N. and Raymond, C.A. (Eds) Eucalypt Plantations: Improving Fibre Yield and Quality. Proceedings of the Cooperative Centre for Temperate Hardwood Forestry-International Union of Forestry Research Organisations Conference. The Cooperative Centre for Temperate Hardwood Forestry, Hobart, Tasmania, Australia.

Borralho, N. 1990. Around the world: Portugal. Commonw. For. Rev. 69: 18–19.

Byrne, M. and Moran, G. 1994. Breakthrough in Genetic Mapping. Onwood, Winter: 1. Division of Forestry, Commonwealth Scientific, Industrial and Research Organisation, Canberra.

Calle, F.R., Furtado, P. and Hall, D.O. 1992. Brazil charcoal. Biomass Users Network News, July–August: 6–7.

Campinhos, E. 1998. Sustainable plantations of high-yield *Eucalyptus* trees for production of fiber: The Aracruz case. In: Boyle, J.R., Kavanagh, K., Winjum, J.K. and Jensen, E.C. (Eds) Planted Forests: Contributions to the Quest for Sustainable Societies. In process.

CEASA (Celulosas de Asturias S. A.) 1994. Intelligent Fibre. World Paper, November–December. CEASA, Madrid.

Cromer, R., Smethurst, P., Turnbull, C., Misra, R., LaSala, A., Herbert, A. and Dimsey, L. 1995. Early growth of eucalypts in Tasmania in relation to nutrition, pp. 331–335. In: Potts, B.M., Borralho, N.M.G., Reid, J.B., Cromer, R.N., Tibbitts, W.N. and Raymond, C.A. (Eds) Eucalypt Plantations: Improving Fibre Yield and Quality. Proceedings of the Cooperative Centre for Temperate Hardwood Forestry-International Union of Forestry Research Organisations Conference. The Cooperative Centre for Temperate Hardwood Forestry, Hobart, Tasmania, Australia.

Davidson, J. 1985. Setting Aside the Idea that Eucalypts are Always Bad. United Nations Development Programme/FAO (Food and Agriculture Organization of the United Nations) Project BGD/79/O17, Working Paper No. 10. FAO, Rome.

DeBell, D.S., Whitesell, C.D. and Schubert, T. 1989. Using N_2-fixing *Albizia* to increase the growth of *Eucalyptus* plantations in Hawaii. For. Sci. 35: 64–75.

Eldridge, K.G. 1995. Eucalypt base populations for selections, pp. 204–207. In: Potts, B.M., Borralho, N.M.G., Reid, J.B., Cromer, R.N., Tibbitts, W.N. and Raymond, C.A. (Eds) Eucalypt Plantations: Improving Fibre Yield and Quality. Proceedings of the Cooperative Centre for Temperate Hardwood Forestry-International Union of Forestry Research Organisations Conference. The Cooperative Centre for Temperate Hardwood Forestry, Hobart, Tasmania, Australia.

Evans, J. 1985. Plantation Forestry in the Tropics. Paper presented at the twelfth Commonwealth Forestry conference. Victoria, British Columbia.

FAO (Food and Agriculture Organization of the United Nations) 1979. Eucalypts for Planting. FAO Forestry Series No. 11. FAO, Rome.

FAO (Food and Agriculture Organization of the United Nations) 1991. Wood and Wood Products. Forestry Statistics Today for Tomorrow, 1961–1989 ... 2010. FAO, Rome.

FAO (Food and Agriculture Organization of the United Nations) 1993. Forest Resources Assessment 1990: Tropical Countries. FAO Forestry Technical Paper No. 112. FAO, Rome.

Fearnside, P.A. 1989. The charcoal of Carajas: A threat to the forest of Brazil's eastern Amazon region. Ambio 18: 141–143.

Foelkel, C. and de Assis, T.F. 1995. New pulping technology and Eucalyptus wood: The role of soil fertility, plant nutrition and wood ion content, pp. 10–13. In: Potts, B.M., Borralho, N.M.G., Reid, J.B., Cromer, R.N., Tibbitts, W.N. and Raymond, C.A. (Eds) Eucalypt Plantations: Improving Fibre Yield and Quality. Proceedings of the Cooperative Centre for Temperate Hardwood Forestry-International Union of Forestry Research Organisations Conference. The Cooperative Centre for Temperate Hardwood Forestry, Hobart, Tasmania, Australia.

Gregersen, H., Draper, S. and Elz, D. 1989. People and Trees – the Role of Social Forestry in Sustainable Development. The World Bank, Washington, DC.

Hendricks, J.J., Nadelhoffer, K.J. and Aber, J.D. 1993. Assessing the role of fine roots in carbon and nutrient cycling. Tree 8: 174–178.

Iuseum, A.N., Correa de Lima, J.P. and Mercado, R.S. 1988. Brazil: The forest sector's participation in international trade, pp. 37–46. In: A. Nagy (Ed) International Trade in Forest Products. A. B. Academic Publishers, Bicester, Oxon.

Jayawickrama, K.J.S., Schlatter, V.J.E. and Escobar, R.R. 1993. Eucalypt plantation forestry in Chile. Aust. For. 56: 179–192.

Kanowski, P.J. 1995. The complex future of plantation forestry, pp. 483–487. In: Potts, B.M., Borralho, N.M.G., Reid, J.B., Cromer, R.N., Tibbitts, W.N. and Raymond, C.A. (Eds) Eucalypt Plantations: Improving fibre yield and Quality. Proceedings of the Cooperative Centre for Temperate Hardwood Forestry-International Union of Forestry Research Organisations Conference. The Cooperative Centre for Temperate Hardwood Forestry, Hobart, Tasmania, Australia.

Kanowski, P.J. and Savill, P.S. 1990. Plantation Forestry. World Bank Forest Policy Issues Paper. Oxford Forestry Institute, Oxford.

Kumar, V. 1991. Eucalyptus in the forestry scene of India, pp. 1105–1116. In: Schonau, A.P.G. (Ed) Proceedings of the International Union of Forestry Research Organisations Symposium on Intensive Forestry: The Role of Eucalypts. South African Institute of Forestry, Pretoria.

Kyotöla, T. 1989. Aracruz ... soon a 1,000,000 t/a pulp mill. Jaakko Poyry Client Magazine, January: 28–29.

Lohmann, L. 1990. Japanese sow gum trees in Thailand to feed pulpmills. Canberra Times (April) 64: 8.

Marien, J.N. 1991. Clonal forestry in Morocco: Propagation and maturation problems, pp. 126–132. In: Schonau, A.P.G. (Ed) Proceedings of the International Union of Forestry Research Organisations Symposium on Intensive Forestry: The Role of Eucalypts. South African Institute of Forestry, Pretoria.

Noble, A. 1992. Nutrition research: Where to now? Eucalypt News 4/92: 3.

Peacock, J. 1995. Improving plantation eucalypts with biotechnology, pp. 416–419. In: Potts, B.M., Borralho, N.M.G., Reid, J.B., Cromer, R.N., Tibbitts, W.N. and Raymond, C.A. (Eds) Eucalypt Plantations: Improving Fibre Yield and Quality. Proceedings of the Cooperative Centre for Temperate Hardwood Forestry-International Union of Forestry Research Organisations Conference. The Cooperative Centre for Temperate Hardwood Forestry, Hobart, Tasmania, Australia.

Poore, M.E.D. and Fries, C. 1985. The Ecological Effects of Eucalyptus. Food and Agriculture Organization of the United Nations (FAO) Forestry Paper No. 59. FAO, Rome.

Pryor, L.D. 1976. Biology of Eucalypts. Edward Arnold, London.

Puri, D.N. 1990. Production potential of degraded lands with multipurpose trees, pp. 184–190. In: Haugen, C., Medema, L. and Lantican, C.B. (Eds) Multipurpose Tree Species Research for Small Farms: Strategies and Methods. Winrock International and International Development Research Centre of Canada, Ottawa, Canada.

Raintree, J.B. and Lantican, C. 1993. Forestry economics and MPTS development, pp. 15–31. In: Thammincha, S., Puangchit, L. and Wood, H. (Eds) Forestry Economics Research in Asia. Faculty of Forestry, Kasetsart University, Bangkok; International Development Research Centre of Canada, Ottawa, Canada.

Rajan, B.K.C. 1987. Versatile Eucalyptus. Diana Publications, Bangalore, India.

Rodrigues, L.C., Silva, A.P. and de Freitas, M. 1989. Eucalypts in Brazil. Aust. For. Ind. J. 55: 22–34.

San Miguel, A. 1988. Short rotation biomass plantations, pp. 540–579. In: Hummel, F.C., Palz, W. and Grassi, G. (Eds) Biomass Forestry in Europe: A Strategy for the Future. Elsevier Applied Science, London.

Schonau, A.P.G. and Herbert, M.A. 1989. Fertilizing eucalypts at plantation establishment. For. Ecol. Manage. 29: 221–244.

Schonau, A.P.G. and R.A.W. Gardener. 1991. Eucalypts for colder areas in southern Africa, pp. 467–479. In: Schonau, A.P.G. (Ed) Proceedings of International Union of Forestry Research Organisations Symposium on Intensive Forestry: The Role of Eucalypts. South African Institute of Forestry, Pretoria.

Shah, S.A. 1988. Experiences with eucalypts in social forestry in India – lessons learnt. Proc. Int. For. Conf. Aust. Bicenten. 5: 13.

Teasdale, R.D. 1995. Optimising the application of biotechnology to forest plantations, pp. 427–429. In: Potts, B.M., Borralho, N.M.G., Reid, J.B., Cromer, R.N., Tibbitts, W.N. and Raymond, C. A. (Eds) Eucalypt Plantations: Improving Fibre Yield and Quality. Proceedings of the Cooperative Centre for Temperate Hardwood Forestry-International Union of Forestry Research Organisations Conference. The Cooperative Centre for Temperate Hardwood Forestry, Hobart, Tasmania, Australia.

Toky, O.P. and Bisht, R.P. 1988. Growing eucalypts in India. Proc. Int. For. Conf. Aust. Bicenten. 3: 22.

Turnbull, J.W. 1991. Future use of Eucalyptus: Opportunities and problems, pp. 2–27. In: Schonau, A.P.G. (Ed) Proceedings of International Union of Forestry Research Organisations Symposium on Intensive Forestry: The Role of Eucalypts. South African Institute of Forestry, Pretoria.

Volker, P.W. 1995. Evaluation of *Eucalyptus nitens* × *globulus* for commercial forestry, pp. 222–226. In: Potts, B.M., Borralho, N.M.G., Reid, J.B., Cromer, R.N., Tibbitts, W.N. and Raymond, C.A. (Eds) Eucalypt Plantations: Improving Fibre Yield and Quality. Proceedings of the Cooperative Centre for Temperate Hardwood Forestry-International Union of Forestry Research Organisations Conference. The Cooperative Centre for Temperate Hardwood Forestry, Hobart, Tasmania, Australia.

Wang, H. 1991. A review of introduction, improvement, silviculture and utilisation of eucalypts in China, pp. 144–153. In: Ministry of Forestry, Scientific and Technology Information Centre (Ed) Development of Forestry Science and Technology in China. China Science and Technology Press, Beijing.

Wingfield, M.J., Swart, W.J. and Kemp, G.H.J. 1991. Pathology considerations in clonal propagation of Eucalyptus with special reference to the South African situation, pp. 811–820. In: Schonau, A.P.G. (Ed) Proceedings of International Union of Forestry Research Organisations Symposium on Intensive Forestry: The Role of Eucalypts. South African Institute of Forestry, Pretoria.

Zheng, H. 1988. The role of Eucalyptus plantations in southern China, pp. 79–85. In: Withington, D. et al. (Eds) Multipurpose Tree Species for Small-Farm Use. Winrock International and International Development Research Centre of Canada, Ottawa, Canada.

Zobel, B.J., Van Wyk, G. and Stahl, P. 1987. Growing Exotic Forests. John Wiley and Sons, New York.

New Forests **17:** 53–70, 1999.

Douglas-fir planted forests

RICHARD K. HERMANN[1] and DENIS P. LAVENDER[2]
[1]*Department of Forest Resources, College of Forestry, Oregon State University, Corvallis, USA;* [2]*Department of Forest Science, College of Forestry, Oregon State University, Corvallis, USA*

Key words: introduced species, growth and yield, productivity, sustainability

Abstract. A combination of superior wood quality and high productivity has made Douglas-fir (*Pseudotsuga menziesii* [Mirb.] Franco) one of the premier timber trees in the world. As such, it is grown as a plantation species in several countries in Europe, South America, New Zealand, and Australia, as well as throughout its extensive natural range in western North America. Decades of experience with the silviculture of young stands have demonstrated that practices such as planting, the use of genetically improved seedlings, precommercial and commercial thinning, and fertilization may dramatically increase the yield of industrial products over that of natural forests. Further, such silviculture is compatible with the production of desired amenities. Vigorous implementation of such practices wherever Douglas-fir is cultivated will increase the world's timber resources, and be an effective strategy for reducing the pressure, occasioned by the world's rapidly increasing population, to harvest the fragile tropical and boreal forests.

Introduction

The poet Rudyard Kipling once wrote, "I keep six honest serving-men (they taught me all I knew); their names are What and Why and When and How and Where and Who." We shall work with these same "men" in our discussion of the role of Douglas-fir (*Pseudotsuga menziesii* [Mirb.] Franco) in plantation forestry.

The Douglas-fir region of the Pacific Northwest is only one of many forest regions in the world where timber harvests are being reduced, or will have to be reduced, to establish or maintain forest reserves. The demand for wood products by the world's population, expected to double to 10 or 11 billion people in the twenty-first century, will put added pressure on remaining natural forests, especially in tropical and boreal regions.

In a 1994 meeting of the Inland Empire Tree Improvement Cooperative, W. L. Libby provided a persuasive argument, even if it was indirect, for planted forests. He pointed out that financial and social costs associated with the reduction or elimination of timber harvests on commercial forest land are well recognized, whereas additional costs associated with foregone harvests have only recently become clear. These costs range from increased pollution to species extinction on undisturbed sites. To link the foregone harvests at some sites to increased productivity of the renewable wood resource at other sites suggests itself as an appropriate way to compensate for these unwanted costs. The intent of our paper is to show how that may be done with Douglas-fir in planted forests at home and abroad.

What

For purposes of this paper, plantations of Douglas-fir will be considered to be areas in which this species was established primarily by either direct seeding or planting. Generally, in managed forests of the Pacific Northwest, young Douglas-fir trees develop in association with natural seedlings of the same species or a range of other coniferous or angiospermous species. The practice of planting a single species has been criticized on the grounds that such action will replace diverse forests with monocultures. However, data from British Columbia (Province of British Columbia 1992) demonstrate that Douglas-fir plantations in Canada support at least the same diversity of coniferous species as do the original forests, because natural regeneration of cedars (*Thuja* spp.), spruces (*Picea* spp.), pines (*Pinus* spp.), and hardwoods results in significant in-growth. And, recently, concern for long-term site productivity has led to methodology for cocultivation of Douglas-fir and the nitrogen-fixing angiosperm, red alder (*Alnus rubra* Bong.). This procedure appears to have substantial promise for regeneration of some sites.

Where

Douglas-fir in western North America

The latitudinal range of Douglas-fir is the greatest of any commercial conifer in western North America. The range resembles an inverted "V" with uneven sides. From the apex in central British Columbia, the western half extends along the Pacific mountain ranges into California for about 2,200 km. The eastern half stretches along the Rocky Mountains of Canada and the United States into the mountains of central Mexico over a distance of nearly 4,500 km. The western half represents the range of the coastal variety (*Pseudotsuga menziesii* var. *menziesii*), and the eastern half, the range of the interior variety (*P. menziesii* var. *glauca* [Beissn.] Franco).

Table 1. Area of Douglas-fir in the United States by variety and year of inventory.

Year	Coastal variety[a] (1,000 ha)	Interior variety[b] (1,000 ha)	Total (1,000 ha)	Source
1936	11,209	4,681	15,890	Mattoon 1936
1952	9,927	2,920	12,847	USDA Forest Service 1958
1977	7,562	4,947	12,509	USDA Forest Service 1982
1987	8,091	6,260	14,351	Waddell et al. 1989

[a] Data from California, Oregon, and Washington.
[b] Data from the Rocky Mountain region.

The area presently occupied by Douglas-fir in the United States is listed as 14.3 million ha. That is 7.3 percent of the country's 196 million ha of nonreserved timberland. The area stocked with Douglas-fir in Canada, approximately 4.5 million ha, is slightly less than one-third of that in the United States. Figures for acreage of Douglas-fir in Mexico are not available.

With the beginning of settlement and commercial logging in the nineteenth century, the area occupied by Douglas fir did shrink, although precise figures of the amount of shrinkage are not available. The decline in the area of Douglas-fir in the United States continued well into the middle of the twentieth century (Table 1) before a reversal of that trend began. Although much of the reduction in the area occupied by Douglas-fir is a consequence of conversion of forest land to nonforest uses, failure to restock the land after harvest or forest fire also played a considerable role. The almost sole reliance on natural regeneration before 1950 often led to extremely variable stocking of harvested areas or their conversion to shrub fields where natural regeneration failed altogether. Planting has made it possible to rehabilitate a major portion of such lands, because planting provides the opportunity to circumvent a major obstacle to natural regeneration, namely a limited supply of seed. Good cone crops occur in Douglas-fir on average at intervals of 6–7 years, and, in regions such as southern Oregon, good seed years may be spaced as much as 12 years apart.

Douglas-fir as an introduced species in Europe

Of all the North American tree species introduced into Europe, none has become as important as Douglas-fir. Its successful introduction is probably the result of a fortunate combination of favorable site conditions and suitable seed sources in early plantations in Scotland, Denmark, and Germany (Matthews 1983). At present, the countries with the largest share of Douglas-fir plantations in Europe are France, Germany, the United Kingdom, and the Netherlands.

France has by far the largest area stocked with Douglas-fir. The French National Forest Inventory showed 220,000 ha occupied by Douglas-fir in 1984 (Bouchon 1984). The annual rate for planting of Douglas-fir was about 10,000 ha from 1960 to 1980. If establishment of plantations continues at this pace, Douglas-fir will occupy approximately 500,000 ha between 2010 and 2020 (Oswald and Pardé 1984). That would allow an annual cut of 6 million m^3. By comparison, the total harvest of softwood in France amounted to 15 million m^3 in 1985 (Pardé 1985).

In the mid-1980s, the Federal Republic of Germany had about 80,000 ha of Douglas-fir (Kleinschmit 1984), and the then still-existing German Democratic Republic had approximately 10,000 ha (Braun and Weissleder 1986) stocked with that species. With an annual planting rate of 5,000 ha (Kleinschmit and Bastien 1992), the area occupied by Douglas-fir probably exceeds well over 100,000 ha. That area is likely to increase further because forest planning programs allot 10–20 percent of the total forest land in Germany to Douglas-fir. That would result eventually in more than 300,000 ha by the middle of the twenty-first century.

The 1982 census of woodlands in the United Kingdom (Locke 1987) counted slightly more than 47,000 ha of Douglas-fir plantations. But, unlike in France and Germany, where it is the most important of the introduced northwest American conifers, Douglas-fir has taken a backseat to Sitka spruce (*Picea sitchensis* [Bong.] Carr.) and lodgepole pine (*Pinus contorta* Dougl. ex Loud.) in the United Kingdom. Sitka spruce and lodgepole pine accounted in 1982 for 40 and 10 percent, respectively, of the total area stocked with conifers in Great Britain, compared to a share of 4 percent for Douglas-fir. However, the Douglas-fir share may increase in the future. "In view of its high timber value, rapid growth, and resistance to butt rot, it is likely to be used on an increasing scale as the emphasis swings away from afforestation of bare ground to replanting of more sheltered valley sites, e.g., those carrying better quality Scots pine [*Pinus sylvestris* L.] and larch" (Lines 1987).

In the Netherlands, 16,000 ha of the country's 300,000 ha of forest land are stocked with Douglas-fir. The area planted with Douglas-fir is projected to grow to 55,000 ha by 2050 (deVries 1990).

Aside from these four countries, plantations of Douglas-fir exist throughout Europe, although their extent varies from country to country. In some countries, the acreage occupied by the species is very small (Table 2).

Table 2. Area occupied by Douglas-fir in Europe.[a]

Country	Area (ha)	Year	Source
France	220,000[b]	1984	Bouchon 1984
	330,000[b]	1997	Champs 1997
Germany	134,000[b]	1986	Hapla 1997; Braun and Weissleder 1986
United Kingdom	47,000[b]	1982	Locke 1987
Netherlands	16,000[b]	1990	deVries 1990
Romania	13,000[b]	1993	Enescu 1993
Belgium	12,000[c]	1993	Nanson 1993
Italy	10,000[c]	1985	Nocentini 1994
Ireland	8,200[b]	1993	Pfeifer 1994
Portugal	7,000[b]	1989	Luis 1989
Spain	5,000[c]	1993	Elena 1993
Czech Republic	3,800[b]	1991	Vancura 1991
CIS (former USSR)	2,000[c]	1985	Rudenko and Derzhnovskaya 1985
Denmark	1,600[c]	1993	Madsen 1993
Slovakia	1,200[b]	1993	Tavoda 1993
Poland	1,000[b]	1993	Mejnartowicz 1993
Hungary	350[b]	1990	Matyas 1993
Austria	100[b]	1973	Rannert 1979

[a] Countries with areas less than 100 ha occupied by Douglas-fir: Finland, Norway, Sweden; Estonia, Latvia, Lithuania; Luxembourg, Switzerland, Slovenia; Croatia, Serbia, Bulgaria.
[b] Inventory data. [c] Estimates.

Douglas-fir as an introduced species in the southern hemisphere

New Zealand has not only the largest share of Douglas-fir plantations in the southern hemisphere, but also the earliest large-scale plantings dating from 1897 (Kirkland 1969). The species, about 96 percent of which grows in state forests, has remained secondary to Monterey pine (*Pinus radiata* D. Don), mainly because of the need for a longer rotation. Private timber companies had little interest in growing Douglas-fir because timber residues were not easily converted to pulp, as was the case with Monterey pine (Anonymous 1994).

The country's plantations of Douglas-fir covered 71,000 ha in 1990, but decreased to 65,000 ha in 1991 (Neumann 1993). That decline is the result of increased harvest to supply the booming log export industry. Whether or not the area of Douglas-fir will further decrease and be replaced by faster growing Monterey pine is not clear at present. However, future prospects for

New Zealand-grown Douglas-fir have received a boost because the declining supply of Douglas-fir from the United States has raised the price of its wood in international markets, thus improving the economics of growing Douglas-fir in New Zealand (O'Loughlin 1983).

Other countries in the southern hemisphere with plantations of Douglas-fir are Chile with 9,000 ha (Instituto Forestal 1986) and Australia with approximately 1,500 ha (Griffin and Matheson 1978). Argentina had less than 100 ha in 1973, but projections were that Douglas-fir plantations would increase to 6,200 ha in the following years (Yacubson 1973). Apparently, that has just remained a forecast.

Why

Two strong arguments can be made for planted forests. Such forests permit increased productivity of the renewable wood resource and thus can contribute toward meeting the global demand for wood products. But, equally important, plantation forestry has a tremendous potential to assist with forest conservation by reducing pressure on native forests (Gladstone and Ledig 1990).

In regard to tropical forests, the latter point is well illustrated by a few figures from Libby's 1994 paper. The productivity of native tropical rain forests, with respect to harvestable wood, averages about 2 m^3/ha/yr. Tropical rain forests cover about 420 million ha and contain an estimated 8 million species of plants, animals, and other life forms. Assuming Dr. Peter Raven's estimate of 25 percent of species at risk of extinction when a regional ecosystem might be 100 percent disrupted, Libby (1994) calculated that, on average, one tropical species becomes extinct for every 210 ha disturbed ($420 \times 10^6/2 \times 10^6 = 210$). That calculation provides a hypothetical basis for determining how much wood from elsewhere would be needed to offset harvest in native tropical rain forests. This can be done by increasing productivity by 2 m^3/ha/yr on 210 ha elsewhere, or by any other combination of productivity increase per unit area that is equivalent to 2 m^3/ha/yr on 210 ha.

Natural Douglas-fir forests in the Pacific Northwest are among the most productive coniferous forests of the temperate regions, and, with intensive silvicultural practices, increases in productivity per unit area will certainly be possible. For example, McMahon (1992) notes that Douglas-fir growing on a good site (site index = 32 m) would be expected to produce a yield of 174 m^3/ha at the end of a 50-year rotation if naturally regenerated. However, a similar site regenerated with genetically improved stock, and subsequently thinned and fertilized, can produce 643 m^3/ha in 50 years. Such an increase on 50 ha would be more than sufficient to save one tropical species, given

Libby's scenario. The need for increased productivity becomes evident in the face of a progressive reduction of forest land available for timber production. Large portions of public forest land in the United States have already been allocated to other objectives, and indications are that this process will continue. That will leave mostly industrial and private timberlands for potential increases in wood production.

Ironically, plantation forestry is most often condemned in Europe, the continent where planting has reestablished forests ravaged by overexploitation in past centuries. Opponents advance doom and gloom predictions on the long-term sustainability of planted forests. Although many examples of plantation failures can be cited, critics of plantation forestry should be reminded of the successes with the practice and the benefits that may be derived from it.

Plantation forestry facilitates, in many instances, the implementation of measures that will result in increased productivity. These measures include the following:

1. Establishment of forests on land potentially suited for growing timber, but devoid of forests because of past shifts in climate or past destructive land-use practices.
2. Replacement of diseased species or species inappropriate for a particular site.
3. The use of genotypes that provide high yields, have desirable wood properties, and are resistant to diseases, insects, and air pollutants.
4. A high degree of control over stocking density from beginning to end of rotation.
5. Rehabilitation of forest lands rendered unproductive by forest fires and/or occupation by shrub and undesirable species.

An outstanding example of forest rehabilitation is represented by the 150,000-ha Tillamook State Forest of northwest Oregon. Much of this area was burned repeatedly in 1933, 1939, and 1945 by intense fires that left a wasteland of charred snags, logs, and brush (Lucia 1984). After intense salvage logging, the area was regenerated over a period of 25 years by a combination of direct seeding and planting. Today, the forest is prized by recreationists (Lucia 1984), and yields an annual wood harvest of about 500,000 m^3, a figure expected to increase to an annual harvest of 1,750,000 m^3 within the next three decades. Given that the United States is a net importer of wood, the harvest from the Tillamook State Forest represents a significant reduction in harvests in other areas. As Oliver (1994) notes, "When wood products are not produced from areas where other ecosystem values are maintained, they will be produced elsewhere in the world where ecosystem values are disregarded – with greater harm to the global environment." Indeed, if the current Tillamook State Forest harvest were to come instead from tropical forests, Libby's mathematics suggest a loss of about 1,000 species/yr. Of course, none of the foregoing would have been possible if, instead of becoming a flourishing forest, much of the Tillamook area had remained shrub-dominated with little value for forest-related uses. Such might be the case if the opposition to salvage harvests and accompanying forest revitalization that exists today had existed 50 years ago. And, interestingly, in spite of the fires and the clear-cut harvests, productivity for much of the Tillamook State Forest is higher now than was predicted on the basis of the original forest stands.

Growth

The coastal Douglas-fir region

Unfortunately, available data are insufficient for accurate predictions of yields under intensive management. Artificial regeneration with seedlings began on a large scale only about three decades ago. Thus, the majority of plantations have not gone through an entire rotation. Variations in extent of site preparation, initial spacing, early vegetative competition, animal damage, thinning regimes, fertilizer schedules, and quality of planting stock, even on comparable sites, contribute to the difficulties of quantifying realizable increases in yields. This necessitates reliance on estimates based on yields from natural stands, data from such trials as the nine cooperative levels-of-growing-stock (LOGS) studies in Oregon, Washington, and British Columbia, or upon models such as DFSIM (Curtis et al. 1981), SPS (Arney 1985), and ORGANON (Hester et al. 1989) for the coastal variety, or PROGNOSIS (Wycoff et al. 1982) and the yield tables derived from PROGNOSIS (Stage et al. 1988) for the interior variety.

Depending on site quality, mean annual increment (MAI) at age 50 ranges from 4 m^3/ha to 14 m^3/ha in unmanaged natural stands of the coastal variety that are fully stocked (McArdle et al. 1961). Estimates of gross yield, that is, inclusion of the volume of trees subject to mortality as a stand progresses in age, indicate the potential for increases of yield in managed stands. Staebler (1955) calculated increases of 15–20 percent, and Curtis (1967), of 45–85 percent over the net yields shown by McArdle et al. (1961).

Some data for yields for managed stands in the Pacific Northwest exist for rotations of less than 50 years. These data suggest that higher yields than occur in unmanaged stands can be realized. For example, in one of the LOGS studies of stands in western Oregon, Marshall (1991) recorded a net MAI of 18.6 m^3/ha at age 45 years. In that stand, 70 percent of the increment in basal area of the control plot had been retained during each of five consecutive thinnings. The MAI (net) for the control plot appeared to have culminated at about 18.9 m^3/ha at age 40 years. Elimination of all natural mortality in that plot would have increased the MAI to 23.6 m^3/ha.

The interior Douglas-fir region

Within the United States interior (longitudes 120° W to 125° W) Douglas-fir region, the species accounts for approximately 881 million m^3 of growing stock. Although this is only about half of the 1.7 billion m^3 of coastal Douglas-fir, interior Douglas-fir is, throughout its range and particularly in the northern Rocky Mountains, an important resource for the forest products industry. Although currently most attention is focused on reducing timber harvest in the Pacific Northwest, declines in the Rocky Mountain region go little noticed.

Nearly three-fourths of the total volume of interior Douglas-fir in the United States is concentrated in the "Inland Empire," that is, in eastern Washington, Idaho, and western Montana. Interior Douglas-fir grows slower and produces less volume than does the coastal variety. However, the sites that interior Douglas-fir occupies are generally better than average in productivity within the species' range. Pfister et al. (1977) estimates ranges of yield capabilities of habitat types where Douglas-fir is "climax" from about 1.4 m^3/ha/yr to over 7.0 m^3/ha/yr, and yield potential to be considerably greater, that is, more than 9.8 m^3/ha/yr on some of the wetter habitat types where Douglas-fir is seral. According to Van Hooser et al. (1991), more than one-fifth of the sites are capable of producing 6.0 m^3/ha/yr, and the overall average for all sites is in excess of 4.5 m^3 annually, compared to 4.1 m^3 for the other Rocky Mountain forest types.

Annual production in the Inland Empire is 16.8 million m^3. This is equivalent to 2.8 m^3/ha/yr, which represents 62 percent of the potential annual growth of 4.5 m^3/ha/yr attainable on those sites where interior Douglas-fir is dominant. As Van Hooser et al. (1991) point out, within the interior Douglas-fir type in Idaho and Montana, just over half of the area is actually fully stocked. An additional 15 percent is either non-stocked or contains conditions that would preclude stocking of any kind. Notwithstanding the fact that yield capabilities are not fully realized, removals until now have been less than growth by interior Douglas-fir. In eastern Washington and Oregon, harvest equaled 60 percent of net growth, and in the southern Rocky Mountains, only 17 percent of net growth has been cut. In the northern Rocky Mountains, for every 1.1 m^3 removed, 1.7 m^3 have been added to the standing inventory.

The 4 million ha in the interior Douglas-fir zone of British Columbia constitute over 90 percent of the range of *Pseudotsuga menziesii* var. *glauca* in Canada. Although this is a commercially important species in the British Columbia Interior, sufficient information about yields is unavailable. Results of a study (Bonnor 1991) conducted in the British Columbia Interior "Dry Belt" show a range of growth rates from less than 2 to more than 9 m^3/ha/yr in 63 permanent sample plots. However, values for most plots cluster around a volume growth of 4 m^3/ha/yr. According to Bonnor (1991), MAI values used by the Ministry of Forests in timber supply analysis reports range from 3.7 to 4.4 m^3/ha/yr for good sites. For medium sites, the range is from 2.0 to 2.7 m^3/ha/yr.

These estimates of productivity are, however, largely predicated on delaying harvest of forest stands until at least the age of biological rotation, i.e., when the curves describing current annual increment (CAI) and MAI cross. Some timberland owners in the Pacific Northwest currently cut trees earlier than "optimum" rotation age, as a result, in part, of current federal government policies (*The Forestry Chronicle* 1995). This practice will likely exacerbate timber shortages early in the next century.

Europe

The Pleistocene glaciations left an impoverished arboreal flora in central and northern Europe. The clearing of about 70 percent of the original forest left on the whole only the poorest sites for the remaining forest. The relatively slow growth on these sites of the four principal coniferous species that had survived the ice ages led to a search for faster growing and also higher yielding species. Trials with northwest American species that began in the second half of the nineteenth century soon indicated that the coastal variety of Douglas-fir could outperform native Norway spruce (*Picea abies* [L.] Karst.) and Scots pine. The interior variety of Douglas-fir, however, turned out to be generally unsuitable for cultivation in Europe.

The first European yield table for Douglas-fir was published in 1937 by Kanzow in Germany. Since then, additional yield tables have been published in Germany (Zimmerle 1952; Schober 1957; Foerst 1980; Kenk and Hradetzky 1984; Bergel 1986), Great Britain (Hummel and Christie 1953; Bradley et al. 1966; Hamilton and Christie 1971), France (Pardé 1956; Decourt 1967), Italy (Bernetti 1965; Cantiani 1965), Sweden (Karlberg 1961), and Portugal (Luis 1989). All of these tables are for coastal Douglas-fir.

Accurate direct comparisons of North American and European yields are not possible. The measure of productivity generally used in the Pacific Northwest is site index, based on height of dominant and codominant trees, either

Table 3. Comparison of estimated yield of 50-year-old Douglas-fir stands.

Source of estimates	Estimated yield (m^3/ha/yr) at site index (m)		
	29	38	47
Foreign countries			
Bruce 1969[a]	8.7	14.0	21.0
United States			
McArdle et al. 1961[b]	5.2	8.4	11.2
Staebler 1955[c]	6.3	9.8	13.3
Curtis 1967[c]	8.4	12.2	16.4

Note: Site index based on age 100 years.
[a] Total production of thinned plantations.
[b] Net yield. [c] Gross yield.

at age 50 or 100 years, whereas European tables rate productivity on the basis of volume produced over periods ranging from 50 to 100 years. In essence, these tables are theoretical thinning guides that list periodic thinning yield and their sum, which is total production. Indicated yields are based on data from sample plots in thinned plantations, which serve as guides to total production that can be expected on various sites.

Bruce (1969) compared yields in natural stands of Douglas-fir in the Pacific Northwest with those in European plantations of that species. He found that "total production of thinned foreign plantations (expressed as cubic feet under bark, including tip and stump) at age 50 years ranged from 75 to 400 ft^3/ac/year (5 to 28 m^3/ha/year). However, most estimates fell between 125 and 300 ft^3 (8.7 and 21 m^3). The average site index of stands producing 125 ft^3 (8.7 m^3) appears to be about 95 (29 m) – the dividing line between sites IV and V. Production of 300 ft^3/year (21 m^3/year) for 50 years is indicated on an average site index of 155 (47 m) – between sites II and III. Site index 125 (38 m) averages about 200 ft^3 (14 m^3) total production. These average estimates of total production are higher than net yield tabulated by McArdle et al. (1961), or gross yields estimated by Staebler (1955) and Curtis (1967) for natural stands in the Northwest" (Table 3). The average of yield estimates indicated by the European yield tables published after 1969 does not differ significantly from those published earlier.

The southern hemisphere

Most data on yield of Douglas-fir in the southern hemisphere come from New Zealand. They indicate that Douglas-fir plantations in New Zealand produce higher yields than do natural forests in the Pacific Northwest or plantations in Europe, except for the best sites in the United Kingdom. A comparison of growth rates of Douglas-fir in the Canterbury high country of New Zealand's South Island with those on the North Island, the United Kingdom, and Canada by Ledgard and Belton (1985) shows the amazing growth potential of the species. The net MAI at age circa 45 years in the low precipitation areas, 600–800 mm annual rainfall, in the Canterbury high country is 14.5 m^3/ha; it is 28.6 m^3/ha in the high precipitation areas, 1,200–1,400 mm annual rainfall. That compares to a net MAI of 23.1 m^3/ha at age 50 years for New Zealand site quality 1; 24.0 m^3/ha at age 50 years for the best plantations in Great Britain; and 18.6 m^3/ha at age 70 years on the best sites in southwestern British Columbia. Only Karioi State Forest in the central North Island of New Zealand has standing volumes as great as those recorded in the South Island high country. A mean standing volume of 1,573 m^3/ha – best plot = 1,949 m^3/ha – was recorded on fourteen unthinned Douglas-fir plots (average age 50 years) in this forest. Average altitude of Douglas-fir stands at Karioi Forest is 901 m. Ledgard and Belton (1985) note that the reason for such outstanding productivity in the high country is not well understood, but is likely to involve the combined effects of an even seasonal distribution of rainfall, long needle retention, and hence high leaf surface areas, high levels of radiation, and marked temperature fluctuations which are conducive to more efficient carbon fixation.

Data on production of Douglas-fir in South America are available for Chile only. The Instituto Forestal (1986) of Chile has produced a map of potential volume growth of the species showing three groups of productivity: 5–10 m^3/ha/yr; 10–15 m^3/ha/yr; and 15–20 m^3/ha/yr. Yield estimates are based on data from experimental plantations.

When

Although the first forest nursery in the Pacific Northwest opened at Wind River in 1912 and reforestation projects, utilizing 1-year-old seedlings, were conducted during the subsequent four decades, the area reforested was generally small, approximately 1,000 ha/yr (Cleary et al. 1978). However, a combination of increasingly efficient fire-control, statutory requirements, and the realization that reforestation was vital to the future of forestry in the Pacific Northwest stimulated significant increases in tree seedling production and planting. In 1950, nurseries in Oregon and Washington produced 16 million seedlings; in 1970, 80 million; and in 1990, 230 million (U.S.D.A. Forest Service 1991). Such data attest to the increasing importance of plantation forestry in the Douglas-fir region of the Pacific Northwest. Data discussed elsewhere in this paper suggest that interest in establishment of Douglas-fir plantations in other parts of the world largely parallels the foregoing.

How

A typical scenario for the establishment and growth of a Douglas-fir plantation involves the following sequence of silvicultural measures:

- Competing vegetation and organic debris are reduced to the extent necessary to permit efficient planting of tree seedlings. Until about 15 years ago, fire was commonly employed to prepare a site for planting. With the advent of smoke control regulations, however, some type of fossil fuel-powered machinery is now more frequently used, with some continued use of fire.
- One- to 3-year-old Douglas-fir seedlings grown from seed appropriate for the site are planted, most commonly, by hand labor. Concerns have been expressed that such methodology may reduce the genetic base of the species, but this hypothesis has not been confirmed by studies of seed production (Adams 1981; Howe 1989; El-Kassaby and Namkoong 1994)
- If the plantation site is characterized by aggressive competing vegetation, it may be necessary to implement a vegetation management program for several years after the seedlings are planted to ensure their maximum survival and growth. Such procedure sometimes involves the use of machinery; more often either hand slashing or herbicides are employed. It is ironic that the latter substances are often banned, for reasons discussed in McGee and Levy (1988), even though Newton and Dost (1984) document data that amply demonstrate that such chemicals have less adverse effects, from both the standpoint of human health and plantation quality, than do alternative techniques.
- On a site-specific basis, a series of precommercial and commercial thinnings are applied in conjunction with judicious pruning and application of fertilizers. Such treatments permit greater plant and animal diversity than would be obtained if stands are maintained with totally closed canopies for decades, and they stimulate maximum yields of merchantable products.

Who

Most seedlings (86 percent) grown in Oregon and Washington are produced by private or industrial nurseries primarily for use on industrial tree farms. About 20 percent of the seedlings are grown in federal nurseries for use in reforesting National Forest and Bureau of Land Management lands, and 12 percent are grown in state nurseries for plantations on both state and private lands (USDA Forest Service 1994). In British Columbia, the great majority of seedlings are planted on provincial forest lands, whereas in countries where Douglas-fir is an exotic species, plantations are generally established by governmental agencies with seed from carefully selected parent trees.

Conclusions

We believe that the material reviewed in this paper supports the following:

- Douglas-fir plantations managed with appropriate silvicultural techniques represent a premier source of forest products. Maintenance of such plantations through a range of rotation lengths from biological rotation to 250 years will provide society with, not only a continuing supply of old-growth ecosystems, but, as well, sources of high-grade timber.
- Designation of the maximum possible productive areas of the Douglas-fir forest type for plantation forestry will simultaneously permit a maximum area of forest in the Pacific Northwest for other forest land uses, as well as further global conservation by reducing pressures that result in species extinction in tropical and other fragile ecosystems.
- Although forest policy is clearly in the public domain, the site-specific methodology necessary to implement policy should be devised by well-educated resource managers using the most current and relevant scientific bases for forestry.

References

Adams, W.T. 1981. Population genetics and gene conservation in Pacific Northwest conifers, pp. 401–415. In: Scudder, G.G.E. and Reveal, J.L. (Eds) Evolution Today. Proceedings of the Second International Congress of Systematic and Evolutionary Biology. Hunt Institute for Botanical Documentation, Carnegie-Mellon University, Pittsburgh, PA.

Anonymous 1994. A brief history of Douglas-fir in New Zealand. N. Z. For. 39(1): 28–30.

Arney, J.D. 1985. A modeling strategy for the growth projection of managed stands. Can. J. For. Res. 15: 511–518.

Bergel, D. 1986. Douglasien-Ertragstafel fuer Nordwestdeutschland. 1985. Allg. Forst. Jagdztg. 157(3/4): 149–159.

Bernetti, G. 1965. Tavola cormometria della Pseudotsuga Douglasii dell'Appenino Toscano. Istituto di Assestamento Forestale della Universita Firenze. Richerche Sperimentali di Dendrometria e di Auxometria 4: 8–17.

Bonnor, G.M. 1991. A growth and yield study of interior Douglas-fir in British Columbia, pp. 269–274. In: Baumgartner, D.M. and Lotan, J.E. (Eds) Interior Douglas-fir, the Species and Its Management. Department of Natural Resource Sciences, Washington State University, Pullman, WA.

Bouchon, J. 1984. Importance des plantations de douglas et épicéa en France. Rev. For. Franc. 36(4): 254–258.

Bradley, R.T., Christie, J.M. and Johnston, D.R. 1966. Forest Management Tables. Forestry Commission Booklet 16. H.M.S.O, London.

Braun, H. and Weissleder, S. 1986. Nachkommenschaftspruefung von Douglasienbestaenden der DDR – ein Beitrag zur Nutzung eigner Reserven. Beitr. Forstwirtsch. 20(4): 180–183.

Bruce, D. 1969. Potential Production in Thinned Douglas-fir Plantations. Pacific Northwest Forest and Range Experiment Station, U.S.D.A. Forest Service, Research Paper PNW-87. Pacific Northwest Forest and Range Experiment Station, Portland.

Cantiani, M. 1965. Tavola alsometrica della Pseudotsuga Douglasii in Toscana. Istituto di Assestamento Forestale della Universita Firenze. Richerche Sperimentali di Dendrometria e di Auxometria 4: 32–73.

Champs, J. de. 1997. Le Douglas en France, pp. 43–49. In: de Champs (Coord) Le Douglas. AFOCEL, Paris, France.

Cleary, B.D., Greaves, R.D. and Hermann, R.K. 1978. Regenerating Oregon's Forests. Oregon State University Extension Service, Corvallis.

Curtis, R.O. 1967. A method of estimation of gross yield of Douglas-fir. For. Sci. Monogr. 13: 1–24.

Curtis, R.O., Clendenen, G.W. and DeMars, D.J. 1981. A new stand simulator for coast Douglas-fir. DFSIM User's Guide. Pacific Northwest Forest and Range Experiment Station, U.S.D.A. Forest Service, General Technical Report PNW-128. Pacific Northwest Forest and Range Experiment Station, Portland

Decourt, N. 1967. Le Douglas dans le Nord-est du Massif Central. Ann. Sci. For. 24(1): 45–84.

deVries, S. M. G. 1990. Use and improvement of Douglas-fir in the Netherlands. Sec. 4.37. In: Proceedings of the Joint Meeting of Western Forest Genetics Association and IUFRO Working Parties S2.02-05, 06, 12, and 14. Weyerhaeuser Company, Centralia, WA.

El-Kassaby, Y.A. and Namkoong, G. 1994. Impact of forest management practices on genetic diversity and its conservation, pp. 205–213. In: Drysdale, R.M., John, S.E.T. and Yapa, A.C. (Eds) Proceedings International Symposium on Genetic Conservation and Production of Tropical Forest Tree Seed. ASEAN-CANADA Forest Tree Seed Centre Project, Muak-Lek, Saraburi, Thailand.

Elena, R. 1993. Instituto Forestal de Investigaciones y Experiencias, Madrid, Spain. Personal communication.

Enescu, V. 1993. Romanian Institute of Forest Tree Improvement. Personal communication.

Foerst, K. 1980. Standort, Wuchsleistung und Ernaehrungszustand aelterer bayerischer Bestaende der gruenen Douglasie (*Pseudotsuga menziesii* [Mirb.] Franco, var. menziesii.) Mitteilungen der Bayrischen Staatsforstverwaltung No. 41. Bayerische Staatsforst Verwaltung, Munich, Germany.

Gladstone, W.T. and Ledig, F.T. 1990. Reducing pressure on natural forests through high-yield forestry. For. Ecol. Manage. 35: 69–78.

Griffin, A.R. and Matheson, A.C. 1978. Early height growth of some provenances from the IUFRO Douglas-fir collection in nine Australian field trials, pp. 217–228. In: Sziklai, O. (Ed) Background Papers and Douglas-fir Provenances (volume II). Proceedings of the IUFRO Joint Meeting of Working Parties S2.02-05, 06, 12, 14. Information Service Branch, B.C. Ministry of Forests, Victoria, British Columbia.

Hamilton, G.J. and Christie, J.M. 1971. Forest Management Tables (metric). Forestry Commission Booklet 34. H.M.S.O, London.

Hapla, F. 1997. How to bring into accord the silvicultural management and the end-users' interests in case of Douglas-fir. In: Zhang, S.Y, Gosselin, R. and Chauret, G. (Eds) Section

V:3-8 Timber Management toward Wood Quality and End-Product Value. CTIA/IUFRO International Wood Quality Workshop Proceedings, Aug. 18–22, Quebec City, Canada. Forintek Canada Corp.

Hester, A.S., Hann, W.D. and Larson. D.R. 1989. *ORGANON*: Southwest Oregon Growth and Yield Model User Manual. Forest Research Lab, Oregon State University, Corvallis.

Howe, G.E. 1989. Genetic effects of even-aged and uneven-aged silviculture, pp. 84–91, 216. In: Murphy, D. (Ed) Proceedings of the National Silviculture Workshop on Silviculture Challenges and Opportunities in the 1990s. USDA Forest Service, Washington, DC.

Hummel, F.C. and Christie, J. 1953. Revised Yield Tables for Conifers in Great Britain. Forestry Commission Forest Record 24. Forestry Commission, London.

Instituto Forestal 1986. Especies forestales exóticas de interés económico para Chile, pp. 139–145. In: Gerencia de Desarrollo, AF 86/32. Corporacion de Fomento de la Produccion, Santiago, Chile.

Kanzow, H. 1937. Die Douglasie. Z. Forst. Jagdwes. 69: 65–93, 113–139, 244–271.

Karlberg, S. 1961. Development and Yield of Douglas-fir (Pseudotsuga taxifolia [Poir.] Britt.) and Sitka spruce (Picea sitchensis [Bong.] Carr.) in Southern Scandinavia and on the Pacific Coast. Bulletin 34. Royal College of Forestry, Stockholm.

Kenk, J. and Hradetzky, J. 1984. Behandlung und Wachstum der Douglasien in Baden-Wuerttemberg. Mitteilungen der Forstlichen Versuchs- und Forschungsanstalt Baden-Wuerttemberg, Heft 113.

Kipling, R. The elephant's child, Stanza 1, pp. 249. In: Scribneajs, C. (Ed) The Works of Rudyard Kipling Vol. XX Just 80 Stories.

Kirkland, A. 1969. Notes on the establishment and thinning of old crop Douglas-fir in Kaingaroa Forest. N. Z. J. For. 14(1): 25–37.

Kleinschmit, J. and Bastien, J.C. 1992. IUFRO's role in Douglas-fir (Pseudotsuga menziesii [Mirb.] Franco) tree improvement. Silvae Genetica 41(3): 161–173.

Ledgard, N.J. and Belton, M.C. 1985. Exotic trees in the Canterbury high country. N. Z. J. For. Sci. 15(3): 298–323.

Libby, W.J. 1994. The role of production forestry in preventing tropical species extinctions. In: Nimbyism, Genetic Improvement, and Plantation Forestry: Growing Wood for the 21st Century. Twenty-first Annual Meeting of the Inland Empire Tree Improvement Cooperative, Post Falls, Idaho.

Lines, R. 1987. Choice of Seed Origins for the Main Forest Species in Britain. Forestry Commission Bulletin 66. H.M.S.O., London.

Locke, G.M.L. 1987. Census of Woodlands and Trees 1979–1982. Forestry Commission Bulletin 63. H.M.S.O., London.

Lucia, E. 1984. Tillamook Burn Country: A Pictorial History. The Caxton Printers, Ltd, Caldwell, ID.

Luis, J.F.S. 1989. Crescimento e competicao em povoamentos de Pseudotsuga menziesii (Mirb.) Franco em Portugal. Ph.D. dissertation. Universidade de Tras-Os-Montes e Alto Duro, Vila Real, Portugal.

Madsen, S.F. 1993. Danish Forest and Landscape Research Institute, Lynby, Denmark. Personal communication.

Marshall, D.D. 1991. The Effects of Thinning on Stand and Tree Growth in a Young, High Site Douglas-fir Stand in Western Oregon. Ph.D. dissertation. Oregon State University, Corvallis.

Matthews, J.D. 1983. The Role of Northwest American Trees in Western Europe. H. R. MacMillan Lectureship in Forestry. University of British Columbia, Vancouver, Canada.

Mattoon, W.R. 1936. Forest Trees and Forest Regions of the United States. USDA Forest Service, Miscellaneous Publication 217. USDA Forest Service, Washington, DC.

Matyas, C. 1993. University of Sopron, Hungary. Personal communication.

McArdle, R.E., Meyer, W.H. and Bruce, D. 1961. The Yield of Douglas-fir in the Pacific Northwest. USDA Forest Service Technical Bulletin 201. USDA Forest Service, Washington, DC.

McGee, A.B. and Levy, E. 1988. Herbicide use in forestry: Communication and information gaps. J. Envirn. Manage. 26: 11–26.

McMahon, J.P. 1992. The place of nutrition in forest management, pp. 1–4. In: Chappell, H.N., Weetman, G.F. and Miller, R.E. (Eds) Forest fertilization: Sustaining and Improving Nutrition and Growth of Western Forests. College of Forest Resources, University of Washington, Seattle.

Mejnartowicz, L. 1993. Polish Academy of Sciences, Institute of Dendrology, Kornik, Poland. Personal communication.

Nanson, A. 1993. Station de Recherches Forestieres, Gembloux, Belgium. Personal communication.

Neumann, A. 1993. A National Exotic Forest Description as of 1 April 1991. Ministry of Forests Report. Ministry of Forests, Wellington, New Zealand.

Newton, M. and Dost, F.N. 1984. Biological and Physical Effects of Forest Vegetation Management. Final Report. Submitted to Washington Department of Natural Resources, Olympia, WA.

Nocentini, S. 1994. University of Florence, Italy. Personal communication.

O'Loughlin, C. 1983. Forest Consultant, Wadestown, Wellington, New Zealand. Letter of 21 April 1983.

Oliver, C.D. 1994. Rebuilding biological diversity at the landscape level, pp. 95–115. In: Danielson, J. and Sampson, N. (Eds) Proceedings of the Conference on Forest Health and Fire Danger in Inland Western Forests. Harman Press, Spokane, WA.

Oswald, H. and Pardé, J. 1984. Die Douglasie in Frankreich: Waldbau und Leistung. Allg. Forstztg. 39(17/18): 438–441.

Pardé, J. 1956. Douglas et tables de production. Annales de l'Ecole Nationale des Eaux et Forêts 15(1): 137–168.

Pardé, J. 1985. Die Douglasie in Ostfrankreich. Allg. Forstztg. 40(4): 62–64, 80.

Pfeifer, A. 1994. Forestry and Wildlife Service, Bray, Ireland. Personal communication.

Pfister, R.D., Kovalchik, B.L., Arno, S.F. and Presby, R.C. 1977: Forest Habitat Types of Montana. Intermountain Forest and Range Experiment Station, Forest Service, U.S. Department of Agriculture, Ogden, Utah 84401, USDA Forest Service, General Technical Report INT-34. U.S. Govt Printing, pp. 778–814.

Province of British Columbia 1992. British Columbia's Forests: Monocultures or Mixed Forests? Ministry of Forests. Queen's Printer for British Columbia, Victoria, British Columbia.

Rannert, H. 1979. Ueber den Anbau fremdlaendischer Baumarten in Oesterreich. Centralbl. Gesamte. Forstwes. 96(2): 86–120.

Rudenko, L.P. and Derzhnovskaya, M.A. 1985. Douglas-fir in the forests of the Carpathians. For. Abstr. 47: 3965, 1986.

Schober, R. 1957. Neubearbeitung des Abschnittes Douglasie der Wiedemannschen Ertragstafel. M. and H. Schaper, Hannover.

Staebler, G.R. 1955. Extending the Douglas-fir yield tables to include mortality, pp. 54–59. In: Proceedings 1954 Society of American Foresters Meeting. Society of American Foresters, Washington, DC.

Stage, A.R., Renner, D.L. and Chapman, R.C. 1988. Selected Yield Tables for Plantations and Natural Stands in Inland Northwest Forests. Intermountain Research Station, USDA Forest Service, Research Paper INT-394. USDA Forest Service, Ogden, UT.

Tavoda, P. 1993. Slovac Forest Research Institute, Zvolen, Slovakia. Personal communication.

The Seattle Times 1995. Hatfield would sell federal timber. For. Chron. 71(2): 176.

USDA (United States Department of Agriculture) Forest Service 1958. Timber Resources for America's Future. Forest Resource Report Number 14. USDA Forest Service, Washington, DC.

USDA (United States Department of Agriculture) Forest Service 1982. An Analysis of the Timber Situation in the United States 1952–2030. Forest Resources Report FRR-23. USDA Forest Service, Washington, DC.

USDA (United States Department of Agriculture) Forest Service 1991. Pacific Northwest Region State and Private Forestry. Pacific Northwest Nursery Directory and Report.

USDA (United States Department of Agriculture) Forest Service 1994. Pacific Northwest Region State and Private Forestry. Pacific Northwest Nursery Directory and Report.

Van Hooser, D.D., Wadell, K.L., Mills, J.R. and Tymcio, R.P. 1991. The interior Douglas-fir resource: Current status and projections to the year 2040, pp. 9–14. In: Baumgartner, D.M. and Lotan, J.E. (Eds) Interior Douglas-fir, the Species and Its Management. Department of Natural Resource Sciences, Washington State University, Pullman, WA.

Vancura, K. 1991. Internal report. Zbraslav, Czech Republic: Czech Forest Research Institute.

Waddell, K.L., Oswald, D.D. and Powell, D.S. 1989. Forest Statistics of the United States, 1987. Pacific Northwest Research Station, USDA Forest Service, Resource Bulletin PNW-RB-168. USDA Forest Service, Portland, OR.

Wykoff, W.R., Crokston, N.R. and Stage, A.R. 1982. User's Guide to the Stand Prognosis Model. Intermountain Research Station, USDA Forest Service, General Technical Report INT-133. U.S.D.A. Forest Service, Ogden, UT.

Yacubson, D. 1974. The problem of forest seeds in Argentina. IUFRO WP S2.01-06 Seed Problems, pp. 10. In: 1973 Proceedings of International Symposium on Seed Processing (volume 2), Paper 28. Royal College of Forestry, Stockholm.

Zimmerle, H. 1952. Ertragszahlen fuer gruene Douglasie, Japaner Laerche und Roteiche in Wuerttemberg. Mitteilungen der Forstlichen Versuchs- und Forschungsanstalt Baden-Wuerttemberg, Heft 113.

New Forests **17:** 71–88, 1999.

Loblolly – the pine for the twenty-first century

ROBERT P. SCHULTZ
Southern Forest Experiment Station, USDA Forest Service, USA

Key words: pine plantations, tree improvement, growth and yield, multiple use management, forest health, sustainability

Abstract. Loblolly pine (*Pinus taeda* L.) was a minor component of the vast natural forests of the southern United States before the region was settled by immigrants. Extensive planting and natural regeneration of cutover forest land and abandoned farmland between 1930 and 1990 made loblolly the leading timber species in the United States. It now predominates on 13.4 million ha (45 percent) of the commercial forest land in the southern United States (between latitudes 28° N and 39° N and longitudes 75° W and 97° W) and directly or indirectly provides 110,000 jobs and $30 billion to the economy of the region. The extreme versatility of loblolly has also provided important environmental contributions to most southern states. These include landscape beautification, erosion control, soil amelioration, excellent wildlife habitat, and outstanding recreational opportunities. Incorporating existing pest management strategies into silvicultural systems can produce substantial and long-lasting insect and disease control with little cost or physical effort. Introductions of loblolly into numerous countries around the world (especially China, Brazil, and Argentina) have proven to be very successful, environmentally sound, and commercially profitable. In some locations, loblolly grows much faster than on sites of similar quality in the southern United States.

Introduction

Loblolly pine (*Pinus taeda* L.) is the major timber species in the southeastern United States. It predominated on 13.4 million ha (45 percent) of the commercial forest land in the southern United States in 1989. In contrast, less than 2 million ha of predominantly loblolly pine forest were scattered throughout the southern United States prior to European settlement of North America. Loblolly pine has thrived in association with the human species.

In the original forests of the South, loblolly pine was a minor component of both the uplands, which were dominated by longleaf pine (*P. palustris* Mill.) or mixed upland hardwoods, and the wet river bottoms and swamps, which were dominated by mixed bottomland hardwoods. However, loblolly was an important component of moist sites that were not subject to regular burning throughout the South.

The combination of intense logging of the original forests in the late 1800s and early 1900s, coupled with expanded fire control and bareroot planting beginning in the 1930s, permitted loblolly pine to become established on an array of upland sites throughout its natural range. Loblolly's easy, quick regeneration and rapid growth soon made it the favored forest tree species throughout the southern United States. It has also become a favored conifer in parts of Asia, Africa, and South America, where nearly one-half billion seedlings are planted annually.

Current U.S. range

Loblolly pine grows naturally in fifteen southern and mid-Atlantic states from central Florida (28° N latitude), northward to Delaware and New Jersey (39° N, 75° W), and westward to east Texas and southeast Oklahoma (97°30′ W). Its range overlaps the natural ranges of longleaf, slash (*P. elliottii* Engelm.), shortleaf (*P. echinata* Mill.), pitch (*P. rigida* Mill.), and Virginia (*P. virginiana* Mill.) pines, but only shortleaf has a wider geographic range than does loblolly.

Loblolly pine reaches merchantable size throughout its natural range, except in southcentral Georgia and the lower Coastal Plain of Florida, Mississippi, and Alabama, where it is important only in localized areas and along the mountain fringes of the mid-South and in the extreme north-eastern part of its range (Figure 1). Large-scale plantings and direct seeding have extended the commercial range throughout much of Tennessee, western Arkansas, and eastern Oklahoma. Even in southern Illinois, plantings have grown at the rate of 3.8 m^3/ha/yr over a 27-year rotation (Gilmore and Funk 1976).

At the southern limit of the species' range, the average monthly temperature variation is only 10 °C throughout the year, whereas temperature variation reaches 25 °C at the northern limit. In addition, monthly rainfall varies from as little as 25 mm in Oklahoma and Texas to more than 200 mm in Florida, but is quite uniform in the northern part of the range. This wide range in habitat provides extensive evidence that loblolly is well adapted to prosper in a wide range of environments.

Historical patterns of loblolly pine culture

Forty years of intense pine harvesting prior to 1930 destroyed 90 percent of the original 48.6 million ha of southern pine forests and left 25 percent of the

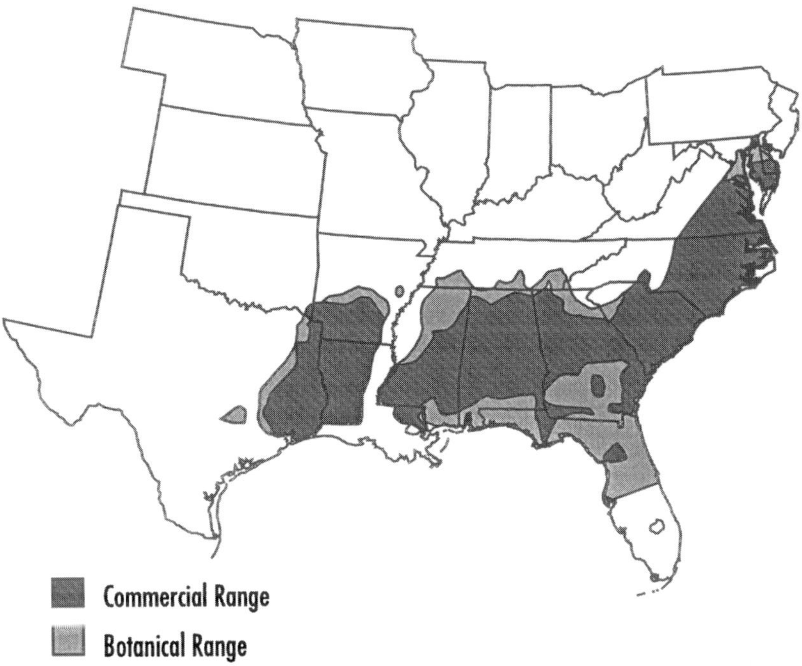

Figure 1. Botanical and commercial range of loblolly pine (*Source*: Goebel 1974).

entire region without natural reproduction or seed trees (Mann 1969). Artificial regeneration, by either planting or direct seeding, was the only means to get this land back into pine production. A few small loblolly pine plantations can be traced back to as early as 1873. By 1931 there were 29,565 ha of southern pine plantations, many of which were loblolly pine. During the early part of the twentieth century, large areas of the loblolly pinery, primarily in the Piedmont, but also on well-drained Coastal Plain soils, were developed on degraded farmlands. Most of these lands had been cropped for extensive periods without the benefit of chemical amendments or water and erosion control (Trimble 1974), and were "worn out" – highly eroded and depleted of nutrients.

Large-scale seedling production began in the mid-1930s when the USDA Forest Service began operating its Ashe Nursery in southern Mississippi. During the Great Depression of the 1930s, the Civilian Conservation Corps planted over 200,000 ha to pines during a 7-year period. These early nursery operations and outplantings proved that loblolly pine could be artificially regenerated on a large scale and that it would grow rapidly on many diverse sites. This information base, coupled with favorable markets for softwood products and state and federal incentive programs, as well as industry's

Table 1. Volume of loblolly pine growing stock on commercial forest lands in the south by State.

State	Volume (million m^3)			
	1960	1970	1980	1989
Alabama	153.4	155.7	167.2	179.7
Arkansas	61.8	76.6	107.2	111.3
Florida	12.3	14.3	16.4	17.3
Georgia	108.8	162.4	184.4	196.7
Louisiana	85.8	134.4	172.3	199.7
Mississippi	57.4	106.1	141.5	152.0
North Carolina	121.7	135.6	146.2	158.2
Oklahoma	1.0	1.7	2.3	3.5
South Carolina	83.7	115.4	160.1	156.9
Tennessee	1.1	5.2	7.6	10.9
Texas	87.0	126.2	144.6	145.2
Virginia	60.8	58.4	63.0	69.0
Total	834.8	1,092.0	1,312.8	1,400.4

commitment to short rotation, high fiber yield forestry, and tree improvement programs, resulted in an exponential increase in loblolly pine seedling production during the mid- to late-1900s.

Loblolly pine growing stock has increased from about 0.6 billion m^3 in the late 1940s and early 1950s (Wahlenberg 1960) to 0.8 m^3 in 1960, 1.1 billion m^3 in 1970, 1.3 billion m^3 in 1980, and 1.4 billion m^3 in 1989 (Table 1). In 1989, loblolly pine comprised one-half of the total southern yellow pine growing stock volume (approximately 1.4 billion m^3) with a total dry biomass of 1 billion metric tons.

A steep increase in loblolly pine growing stock of about 0.2 billion m^3 per decade, from 1950 to 1980, was reduced by more than one-half during the 1980s. Factors contributing to this decline include failure to establish loblolly pine after logging, declining forest area and increasing mortality in older stands on public and non-industrial private forest lands (Sheffield et al. 1985; Williams and Gresham 1986).

Monetary benefits to society from past investments in loblolly pine regeneration have been substantial. Foster (1982) estimated that for every \$100 of public moneys spent to regenerate pines, society will benefit from a \$3,800 reduction in the total price of the wood products it consumes in the year 2020 – a compound rate of return that exceeds 10 percent after inflation. However,

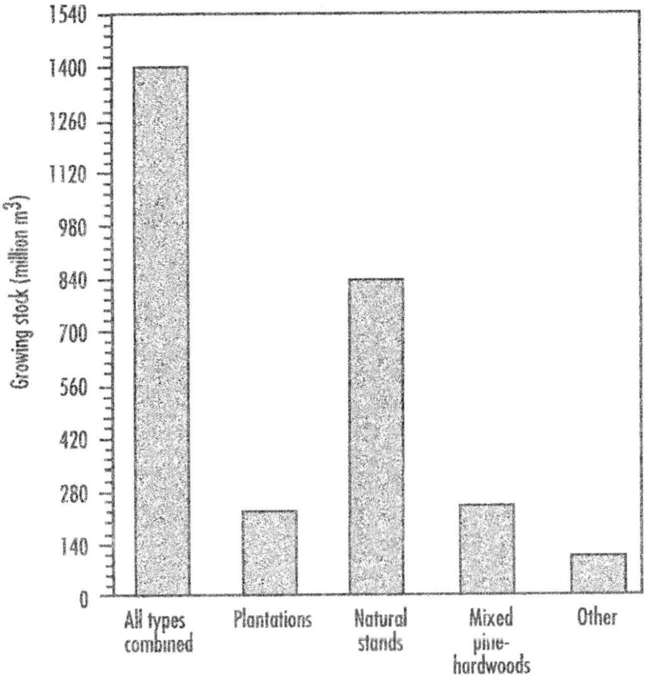

Figure 2. Loblolly pine growing stock by forest management type in 1989.

regeneration costs and investment schedules must be analyzed carefully over an entire rotation on the basis of site productivity and expected product values. The present annual value of loblolly pine to the southern economy is about $30 billion, including the direct or indirect employment of 110,000 people.

Current growing stock, ownership, and regeneration

In the United States, loblolly pine makes up over one-half of the total southern yellow pine volume. Fifty-seven percent of the total loblolly volume is located in the Midsouth, Texas to Alabama, and 43 percent in the Atlantic Coastal states. Of the 1.4 billion m^3 of growing stock in 1989, 59 percent was in natural stands and 16, 18, and 7 percent in plantations, mixed pine-hardwoods, and other types, respectively (Figure 2). The largest concentration of loblolly pine is in central and northern Louisiana, southern Arkansas and eastern Texas, where each of 57 contiguous counties and parishes has over 2.8 million m^3 of growing stock.

The majority of the merchantable loblolly pine growing stock (12.7 cm at diameter breast height [dbh] and above) is in the 20–41 cm diameter

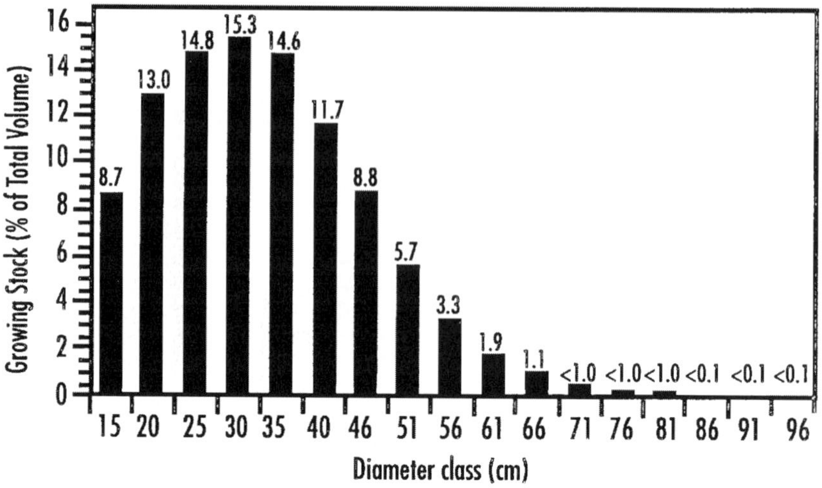

Figure 3. Loblolly pine growing stock as a percentage of total volume by diameter distribution in 1989.

classes (Figure 3). About 0.9 billion m^3 (or 63 percent) of this growing stock volume belongs to approximately 3 million nonindustrial private forest (NIPF) landowners, each of whom holds an average of 13 ha (Rosson and Doolittle 1987). Industry owns an additional 27 percent of the growing stock, and 10 percent is in public ownership (Figure 4). Forest industry leases and manages, or provides landowner assistance on nearly 3.6 million ha of NIPF lands across the South (Meyer et al. 1986; Smith 1989), and many of these lands are either pure loblolly pine, or mixed stands with loblolly pine as a primary or secondary species.

Across the South, between 1952 and 1985, the amount of naturally regenerated pines on NIPF holdings steadily decreased from about 36 percent to about 21 percent. Although about one-half of this loss was offset by planting pine (mainly loblolly), the remainder of the unstocked sites reverted to natural upland hardwoods. A similar decrease occurred on industrial holdings (a reduction from 42 to 20 percent); however, all of the loss was made up by planting pines (Alig et al. 1986). At least 75 percent of the unstocked or poorly stocked forest land in the South belongs to NIPF owners (Birdsey and Pitcher 1986; Rosson and Doolittle 1987). Seventy percent or more of loblolly pine stands are within 0.3 km of a road (Rudis et al. 1984) making access easy for harvesting, which is common and frequent, especially on privately owned lands. Partial cuts and natural regeneration prevail on nonindustrial lands, whereas clearcutting and planting are routine on industry holdings.

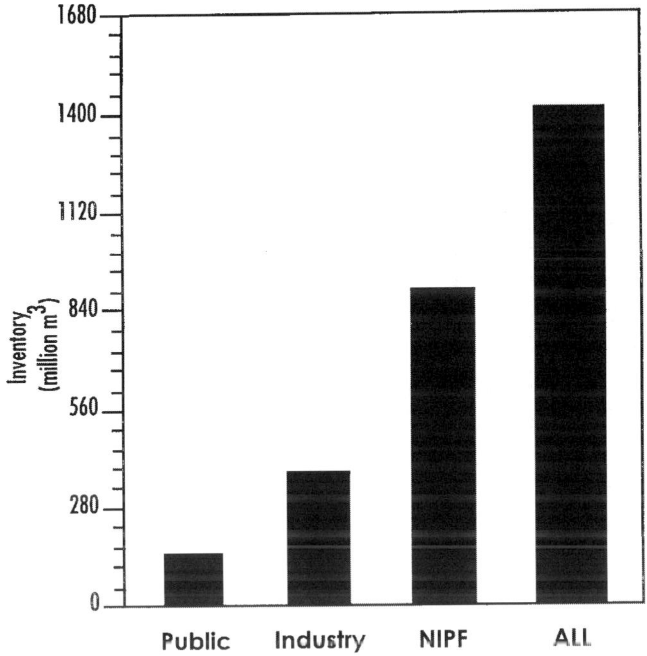

Figure 4. Loblolly pine inventory by ownership class in the Southern United States in 1989.

In 1974 loblolly pine seedling production was 337 million (Abbott and Fitch 1977). In 1980 it was almost 1 billion. This was 60 percent of all nursery stock produced in the U.S.A. and 75 percent of that in the South. Between 1985 and 1993 annual production and planting of loblolly pine bare-root seedlings across the South ranged from 1 to 1.5 billion, about 80 percent of all pine planting there. More than 85 percent of these seedlings were from genetically improved sources (Lantz 1988).

Successful artificial regeneration of loblolly pine requires systematic planning and organization to merge nursery and field procedures and to get rapid seedling development. Seed collection and processing, nursery seedling development, and outplanting or direct seeding must fit into a careful scheme that includes evaluation of site quality and site preparation options, and matching seedlings to a specific environment. For maximum volume growth on most sites, at least 1850 healthy seedlings per hectare should be planted where competing vegetation has been eliminated, but where topsoil and organic matter have not been moved from the immediate rooting zone. With bare-root seedlings early mortality can average 30 percent (Schultz 1975).

Containerized seedlings are a relatively recent addition in loblolly pine regeneration, and, if produced in quantities of 3 million or more, can be cost competitive with bare-root stock. They can be grown any time of the year to meet unexpected demands or to extend the planting season for excessively dry or wet sites. Container loblolly seedlings often have better survival than bare-root stock.

Direct seeding is much less expensive than planting and provides quick and reliable regeneration of large areas of difficult terrain or following wildfire. Although it does not provide as high assurance of success as planting, seeding may be the best or only viable option for some situations such as rough mine spoils or very rocky sites.

Tree improvement

Loblolly pine tree improvement is having a major impact on southern forestry. Vast areas of superior trees are now established across the South, with anticipated growth increases of at least 7–13 percent in volume at rotation age. Significant gains during the first generation of genetic improvement have also been made in terms of wood density, tree form, wood quality, and fusiform rust resistance. Progeny of rogued first-generation orchards may result in stand value gains of 18–32 percent at harvest time, depending on site quality, management type, rotation age, and product marketed. Unless improved seedlings are specifically designed to excel with vegetative competition or other adverse conditions, some soil or vegetative manipulation must be carried out to obtain the best growth rates.

Future challenges and opportunities are now concentrated in second- and third-generation selections and in advanced cloning techniques, with tissue and individual cell propagation greatly speeding the process of genetic improvement. Growth and value increases from these selections should surpass that of first-generation tree improvement. Hybrids between loblolly and other southern pines also offer opportunities for improving growth under specific environmental conditions.

Management, growth and yield

It is practical to manage loblolly pine plantations in rotations of 25 to at least 120 years, depending on objectives. However, some control of competing hardwoods, shrubs, vines and herbaceous plants is necessary during the first several growing seasons in order for young loblolly to compete for resources and to become well-established.

The most cost effective method of hardwood control is prescribed burning (Franklin 1985). Control of only large hardwoods can double pine production and increase financial returns by more than 200 percent (Balmer et al.

1978). Herbaceous weed control in new plantations can be economically advantageous if rotation length is reduced by only one year (Busby 1989). However, controlling woody competition is generally more cost effective than controlling herbs (Dangerfield and Merck 1989).

Stand density, site quality, competing vegetation, and tree improvement each can have significant effects on loblolly pine tree growth and yields. Stocking levels needed to maximize profits are much lower than levels that optimize physical production. On average sites (site index = 24 m, base age 50 years), where integrated products are grown, maximum merchantable yield can be obtained with initial establishment densities of 1,730–1,975 trees/ha and by maintaining basal area stocking between 20 and 23 m^2/ha.

Commercial thinning can begin as early as age 10 years on high sites and continue at intervals of from 5 to 7 years. Commercial thinning before age 25 may reduce total yields somewhat, but will increase average tree diameter and product value. To maximize the amount of high-quality sawtimber in crop trees, the first thinning should be at about age 15, when trees no longer produce juvenile wood. After about age 20, well-managed stands can produce 5.6–7.0 m^3/ha/yr of high quality sawtimber for many years. By age 30, total yields can be from 294 to 378 m^3/ha, with merchantable yields of 252–327 m^3/ha, much of which will be sawtimber.

Phosphorus (P) fertilization can greatly increase seedling growth on wet Coastal Plain sites. Nitrogen (N) fertilization alone or together with P fertilization can increase the growth of loblolly from seedling stage to maturity on many soils in the Coastal Plain, Piedmont, and Mountain provinces throughout the South. Application of as little as 56 kg P/ha, or 56 kg P/ha and 40 kg N/ha may result in optimum growth of newly established trees. Increases in tree volume of 5–89 percent for at least 3–5 years can occur when 56–168 kg N/ha are applied to pole-sized loblolly stands on clay soils.

Multiple use management

In addition to wood products, loblolly pine plantations and individual trees have many uses. Loblolly provides diverse wildlife habitat, cattle forage, erosion control, water quality improvement, land reclamation and rehabilitation, agroforestry, aesthetics, and a wide array of recreational opportunities.

Wildlife

Loblolly pine and associated grass, herb, shrub, and tree species are important to the well-being of many of the 400 species of vertebrates – including amphibians, reptiles, birds, and mammals – across the South. Interspersion of forest types provides the best all-around wildlife habitat. Loblolly pine plantations of modest sizes and variable ages should be interspersed with hardwood stands, and/or mixed pine-hardwoods, and narrow strips of forest cover should be maintained along all permanent streams (streamside management zones). Mixed pine-hardwood stands provide good habitat for many wildlife species. Hardwoods, in both the overstory and the understory, are especially valuable in promoting diverse populations of wildlife species. Extensive unbroken tracts of pure loblolly do not provide good habitat for most wildlife.

In loblolly pine plantations, the number of wildlife species and individuals increases as vertical layering increases. To provide a high level of both plant and animal diversity, adjacent loblolly pine plantations should be small, irregular, and separated by a minimum of 7 years in age. Uncut, or sparsely cut, management zones (50–100 m wide) should be maintained along streams to improve wildlife and fish habitat, increase water quality, and improve the aesthetic quality of clearcuts. Early succession, following harvest cutting, is especially rich and diverse in its vegetative composition and variability, and makes good-to-excellent habitat for a multitude of game and nongame wildlife species. The effect is most pronounced the first 3–5 years after clearcutting and site preparation.

Cultural treatments, such as burning, thinning, pruning, and fertilization, all increase abundance of wildlife food and cover. The diversity and quantity of good wildlife habitat decrease as a stand ages and the overstory crown cover closes. Cultural treatments can stimulate renewed use by wildlife. Increases last only about 1 year following burning or fertilizing, but may last as long as 3 years following combinations of burning, thinning, and fertilizing. Open, regularly burned stands are crucial habitat for some species, such as bobwhite quail (*Colinum virginianus*), gopher-tortoise*(Gopherus polyphemun*), eastern indigo snake (*Drymarchon corais couperi*), and the red-cockaded woodpecker (*Picoides borealis*). Older loblolly stands provide high-energy seeds in the fall that are especially important to the well-being of birds and rodents.

Grazing

An estimated 21.8 million ha of forest range are available for cattle grazing in the loblolly-shortleaf pine-hardwood ecosystems, stretching from east Texas to northern Virginia. If planned and managed carefully, integrating wildlife, livestock, and loblolly pine production offers an opportunity to increase economic returns to a landowner, especially during the first 10 years of a plantation, when forage levels are high. Cattle damage to pine seedlings can be minimized by decreasing grazing or excluding cattle during winter and spring until sufficient forage is produced. If cattle grazing is planned throughout a pine rotation, fewer trees/ha should be planted and, once trees reach 15 cm dbh, stands should be thinned at 5-year intervals to maintain open conditions. Other techniques include judicious burning to move livestock continuously, or supplemental feeding, which concentrates feeding in areas not used for timber production. Some sacrifice in wood yield will occur in forage-timber operations because lightly stocked stands will not produce high volume yields. However, large, high-quality sawlogs can be grown in combination with continuous beef production.

Water quality and erosion control

Loblolly pine stands are very effective in protecting water quality, but will reduce water yields because of high evapotranspiration. Soil erosion is generally less than 180 kg/ha/yr in undisturbed watersheds, most from stream channels. Harvesting and site preparation cause temporary, but sometimes very high, increases in storm flow and peak discharge rates. Best management practices (especially revegetating severely degraded sites; constructing dips and bars along forest roads to control water; seeding, fertilizing, and mulching unprotected areas; and leaving residual trees in streamside management zones) may cause a direct loss of 3–5 percent in gross timber sales revenue (Lickwar et al. 1990), but should be cost effective in terms of improved water quality, aesthetics, and wildlife protection.

Loblolly pine has been used throughout the South to control erosion. The Piedmont region of the Southeast includes many thousands of hectares of overworked, abandoned farms that are characterized by little or no topsoil, low fertility, poor soil physical properties, and continuous erosion. Planting loblolly at about 2,200 trees/ha can effectively stabilize these soils and bring them back to a productive state, especially when some form of site amelioration is used. Subsoiling is effective in promoting tree growth on compacted sites. The addition of dried sewage sludge can also improve pine growth on some of these sites (Berry and Marx 1976; Berry 1979).

Even without fertilization, planting loblolly may be very profitable on eroded slopes. For example, 2 ha of eroded submarginal hillside in the North Carolina Piedmont were planted or direct seeded to loblolly and shortleaf pines in 1927. Thirty-seven years later, total yield (standing plus removals) was 390 m^3/ha or 10.6 m^3/ha/yr. In addition, erosion was halted and the soil was improved (Maki 1964).

Loblolly pine is an excellent species for stabilizing new roadcuts, borrow pits, and lands denuded by air pollution. Loblolly can stabilize soil faster and grow better than other conifers and most hardwoods on many types of mine spoils. For example, planted loblolly seedlings grow well on a mixture of overburden fill material in lignite mine pits in Louisiana, so there is no need to keep topsoil separate during the mining process (Burton and Tiarks 1986).

Eroded loess soils: The Yazoo-Little Tallahatchie story

In the 1800s, exploitive agriculture on rolling erosive soils in the hills of western Tennessee and western Mississippi (the Yazoo-Little Tallahatchie river basins) changed an area of pine-hardwood forests to the most severely eroded portion of the eastern United States. Millions of hectares of uplands suffered extensive loss of productivity. Debris from eroded hillsides buried rich bottomlands, and extreme runoff caused frequent and serious flooding. Research was begun in the 1940s to stop the soil loss and to develop means of rejuvenating the land. Planting proved to be the best method for establishing trees on eroded sites because seeds are too easily washed from slopes by rain. Among all the conifers planted on eroded loess ridges, only loblolly pine survived well and grew rapidly. Over 918 million loblolly pine seedlings were eventually planted on more than 338,000 ha, between 1948 and 1982, demonstrating that severely degraded sites can be returned to a productive state by proper handling and careful planting of quality seedlings of the appropriate species (Williston 1988).

Agroforestry

The production of agricultural crops and loblolly pine simultaneously on the same site is feasible for several years in new plantations. Trees use very little space during the first 2–3 years, so crops, such as soybeans, tomatoes, watermelons, and corn, can be grown between the rows of pines. This type of operation provides immediate economic returns that defray the costs of plantation establishment. To make such a system work, extensive changes in plantation spacing and management operations are necessary. For example, a spacing of 5–8 m between tree rows provides better opportunities for culturing several rows of annual crops than does a customary spacing of 3 m between rows. Some loss in wood productivity can be expected, but financial analyses indicate that intercropping can be more profitable, under some conditions, than growing pines alone (McNeel and Stuart 1984).

Planting one or two rows of closely spaced (1.0–1.2 m apart) trees on one or more sides of a field can provide substantial protection for agricultural crops where winds are a continuous problem (Maki and Allen 1978). Some loblolly pine genotypes may be more wind-tolerant than others. Breeding for wind resistance could ultimately produce trees that are especially suited to windbreak development (Telewski 1986).

Visual impacts

Landscape management, or emphasis on visual effects, has only recently become an important part of forest management. It is becoming increasingly important for managing loblolly pine even in the Coastal Plain, where opportunities for species diversity in the overstory are limited and few spectacular changes in the landscape exist.

The greatest visual impact occurs during the harvest and regeneration period – a cutover area covered with debris next to a tall stand of trees can be displeasing. As distance from a cutting unit increases, visual differences and concomitant displeasure gradually fade. Generally, the scars left from harvesting and intensive site preparation are hidden by rapidly growing vegetation within 2 years. To promote visual appeal, clearcuts should be preplanned to merge with other stand types and ages, should conform to the land contour, and should include uncut streamside management zones that lessen the stark changes. Site preparation treatments that minimize initial disruption of soil and understory vegetation and speed site recovery and tree growth are best from aesthetic and recreational viewpoints. Some large, old trees and standing snags may be kept in cutover areas for their visual effects. Similarly, it is aesthetically pleasing to view dogwoods (*Cornus florida* L.) or other attractive hardwood trees and shrubs in the understory or midstory of pine stands along roadways or in special use areas.

The contrast between clearcuts and adjacent stands can also be made to appear less harsh by thinning along the edges of adjoining stands and maintaining small trees and shrubs in this transition zone. When harvesting, leaving vegetation in and around ponds will provide diversity within cutting areas and minimize the overall visual effect of tree removal. Clearcuts along skylines and at higher elevations should be avoided. A shelterwood or selection cut can be used to retain a forest landscape in front of openings, especially along well-traveled roads. Alternatively, uncut buffer strips can be retained along highways to minimize the effect of clearcutting. Once a new stand is established treatments such as thinning and prescribed burning do not normally pose significant visual problems. Prescribed burns, properly timed and implemented, result in extremely short visual impact periods. Thinning normally presents little change in color, form, or texture in the overall landscape. Small openings should be promoted in uneven-aged stands to provide contrasting effects. Long rotations generally increase the aesthetic appeal of a stand because large, old trees are especially desirable.

Hull and Buhyoff (1986) compared the visual appeal of four loblolly pine management practices and concluded that natural stands were of highest scenic quality, followed by unthinned planted stands and heavily thinned planted stands. Lightly thinned stands were the least scenic. The researchers concluded that scenic beauty should not be limited to a single year of a stage in the life of a stand, but rather summarized over an entire rotation. This would improve the visual assessment of newly harvested stands that are unduly criticized for 2–3 years of very low visual quality. Similarly, lengthening a rotation will increase the scenic beauty available over the life of a stand.

Forest health

Damage from pests and environmental stresses have become increasingly severe in southern pine forests over the past 50 years. This is principally the result of human changes in ecosystems from mixed species to rapidly growing even-aged stands of a single species, such as loblolly pine. Research has shown that such development has altered many natural balances that previously kept pathogenic organisms in check in ecosystems.

Major diseases affecting loblolly pine include fusiform rust (*Cronartium quercuum* [Berk.] Miyabe. ex Shirai. f. sp. *fusiforme*) and annosus root rot (*Heterobasidion annosum* [fr.] Bref.). Both predominate in plantations. Annosus root rot damage is greatest following thinning in pole-sized stands. It usually infects newly cut stumps and destroys roots in pockets surrounding infection courts. It can be effectively controlled with good stand management. Fusiform rust commonly forms galls on the stems of nursery seedlings and on both stems and branches of planted seedlings and saplings. Growing galls often badly deform and weaken loblolly pine stems, thus causing early mortality. Newly regenerated trees are less susceptible to fusiform rust under a pine overstory than in open, planted areas, so shelterwood or all-age management may be preferable on some high risk sites. Control of fusiform rust in loblolly pine stands has improved significantly over the past decade. Systemic fungicides can now protect nursery seedlings from rust. Disease resistant seedlings, especially from east Texas and south Louisiana, have reduced infection in plantations by more than 50 percent.

The southern pine beetle (*Dendroctonus frontalis* Zimm.), the black turpentine beetle (*D. terebrans* Oliver), and pine engraver beetles (*Ips* spp.) are the most destructive insects of loblolly pines. In most years, southern pine beetles account for more than 80 percent of the total economic loss caused by insects. Their success in killing trees can be partly attributed to their symbiotic relationships with fungi. Beetle-induced fungi are thought to be important in overcoming host resistance and are also important for larval nutrition. Beetles readily attack loblolly, whether in pure stands or in mixtures with hardwoods. When stands are rated for potential beetle hazard, the most susceptible are generally older stands with high stocking densities.

However, as much as 50 percent of the bark beetle infestations in loblolly are associated with lightning-damaged trees, attributable to numerous summer electrical storms.

Insects other than bark beetles account for less than 5 percent of the total annual insect losses of loblolly pine. New loblolly pine regeneration is most seriously affected by pales weevil (*Hylobius pales* [Herbst.]), whereas primary damage in sapling stands is caused by the Nantucket pine tip moth (*Rhyacionia frustrana* Comstock).

Incorporating existing pest management information into silvicultural systems can produce substantial and long-lasting benefits with little additional cost or physical effort. This can be done by careful planning and a knowledgeable assessment of the risk of attack for each stand or area.

International importance of loblolly pine resources

Loblolly pine is now a valuable forest resource in Africa, Asia, and South America. Some 400 million seedlings are presently being planted annually on these continents. From 200 to 300 million seedlings/yr are planted in China, and more than 100 million/yr are planted in Brazil. Smaller but locally important numbers of loblolly are planted annually in Argentina, Uruguay, South Africa, and Zimbabwe. Most of these countries have made long-term commitments to continuous loblolly pine regeneration and management. Trial plantings have been made in more than a dozen other countries.

The keys to successfully growing loblolly pine in exotic locations are tree improvement and proper site selection. In some locations in Brazil and Argentina loblolly stands are in their second rotation and growth rates can reach 25–39 m^3/ha/yr; these rates are three times greater than those of similar stands in the United States. Outstanding growth and vigor of planted loblolly pines have also been achieved on degraded sites, old fields, and pastures in many locations. Careful control of planting sites has also resulted in superior growth for a variety of products, including cattle, annual crops, and perennial crops, as well as pulpwood and sawtimber (e.g., Schreiner and Baggio 1984). Fertilization is generally not needed for rapid growth on most sites, and pest control measures are not needed, except where leafcutting ants (*Atta* spp. and *Acromyrmex* spp.) are a problem.

Although loblolly pine is very adaptable, it has not been used to its full potential in many exotic locations. Many plantings have failed because the wrong seed sources were used or because plantings were made in environments for which the species or the seed sources were not adapted. Seed sources from the southernmost United States perform especially well in subtropical climates. In some cases, hybrids may outgrow pure loblolly in harsh environments. For example, pitch pine x loblolly pine hybrids grow faster than the pitch pine female parent and are more cold hardy than the loblolly male parent in South Korea. (Hyun and Ahn 1959; Hyun 1972, 1974; Han et al. 1987a, 1987b). As genetically improved trees become more available and the geographic diversity of the species is more fully recognized and used, interest in loblolly as a superior exotic pine should continue to increase.

Summary

Loblolly pine is a major forest tree species and predominates on 45 percent of the commercial forest land in the southeastern United States. It is an ideal tree for forest management and site restoration. It is hardy, versatile, and grows rapidly on diverse sites. It has proven to be especially valuable in rejuvenating sites that have been physically abused, and even destroyed, by human activities. It seeds profusely, is easy to plant, provides large yields/hectare, and can be used for many different marketable products at a relatively early age. On good sites, 15–25 cm diameter trees can be removed by thinning at 10–15 years of age. Final harvest can occur anywhere between ages 25 and 120 years, depending on management objectives. Loblolly pine is extremely important to the overall economy of the southern United States, directly or indirectly providing 110,000 jobs and $30 billion to the economy.

Loblolly pine planted forests provide excellent wildlife habitat and offer outstanding recreational opportunities when stands of many ages are grown in close proximity. Growing loblolly for several successive rotations on an area will not reduce site productivity as long as tree branches and needles are not removed from the site during harvest, care is taken to minimize soil compaction and erosion, and surface soil is not removed from planting sites during site preparation.

Introductions of loblolly into other parts of the world at similar latitudes to its native range (especially in China, Brazil, and Argentina) have proven very successful and commercially profitable. In some locations trees grow much faster than on sites of similar quality in the southern United States.

References

Abbott, H.G. and Fitch, S.D. 1977. Forest nursery practices in the United States. J. For. 75: 141–145.

Alig, R.J., Knight, H.A. and Birdsey, R.A. 1986. Recent Area Changes in Southern Forest Ownerships and Cover Types. Research Paper SE-260. Southeastern Forest Experiment Station, USDA Forest Service, Asheville, NC.

Balmer, W.E., Utz, K.A. and Langdon, O.G. 1978. Financial returns from cultural work in natural loblolly pine stands. South. J. Appl. For. 2(4): 111–117.

Berry, C.R. and Marx, D.H. 1976. Sewage sludge and *Pisolithus tinctorius* ectomycorrhizae: Their effect on growth of pine seedlings. For. Sci. 22(3): 351–358.

Birdsey, R.A. and Pitcher, J.A. 1986. Management options for poorly stocked stands in the south central United States. South. J. Appl. For. 10(2): 73–77.

Burton, J.D. and Tiarks, A.E. 1986. Available nutrients and early growth of woody plants vary with overburden material in lignite minesoils of Louisiana, pp. 79–85. In: Proceedings of the National Meeting of the American Society of Surface Mining and Reclamation. American Society for Surface Mining and Reclamation, Princeton, WV.

Busby, R.L. 1989. Economic returns using sulfometuron methyl (Oust) for herbaceous weed control in southern pine plantations, pp. 359–364. In: Proceedings, 5th Biennial Southern Silvicultural Research Conference, 1988. Memphis, TN. Gen. Tech Rep. SO-74. U.S. Department of Agriculture, Forest Service, Southern Forest Experiment Station, New Orleans, LA.

Foster, B.B. 1982. Taxpayers gain from southern pine regeneration programs. South. J. Appl. For. 6(4): 188–193.

Franklin, E.C. 1985. How to manage your woodlands profitably. For. Farm. 44(5): 14–15.

Gilmore, A.R. and Funk, D.T. 1976. Shortleaf and loblolly pine seed origin trails in southern Illinois: 27-year results, pp. 115–124. In: Proceedings of the 10th Central States Forest Tree Improvement Conference. Department of Forestry and Natural Resources, Purdue University, West Lafayette, IN.

Goebel, N.B. 1974. Growth and yield of planted loblolly pine. In: Proceedings, Symposium on Management of Young Pines 1974 October 22–24; Alexandria, LA/December 3–5, Charleston, SC, Asheville, NC; USDA Forest Service, Southeastern Forest Experiment Station 264–275.

Han, Y.C., Hong, S.H., Sohn, S.I., Ahn, J.K. and Byoun, K.O. 1987a. Studies on late growth performance of *Pinus rigida* × *Pinus taeda*. Res. Rep. Inst. For. Genet., Korea Republic 23: 11–15.

Han, Y.C., Lee, K.Y., Choi, S.K., Ryu, K.O., Chang, S.O., Song, H.U., Noh, C.C., Chang, S.K., Kim, K.S. and Youn, K.S. 1987b. Growth performance of 24-year-old loblolly pine (*Pinus taeda* L.) in five locations. Rep. Inst. For. Genet., Korea Republic 23: 96–100.

Hull IV, B.R. and Buhyoff, G.J. 1986. The scenic beauty temporal distribution method: An attempt to make scenic beauty assessments compatible with forest planning efforts. For. Sci. 32(2): 271–286.

Hyun, S.K. 1972. The possibility of F utilization in pine hybridization, pp. 1–10. In: Proceedings of the IUFRO genetics-SABRAO Joint Symposia. Government Forest Experiment Station, Tokyo, Japan.

Hyun, S.K. 1974. Pine hybridization and performance of hybrids in Korea. Indian J. Genet. 34A: 375–386.

Hyun, S.K. and Ahn, K.Y. 1959. Principal characteristics of × Pinus rigitaeda. Res. Rep. Inst. For. Genet., Korea Republic 1: 35–40.

Kline, W.N. and Kidd, F.A. 1986. Economics of brush control in loblolly pine plantations, pp. 324–334. In: Weed Science and Risk Assessment: Proceedings, Southern Weed Science Society 39th Annual Meeting. Southern Weed Science Society, Champaign, IL.

Lantz, C.W. 1988. Overview of southern regeneration, pp. 1–7. In: Proceedings of the 1988 Southern Forest Nursery Association. Southern Forest Nursery Association, Charleston, SC.

Maki, T.E. 1964. From poverty grass to pine timber. Res. Farm. 23(1): 4–5.

Maki, T.E. and Allen II, L.H. 1978. Turbulence characteristics of a single line pine tree windbreak, pp. 81–92. In: Proceedings of the 37th Annual Meeting of the Soil and Crop Science Society of Florida. Soil and Crop Science Society of Florida, Gainesville, FL.

Mann, W.R., Jr. 1969. Techniques and progress in regenerating southern pines. For. Prod. J. 19(8): 10–16.

McNeel, J.F. and Stuart, W.B. 1984. Feasibility of Agri-Silviculture for Pine Plantations in the South. Paper 84-1609. American Society of Agricultural Engineers, St. Joseph, MI.

Meyer, R.D., Klemperer, W.D. and Siegel, W.C. 1986. Cutting contracts and timberland leasing. J. For. 84(12): 35–38.

Rosson, J.F., Jr. and Doolittle, L. 1987. Profiles of Mid-South Nonindustrial Private Forests and Owners. Resource Bulletin SO-125. Southern Forest Experiment Station, USDA Forest Service, New Orleans, LA.

Rudis, V.A., Rosson, J.F. and Kelly, J.F. 1984. Forest Resources of Alabama. Resource Bulleton SO-98. Southern Forest Experiment Station, USDA Forest Service, New Orleans, LA.

Schreiner, H.G. and Baggio, A.J. 1984. Intercropping of corn (Zea mays L.) with *Pinus taeda* L. planted stands located in Southern Parana, Brazil. Boletin de Pesquisa Florestal, Curitiba 8/9: 26–49.

Schultz, R.P. 1975. Intensive culture of southern pines: maximum yields on short rotations. Iowa State J. Res. 49(3): 325–337.

Sheffield, R.W., Cost, N.E., Bechtold, W.A. and McClure, J.P. 1985. Pine Growth Reductions in the Southwest. Resource Bulletin SE-83. Southeastern Forest Experiment Station, USDA Forest Service, Asheville, NC.

Smith, D.W. 1989. Industry-sponsored landowner assistance programs. For. Farm. 48(5): 41.

Telewski, F.W. 1986. Physiological responses of trees to mechanical perturbation: Implication for genetic evaluation for wind tolerance, pp. 251–252. In: Proceedings of the International Symposium on Windbreak Technology. Great Plains Agricultural Council Publication 117. Cooperative Extension Service, Montana State University, Bozeman, MT.

Wahlenberg, W.G. 1960. Loblolly Pine: Its use, Ecology, Regeneration, Protection, Growth and Management. School of Forestry, Duke University, Durham, NC. 603 p.

Williams, T.M. and Gresham, C.A. 1986. Factors important to loblolly pine productivity on the southeastern Coastal Plain of the United States, pp. 177–186. In: Agren, G.I. (Ed) Predicting Consequences of Intensive Forest Harvesting on Long-Term Productivity. Report 26. Institutiionen for Ekologi och Miljovard, Sveriges Lantbruksuniversitet, Uppsala, Sweden.

Williston, H.L. 1988. The Yazoo-Little Tallahatchie Flood Prevention Project: A history of the Forest Service's role. Forestry Report R8-FR 8. Southern Region, USDA Forest Service, Atlanta, GA. 63 p.

New Forests **17:** 89–93, 1999.

Planted forests: poplars

PAUL E. HEILMAN
Western Washington Research and Extension Center, Washington State University, Puyallup, WA, USA

Key words: hybrid poplars, poplar culture

Abstract. Poplars (*Populus* spp.) have been planted in Europe and Asia since very early times. Known in the Near East as the "blessed tree", poplars have been the primary timber producer in regions lacking natural forests in the northern hemisphere. Now, though, we are seeing poplar plantations becoming part of the forest resources even in the abundantly forested Pacific Northwest. Though still a minor contributor to the world's timber supply, the area of land planted to poplars is, nevertheless, increasing rapidly, particularly in China, South Korea, and the United States. The 1992 report from the International Poplar Commission listed nineteen countries with at least 10,000 ha of planted poplars and seven with more than 100,000 ha. Much of the success of poplar plantations results from the breeding of fast-growing and disease-resistant poplar hybrids. Accordingly, at least eleven countries support poplar breeding programs. Uses of poplar wood range from peeled poles for rafters and other elements of construction in agrarian economies to the manufacture of paper, plywood, oriented strand board, and engineered lumber in industrial nations. Interest has also developed in poplar plantations as a renewable source of energy.

Introduction

Poplar (*Populus* spp.) culture dates to very early times. In countries of Europe, poplar growing has a relatively short history, but was firmly established early in the eighteenth century when the desirability of hybrids between European and North American cottonwoods was recognized (FAO 1980). In this paper, I present a picture of world poplar culture, describe some botanical and genetic characteristics of poplars, list the uses of the wood, and finally present a short overview of poplar culture as it is being practiced in the Pacific Northwest.

What is poplar?

Poplar is a general term for the trees of the genus *Populus*. Ecological preferences of the species in the genus can vary from moist, riparian sites to

Table 1. Sources of *Populus* hybrids. Many successful hybrids have come from crossing eastern cottonwood (*Populus deltoides*) with the following species.*

P. nigra L.	*P. trichocarpa* T. & G.	*P. maximowiczii* A. Henry
P. ciliata Royle	*P. szechuunica* Schneider	*P. yunanensis* Dode
P. balsamifera L.		

*Breeding programs have also developed hybrids among the above listed species as well as among aspens and between aspens and white poplars.

well-drained uplands. Some species, like *P. trichocarpa* Torr. & Gray (black cottonwood), can tolerate a wide range of sites. The genus contains twenty-nine different species (Eckenwalder 1996). In addition, the genus contains a large number of hybrids and numerous named and unnamed hybrid clones. For the most part, poplars are native to northern and temperate regions of the northern hemisphere down to about latitude 30°N. Nevertheless, poplars have been widely planted beyond their natural range in both the southern and the northern hemispheres.

The genus *Populus* is in the family Salicaceae, which includes willows (*Salix* spp.), poplars (or cottonwoods), and aspens. A characteristic of this family is that the plants are dioeceous, meaning that they have either male or female flowers. Flowers of both males and females are arranged as catkins. The seeds are very small and are shed attached to fine cottonlike fibers, which allow them to be carried long distances by wind. Although large numbers of seed are produced, vegetative propagation from stem and branch segments, and from roots in some species, is an important means of both natural and artificial reproduction.

The genus *Populus* is divided into five sections, with three being the most important from the standpoint of poplar cultivation (FAO 1980). These three sections are Leuce, which includes the aspens and white poplar (*P. alba* L.); Aigeros, which includes the "true" poplars and is represented by the two species, Lombardy poplar (*P. nigra* L.) and eastern cottonwood (*P. deltoides* Bartr.), which together with their hybrids make up over 90 percent of all cultivated poplars of the world; and finally Tachamahaca, the balsam poplars. In this section is the Pacific Northwest native, *P. trichocarpa*, which has been featured in our poplar breeding and plantation research effort of the past 16 years (Stettler and Heilman 1984).

Poplar culture

Many countries with limited natural forests use poplar from plantations as an important source of timber. Therefore, it is not surprising that in the Near

East, for instance, they are known as "blessed trees" (FAO 1980). However, poplar plantations are not limited to such areas. Even here in the conifer-rich Pacific Northwest, large-scale plantations of poplar hybrids have begun to appear.

Relative to other planted forest trees, eucalypts (*Eucalyptus* spp.) and pines (*Pinus* spp.), for instance, the world's area of poplar plantations is small. According to the International Poplar Commission (1992), seven countries reported more than 100,000 ha of poplar plantations. These countries are China, France, Germany, Hungary, Romania, Turkey, and Yugoslavia. Five countries reported between 30,000 and 100,000 ha, and seven reported from 10,000 to 30,000 ha of poplar plantations. The International Poplar Commission (1992) also reported that eleven countries have poplar breeding programs. These countries are Australia, Belgium, Canada, China, France, Germany, Italy, the Netherlands, New Zealand, Turkey, and the United States. Part of the reason for such interest in breeding is the ease of vegetative propagation of poplar. With stem cuttings, superior hybrids can be easily and quickly cloned and, within a very short time, made available for planting on a large scale.

Although hybrids constitute the primary interest for poplar plantations, in some situations clones of the pure species can be favored. A notable example is eastern cottonwood, which can be the best choice in the warm and humid climates of subtropical lower latitudes. In India, for example, clones of *P. deltoides* can be the best choice. Other examples are individual selections from *P. nigra*. The widely planted Lombardy poplar is a clone of the species *P. nigra* used for ornamental and windbreak planting. Clones with similar columnar form in the species *P. alba* are also widely planted.

Perhaps as much as 90 percent of the poplar hybrids employed in plantations are derived from the species *P. deltoides* mated with the following species: *P. nigra*, *P. trichocarpa*, *P. maximowiczii* Henry, *P. ciliata* Royle, *P. szechuunica* Schneider, *P. yunanensis* Dode, and *P. balsamifera* L. (balsam poplar). Breeding programs have also developed hybrids among the above-listed species, as well as among aspens and between aspens and white poplars. Most such hybrids are from crosses with *P. nigra* or European black cottonwood. Another successful cross with *P. deltoides* is that with *P. trichocarpa*, our own native cottonwood. One of the real advantages of these hybrids over the pure *P. deltoides* parental species is the substantially better rooting and survival from cuttings. Another advantage is higher productivity, which is the result of both faster tree growth and the ability of the hybrid to tolerate closer spacing in plantations. In the Pacific Northwest, the *P. trichocarpa* × *P. deltoides* cross has produced most of the better hybrids.

However, crosses of *P. trichocarpa* with *P. nigra* and *P. maximowiczii* show considerable promise.

Advantages and risks of poplar culture

Several features of poplar make these trees desirable for plantations. The first is the ease in producing planting material. Stool beds for cutting production are easily established from cuttings. In the dormant season, 1-year-old shoots from nursery plants are cut into segments 15–20 cm long for planting stock. Alternatively, extra shoots and branches from production plantations can be pruned and used as sources of cuttings. Larger material, up to 7 or more cm in diameter, can be desirable in special situations, and will survive and grow surprisingly well. Also, whips consisting of the entire 1-year shoot up to 3 m in length or more can be planted in the ground, with high survival rates. Such large material is useful in wild sites where brush and animal browsing can cause problems.

Another advantage of poplars is the ease of planting. On adequately prepared sites, where rows are marked, it is possible for one person to plant 5,000 or more cuttings/day. In fact, a worker in eastern Washington planted about 10,000 cuttings in 1 day. On appropriate sites and with proper site preparation and weed control, plantations are very productive. The final advantage is that relatively short harvest cycles are possible. For pulp, rotations can be as short as 4 years, but practice is dictating 6–7 years for this purpose. For solid wood products, harvest cycles can range from 10 to 20 years or longer, depending on the rate of growth and the product to be manufactured (Heilman et al. 1995).

Uses of poplars vary between agrarian and industrial societies. In agrarian societies, peeled but otherwise unprocessed poles are used for rafters, posts, and beams in construction (FAO 1980). Poplar wood is also used for fuel, although other species are often preferred. Finally, branches and leaves can provide feed for livestock. Processing in industrial societies gives paper, lumber, veneer, plywood, and engineered wood products, including oriented strand board.

Despite the above advantages, poplar growing is not without risk (Dickmann and Stuart 1983; Heilman et al. 1995). Grass and other competing vegetation in young plantations can significantly reduce growth and productivity. Grazing animals, rodents, beavers, and porcupines find poplars quite tasty, as do many species of insects (Ostry et al. 1989). Also, poplar can be susceptible to a variety of fungal and bacterial diseases (Ostry et al. 1989). Developing clones resistant to such agents provides continuing challenges for poplar breeders. Growers must be aware of these risks and manage plantations to minimize their impacts (Dickmann and Stewart 1983, Heilman et al. 1995).

Conclusions

Poplar plantations provide a source of wood for nonforested areas of the world, and also can be important in supplying wood in areas where natural forests have been the primary source. With public sentiment in favor of preserving most remnants of natural forests, it is important to find alternatives for wood production. Poplar plantations offer one option in supplying wood because these plantations can be integrated with other land uses, including agriculture, in ways that both enhance agricultural uses and protect environmental values. That poplar culture has been associated with agriculture since antiquity should give us hope that, not only is poplar culture sustainable, but possibly agriculture, too, is more sustainable when poplar growing is included with it.

References

Dickmann, D.I. and Stuart, K.W. 1983. The Culture of Poplar in Eastern North America. Hickory Hollow Associates, Dansville, MI.

Eckenwalder, J.E. 1996. Systematics and evolution of Populus, pp. 7–32. In: Stettler, R.F., Bradshaw, H.D., Heilman, P.E. and Hinckley, T.M. (Eds) Biology of Populus and Its Implications for Management and Conservation. NRC Research Press, O Hawa, Ontario, Canada.

FAO (Food and Agriculture Organization of the United Nations). 1980. Poplars and Willows in Wood Production and Land Use. FAO, Rome.

Heilman, P.E., Stettler, R.F., Hanley, D.P. and Carkner, R.W. 1995. High-yield Hybrid Poplar Plantations in the Pacific Northwest (revision). Pacific Northwest Regional Extension Bulletin PNW 356. Washington State University Cooperative Extension, Pullman, WA.

International Poplar Commission. 1992. Synthesis of national reports on activities related to poplar and willow areas, production, consumption and the functioning of national poplar commissions. Note from the Secretariat, FO:CIP:Misc/92/1, Nineteenth Session, 23–25 September 1992. Zaragoza, Spain.

Ostry, M.E., Wilson, L.F., McNabb, H.S., Jr. and L.N, Moore. 1989. A guide to Insect, Disease and Animal Pests of Poplars. USDA Forest Service Agriculture Handbook 677. USDA Forest Service, Washington, DC.

Stettler, R.F. and Heilman, P.E. 1984. Short rotation poplar culture in the Pacific Northwest: Components of a genetics program, pp. 217–232. In: Perttu, K.L. (Ed) Ecology and Management of Forest Biomass Production Systems. Report 15. Department of Ecology and Environmental Research, Swedish University of Agriculture Science, Uppsala, Sweden.

New Forests **17:** 95–109, 1999.

The need for planted forests and the example of radiata pine

WILLIAM R.J. (WINK) SUTTON
Consultant, Plantation Focus Limited, 24 Kerswell Terrace, Roturua, New Zealand

Key words: wood supply, Monterey pine, New Zealand forests, sustainability

Abstract. Ten thousand years ago our 5 million hunter-gatherer ancestors used, over the period of a year, about 200 species of local food items. Now, the nearly 6 billion global population gets 70 percent of its food from nine species of plants, one species of bird, and two species of mammals. This is despite global access to about 50,000 edible plants, as well as large numbers of bird and other animal species. We have concentrated our agricultural efforts on those species which are the most desirable, the most convenient, the most productive, and the most easily stored. Of equal importance, the species must be responsive to both genetic improvement and to management manipulation (i.e., domestication). Globally, very little of our food, except fish, now comes from wild sources.

Although we use at least as much wood as we do food, wood supply (which still comes mostly from natural forests) is only now becoming limiting. There are no energy-efficient or environmentally acceptable substitutes for wood. The most likely means we have of satisfying the increasing global demand for wood is to obtain more of our wood from deliberately planted forests. Management of planted forests will increasingly parallel intensive agriculture. One consequence is that, of the 1,000 tree species we now use globally, we may eventually get most of our wood from four or five species. One of these tree species will be radiata pine (*Pinus radiata* D. Don). Radiata grows fast on a wide range of sites, has desirable wood properties, and responds well to both genetic improvement and stand management. The planted forest technology for radiata pine is probably the most advanced of any potential tree species. The evolution and success of radiata pine in New Zealand has major implications and lessons for the rest of the world.

The evolution of agriculture

About 10,000 years ago there were about 5 million people on earth. There was probably little or no tilling of the land, no deliberate planting of crops, and no husbandry of animals. There was no agriculture as we know it today. There was little or no storage of food. Our ancestors were hunter-gatherers. They survived on the naturally occurring plants and animals within the local forests. Although the available food varied enormously among places, the total number of food sources appears to have been similar. North American

Indians, Australian aborigines, and Kalahari bushmen all used from 200 to 250 species of food items (plants, birds, and other animals) over the course of a year.

The population of the world has now increased over a 1,000-fold – to nearly 6 billion. We no longer have to rely on locally available natural foods. Most of us now have potential access to the entire food biodiversity of the earth. For the plants alone that means we have potential access to about 50,000 species of edible plants (United Nations estimates). However, we have been very selective. Worldwide we now use only 200–250 plant species in any significant way for food production (i.e., one in approximately 200 of the plants known to be edible!). In reality we have been even more selective. According to that same United Nations report, 70 percent of our plant food comes from just nine plants (probably wheat, maize, rice, potato, barley, cassava, soya bean, sugar cane, and oats). This means that most of our plant food comes from less than one in every 5,000 of the world's edible plants!

This same incredible selectivity is found in our use of birds and mammals. Most of the world's bird meat comes from just one bird, the chicken. Of the mammals, pigs and cows dominate. Sheep and goats are less important. Only with fish (the source of which is still largely from the wild) do we not have this dominance of a few species.

On first consideration the present situation appears to be a major paradox. We now have access to the total biodiversity resource of the world, and yet we are becoming more and more reliant on fewer and fewer species. However, it may not be such a paradox. When our hunter-gatherer forefathers had to use only what was available locally, they were forced to use almost everything that was edible and usable. Now we can be far more choosy. We have concentrated our advanced agricultural practices on those few foods which are highly desirable, very productive, and pleasant to eat. Also, the food must be easy to harvest, distribute, store, and prepare. Fern roots provide an example of this change. These plants were an important food item for some earlier cultures. They required a great deal of preparation, contained a high proportion of indigestible fibers, and had little flavor. In large quantities fern roots are poisonous. They were only eaten because more desirable foods were not available. When more convenient, more productive, and better tasting foods, like the potato, became available, no one continued to eat fern roots.

To be a success in agriculture a food species had to be more than just convenient and desirable. A food species also had to be responsive to both genetic improvement and domestication. Responsiveness to genetic improvement provides the ability to:

- Enhance desirable aspects (such as flavor).
- Reduce inconvenient, unpleasant, or undesirable aspects (such as sourness, poisons, and shattering – the tendency for naturally occurring cereal seed to break away easily from the seedhead).
- Increase overall yields, as well as the proportion of desirable parts.
- Improve uniformity of the species (e.g., in terms of the timing of maturity, as well as the size of the product).
- Breed increased resistance to disease and insect attack.
- Increase fertilizer response.
- Hybridize and/or breed new varieties.

Clonal propagation and genetic engineering now offer further potential for greater and faster genetic improvements.

The other attribute required is the ability to be domesticated. This ability tends to be overlooked, especially because it is easily confused with the ability to respond genetically. It should not be. By domestication, we mean the ability to respond to management. Chickens and pigs are good examples. Their ability to be domesticated, at least in part, explains why these two animals have been so successful. The advantage is not just the ability to respond genetically (which of course they both do), but also includes the abilities to:

- Reproduce quickly and easily (and at any time of the year).
- Be easily herded.
- Live in confined conditions.
- Grow rapidly and efficiently on readily available food sources.

For the plant food species, successful domestication requires the abilities to:

- Reproduce quickly (either from seed or vegetatively).
- Grow on a wide range of sites.
- Respond to management (e.g., site preparation, fertilizer, and tending).

There are probably hundreds of species of plants, birds, and mammals that may be very responsive to genetic manipulation. The diversity of cultivars in our parks and flower gardens demonstrate this. So also do the many different breeds of our pets. However, the reason we have been so selective when it comes to food production is that we emphasize the ability of a species to be domesticated.

Our agriculture and animal husbandry have become so successful that now almost all the world's food (except fish) comes from managed crops, birds, and mammals. This modern agriculture has meant that earth can support a global population of nearly 6 billion. We enjoy an unprecedented access to food and a freedom from threat of famine.

The importance of wood

Everyone recognizes our need for food, and hence the need for agriculture. In contrast, very few recognize the importance of wood in the global economy. Wood is still the major energy source for the majority of the world's population, and the burning of wood as a fuel is still the major use of wood. The other wood uses are all industrial. Of these the sawlog for the production of sawntimber (and plywood) is the most important. Some wood is still used in the round for poles, posts, and pit props.

The pulp and paper industry attracts a great deal of interest, but my analysis of world wood demand is that the industry is not directly a major user of wood from "nonplanted" forests. Wood supplies for the pulp and paper industry come mostly from the following sources:

- Small wood and the lower value tops of trees left after the removal of higher quality sawlogs.
- Wood residues from the sawmilling and plywood industry.
- Pulpwood plantations.
- Natural forests managed like planted forests.

The pulp and paper industry uses less than 20 percent of the global industrial wood harvest, and most of this is less desirable or unacceptable for other uses.

Statistics from the Food and Agriculture Organization of the United Nations (FAO 1995) have the world using about 3.5 billion m^3 of wood. To give that huge volume a more meaningful expression I usually exhibit a wedge of wood with the volume of 1.8 liters. That wedge represents the average global daily per capita consumption of wood (which includes fuelwood use). Every day the world consumes 5.6 billion times that amount of wood. (Further details are given in Appendix 1.) Wood usage in the world is uneven. Advanced economies tend to use more wood than the global average. The United States, for example, uses about three-and-a-half times the global average (Bowyer 1992). Canada and New Zealand have per capita consumption levels just a little less.

This raises a major issue. Most of our food comes from annual crops, and, therefore, very short time horizons are needed to plan for an increased food harvest. Wood, of which we use a great volume, requires decades in which to grow. Also, all wood substitutes (steel, cement, plastics, and aluminum) are far less environmentally friendly than wood, and require a great deal more energy (Koch 1992; Sutton 1993). Many wood substitutes also involve mining, pollution, and the atmospheric release of fossil carbon from fossil fuels and limestone (as in the case of Portland cement).

Why has wood been almost totally ignored?

Why has wood, especially when it is so vitally important to us, been almost totally ignored in both national and international planning? For an explanation let us go back to the hunter-gatherer era. In the absence of agricultural clearance, the earth's surface was mostly forest covered. Forests dominated because trees have evolved as very competitive, extremely efficient, and long-lived organisms. Take any land where the climate is not at the extreme of very cold, very wet, or very dry. Keep the area relatively free of both grazing animals and fire for a few years, and soon there will be a forest cover. That forest cover will remain for a long time. Forests will recover quickly if damaged through fire, wind, insects, and even volcanic eruption. An excellent example is the recent (1980) volcanic eruption of Mt. St. Helens in Washington state, United States. In an instant everything above the ground over an area of 40,000 ha was killed. Fifteen years after the eruption, much of the forest has come back, even without human help. Forests can even recover after ice ages. This ability of tree growth to dominate, and to recover after any catastrophe, has meant that forests have dominated the earth for at least 100 million years.

Ten thousand years ago the hunter-gatherer had to rely on what was produced naturally in the forest. Natural forests around the world produce very little in the way of food. My estimate is that the food yield from reasonably productive natural forests is between 10 and 20 kg/ha. That includes the fruits, nuts, shoots, leaves, and roots, and the birds and other animals that feed on the same food. Some forests would provide even less food. Also, the period of availability for most of the food products is short – a few days in the case of some berries to little more than a month for some nuts. The food must be harvested or it is lost forever. Nature rarely stores food material in a form useful to humans. The exceptions are the tissues of birds and mammals, some seeds, and plant storage organs (e.g., tubers).

We know little about how much food our hunter-gatherer ancestors needed to survive, but there must have been times during the year when food was very limited. There were, of course, other times when there was plenty of food, but with virtually no means of preserving it. However, let us assume that we need about 2 kg of vegetable food/day (or about 750 kg/yr). Much of this would not have been nutritious. High-energy foods, such as cereals and potatoes, come from agricultural improvements. At a food production rate of 15 kg/ha it would require about 50 ha of forest to provide a person with enough food for a year. However, this probably understates the requirements. The critical period is that of winter and spring. There was no storage of food, except in the flesh of animals and some seeds or tubers. The area of forest required to sustain food availability through the critical period could be twice as much – say 100 ha/person.

The situation for wood supply is entirely different. Relative to their production of food, forests produce huge volumes of wood. Seven and a half m^3/ha for all wood production would be about the global annual average. Wood, once grown, is stored in the stem of the tree. That wood is available any time that it is required. A 100-year-old tree contains 100 years of wood growth, and all of this is available to the hunter-gatherer whenever it is required.

I know of no estimates of the wood needs of the hunter-gatherer. However, let us assume that they needed (mostly for fuel) twice the current global average wood consumption. In that case they would need about 1.5 m^3 of wood/person/yr. At an average growth increment of 7.5 m^3/ha/yr, the amount of forest required to sustain the wood needs of one person is only 0.2 ha.

We can now compare the relative areas of natural forest required to sustain a human being in both food and wood requirements. These areas are 100 ha for food, and 0.2 ha for wood. Although there is little evidence of the actual amounts needed, it is likely that these estimates are of the right order of magnitude. Therefore, without agriculture, our hunter-gatherer forefathers needed about 500 times more forest area for their food needs than they needed for their wood requirements.

Given this enormous difference in relative productivity, it is obvious that food supply for human communities became limiting long, long before any wood supply. However, the Agricultural Revolution has now given us the ability, in the 1990s, to grow food on a scale undreamt of, even 100 years ago. In the past the vast difference in relative productivity may have made it theoretically improbable, but the vast improvements in agricultural productivity could result now in the global wood supply becoming limiting. The two major reasons are that:

- Most of the world's forests have been destroyed for conversion to agriculture.
- Vast areas of the remaining natural forests (which supply most of the world's wood) are being set aside for other values (e.g., recreation, biodiversity, scenic, and wildlife).

The increasing need for wood – how this might be satisfied

We are going to need increased volumes of wood. The need for wood for energy in the developing world is well recognized. The energy needs of the developed world could also be partially satisfied by using wood as an energy source. Most of the wood used in developed countries is, however, for industrial purposes (sawntimber, plywood, roundwood, reconstituted boards, and paper). The world demand for wood products can be expected to increase significantly. If developed countries are sincere about their commitment (under the 1992 Climate Change Convention) to limit atmospheric emissions of fossil carbon, then wood consumption must increase. All wood substitutes are major energy users. A major shift away from wood and an increased use of wood substitutes would increase overall energy requirements. This would almost certainly increase the use of fossil fuels (and, therefore, increase the release of fossil carbon into the atmosphere). Unless there is a major breakthrough in the development of a cheap, environmentally friendly, and renewable energy source, large-scale substitution for wood will not be possible. More use of wood, both as a fuel and as industrial wood, is a solution. This reduces the need for fossil fuels.

The rising expectations and improving living standards of developing countries, especially in Asia, also seem certain to increase wood use (Sutton 1993). Appendix 2 provides a vision of what the ultimate demand might be for industrial wood. The potential global wood demand for industrial purposes alone could be more than ten times that used in the early 1990s!

The enormous difference between the yields of wood and food from natural forests has meant that, at the global level, human civilization has not yet been constrained by an overall wood shortage. Earlier wood shortages have all been either economic or local, but we are approaching the limits of wood harvest from the natural forest. The use of natural forests for wood production is being increasingly opposed by environmental and preservationist groups; the pressure is to retain the remaining natural forests of the world in their natural state. The pressures are greatest to stop the harvest of old-growth forests. Forest managers may be convinced that wood harvest levels can be maintained, or even increased, in the remaining natural forest. However, the public perception now is that the remaining natural forests are under serious and increasing threat.

We have a dilemma. The use of the world's remaining natural forests for wood harvest is being increasingly opposed, yet wood is the most environmentally friendly product we have. Our global economy will require increasing volumes of wood. The only solution to this dilemma is to increasingly shift the wood harvest from the remaining natural forests to deliberately created planted forests.

This solution will require a major rethinking of the objectives of forest management. In most countries (and certainly those countries harvesting most of their wood from natural forests), the prevailing forest management objective is multiple use. Wood harvest is only one of many objectives. Environmental protection, biodiversity enhancement, endangered species protection, recreation, and hunting are often given consideration equal to the need for wood. There are additional pressures to use only local indigenous species, to achieve age diversity, and even to avoid clearcutting as a harvesting system. The management objectives are multiple. Unfortunately such management objectives are not only expensive (and that raises the question of who pays), but also increasingly difficult to administer. How do we determine the balance between the competing objectives (some of which are almost incompatible)? One outcome seems inevitable. Natural forests, under multiple-use management objectives, will have greater and greater difficulty in providing the volumes of wood that will be required by the global population. Wood from such systems will be increasingly expensive.

The solution of increasingly moving to planted forests for our wood production has many attractions. It leaves the remaining natural forests to be managed for their nonwood-producing objectives, e.g., protecting wilderness, protecting the environment, enhancing biodiversity, and meeting demands for recreation and hunting. These demands will be easier to satisfy because there will be no conflict with the need to harvest wood. Planted forests, although managed primarily for wood production, will still be responsibly managed with respect to the environment. However, management will not have to meet the pressures for biodiversity, for species and age diversification, and of restrictions resulting from demands to use only indigenous species. All of these demands and more are expected when managing natural forests.

The agricultural parallel

As the pressures for larger volumes of wood at competitive prices increase, the management of planted forests will increasingly parallel the developments of modern agriculture. That will almost certainly result in productivity, wood quality, and economic gains that could never have been possible when wood came from natural forests.

A major and long-term consequence of this management change will be the eventual use of very few tree species. Planted forests will no longer be restricted to what nature provides. At present we use at least 1,000 tree species globally for wood production. They are what nature has given us. We may well find, just as we have in agriculture, that, as our experience increases, very few tree species will prove pre-eminently successful in managed planted forests. It is still very early in this process, but we already know enough to predict the characteristics of those few tree species that will eventually be used in most planted forests. The winning species will have all of the following characteristics:

- Fast growing on a wide range of sites.
- Relatively free of disease and insect problems.
- Responsive to genetic improvement.
- Responsive to management.
- Able to provide wood that is both desirable for and versatile in a wide range of end uses.

A key factor is going to be economics. Planted forests require a great deal of capital (planted forests are probably the most capital-intensive industry in existence). Traditionally, governments have been major forest investors. In the future, pressures for increased social spending and for tax reductions will mean that governments will be increasingly reluctant to invest in forestry. Investments will increasingly come from the private sector. Investors in any planted forests will increasingly be attracted to those ventures that offer the highest returns and that have the lowest risk. As in agriculture, investment returns will decide the winning tree species.

My prediction is that the eventual winners will include, or be similar to, the five species (or species groups) specifically considered in this volume, Douglas-fir (*Pseudotsuga menziesii* [Mirb.] Franco), the southern pines (*Pinus* spp.), radiata pine (*Pinus radiata*), eucalypts (*Eucalyptus* spp.), and poplars (*Populus* spp.). I will now discuss the success story of the species I am most confident will be a winner – radiata pine. I will refer mostly to New Zealand's experience. It was that experience that stimulated many of the ideas and concepts advanced in this paper.

Radiata pine and its success in planted forests

Radiata pine (or Monterey pine) occurs naturally on the coast of California. Because of the small area of natural stands, and because of the limited number of areas where it can grow locally, it is a minor species in the wild. Since the middle part of the last century, radiata pine has been introduced all over the world. It grows only in temperate regions with a maritime climate. This is because radiata pine is killed by the occasional severe frost which can occur in temperate regions with a continental climate, and because radiata does not function well in tropical regions (except at very high altitudes). Where the climate is satisfactory, radiata is very tolerant of site (e.g., sand to clay, acid to alkaline, high fertility to low fertility, and sea level to 1,000 m).

Radiata has all-around desirability. It is excellent for pulping, the manufacture of reconstituted board products, and plywood. It is not a high-quality structural timber, but it is adequate for most purposes. Radiata is an ideal packaging material. It is probably at its best as a clearwood, for radiata wood has excellent finishing properties and trees have the ability to grow quickly to a large size after timely pruning (which eliminates the lower branches). Radiata has poor natural durability, but it is one of the easiest timbers in the world to effectively treat with wood preservatives of almost any kind.

Radiata is now established in most areas of the world in which the climate is suitable. The 1995 estimates of the area (1,000 ha) of radiata pine successfully planted in the world are as follows: Chile, 1,381; New Zealand, 1,260; Australia, 720; Spain, 240; South Africa, 55; and others, 25. This gives a total of 3,681,000 ha.

Radiata grows fast. Research growth plots have recorded growth rates of over 40 m^3/ha/yr. Planted radiata forests, even for high quality sawlogs, are now expected to average at least 20 m^3/ha/yr or better. These growth rates are for rotations of from 20 to 30 years. Radiata may not be the greatest volume producer known, but it is able to sustain its very high growth rate over several decades.

Radiata pine has shown considerable response to genetic improvement. Tree straightness and form can be improved. Branch habit can be changed. Growth rates can be enhanced. Wood density and disease resistance can also be manipulated. Radiata is generally a good seed producer, and the seed remains viable for many years. Vegetative propagation (cuttings) and, now, tissue culture and embryogenesis techniques allow for the rapid multiplication of improved and proven tree breeds.

The ability of radiata to respond positively to management is a major attribute that few yet recognize. Radiata responds to site preparation, to initial spacing, and especially to the practices of pruning and thinning. This responsiveness to pruning and thinning, combined with the ability to grow rapidly for several decades, means that radiata can achieve a large diameter in a relatively short time. Trees grown on regimes that reduce stocking to relatively low numbers at an early age can achieve mean diameters at breast height of 60 cm in a little over 20 years! Radiata can achieve a large diameter more quickly than almost any other tree species in the world. These diameters can be achieved even after the tree has been early pruned to 6 m.

Radiata is not a newcomer. In New Zealand, the species was first introduced in the late 1840s and was first used commercially in a stand planted in 1857. That stand was felled and utilized in 1877. New Zealand has, therefore, just under 120 years of utilization experience with the species. Since the mid-1950s there has been an applied, highly focused, and well-funded research program on the species. It is probable that as many as 10,000 scientist/technician years of research effort have gone into work with radiata over the last 50 years.

Some critics of New Zealand's planted forests have questioned the practice of even-aged monocultures. Observations in the original California stands also show that radiata grows as an even-aged monoculture. Current radiata management, therefore, parallels the growth characteristics of natural stands. Since radiata's first introduction, New Zealand has been through at least four, or maybe five, rotations of experience. This has not always been on the same site, but there are some sites that are now well into their third rotation. On almost all of these sites there has been no measurable loss in site productivity or any site degradation.

Plantation growers, of course, are well aware of the biological risks. New Zealand, for example, has strict quarantine regulations. New Zealand monitors, not only all radiata developments overseas, but also most stands in New Zealand. All of this is supported by a viable research group. Experience to date shows that the intensive management practices used in New Zealand (especially wide initial spacing, early pruning and thinning to a relatively wide spacing, and relatively short rotations – less than 30 years) are a considerable help in reducing biological risks.

Despite some ill-informed and emotive claims to the contrary, New Zealand planted forests are not biological deserts. As stands age, tending and wide spacing encourage an understory that often contains many indigenous plant species. The stands support a surprisingly large number of birds, both indigenous and introduced. Streams in planted forests host abundant indigenous aquatic life.

New Zealand's planted forests (over 90 percent of which are radiata pine) now provide more than 98 percent of New Zealand's annual 15 million m^3 wood harvest. More than half of this is exported, and wood is now one of New Zealand's major exports. This is a change from a century ago when almost all of the wood used in New Zealand came from indigenous forests. This significant shift has allowed New Zealand to set aside as protection forests (for biodiversity and other aspects) most of the remaining indigenous forests. New Zealand can, therefore, rightly claim that the planted forests have aided biodiversity protection (Sutton 1995a). Recognition of the importance of planted forests is now formalized in the Forest Accord signed in 1991 among most environmental groups (Greenpeace was the most notable exception) and all of the major industrial planted forest owners and wood users in the country (Sutton 1995b).

New Zealand has benefited greatly from planted forests. These forests have made a significant contribution to employment, especially in rural regions where they provide good earnings and meaningful employment. The establishment of planted forests has diversified both land use and the economy away from an almost total dependence on farming (sheep and dairy). On harvest, planted forests provide wood for a sizable industry, much of which is exported. In 1995, about 13 percent of New Zealand's total exports came from planted forests, yet the forests account for less than 5 percent of New Zealand's total land area.

Radiata's emergence as a winner took a long time

New Zealand's current concentration on radiata pine is relatively recent. Over the last century as many as 400 tree species have been tried. Even by 1920, over 200 species had been given trial by the State Forest Service. Initially, radiata was very largely ignored, especially by the state. This was because it grew fast, and there was a general misconception at that time that fast growth automatically meant poor quality.

Over that same period the most promising mixtures of species were also given serious trials. Almost without exception every one of those trials failed. In almost every case one species dominated. It is extremely difficult to find any combinations of species that are worthy of serious consideration in commercial forestry. The failure of almost all other species (and mixtures of them) to live up to their initial promise has resulted in the dominance of radiata – not, as we are so often accused, because we ignored other species or species mixtures.

A major advantage of having only one dominant commercial tree species is that research (genetic, silvicultural, growth, and yield) can be very concentrated. Had New Zealand split its research effort over five or six species, it is unlikely that it would have developed the current efficient planted forest management regimes or made all of the genetic improvements it has. Nor would New Zealand have developed the sophisticated management, yield, and financial models that are now a major factor in the management of its radiata forests. Of course, such a concentration on one species has meant that all other species have been less researched. That effort also reinforced the dominance of radiata pine. I suspect that the same division of research has gone into the other agricultural winners as well. The species that demonstrates all the characteristics of a winner receives the lion's share of research effort. That effort only serves to reinforce the species as a winner.

Conclusions

The concepts advocated in this paper are the result of my attempt to explain the New Zealand experience, and to consider what implications that might have for the rest of the world. Over the last 100 years New Zealand has seriously tested several hundred promising tree species, as well as many tree mixtures. The result was not a group of winners with similar potentials. The result was a domination of one species, radiata pine. Radiata now dominates the planted forests of New Zealand, Chile, and Australia (except Queensland). When we evaluate what has happened in agriculture, we find this same dominance of a very few species.

The global implications of the New Zealand experience with planted forests are that the world will obtain more and more of its wood from deliberately planted forests and less from indigenous forests, even if managed. To increase productivity and to be economically viable, these planted forests will be managed like agricultural crops. It is highly probable that very few species will prove suitable. It could well be that less than five species will eventually supply most of the world's wood.

A shift in wood harvest from natural forests to planted forests has some major advantages, including the following:

- Remaining natural indigenous forests will be conserved.
- As the management of planted forests improves, productivity will increase. This will reduce wood costs, as well as reduce the area of forests needed.
- The intensification of management offers the potential for more uniform tree crops. In the longer term, there is the prospect of wood quality and productivity improvements few now could even contemplate.

Greater recognition of the contribution that planted forests could make may resolve some (many) of the current controversies on forest utilization. The shifting of the wood harvest to planted forests will reduce the need to harvest wood from the world's remaining natural forests. It will also ensure an increasing and sustainable global supply of wood from environmentally friendly sources. These planted forests, some of which will be with the proven winner, radiata pine, will be making a significant and sustainable contribution to society and to the global environment.

Appendix 1

World wood consumption: A graphic illustration

The conversion of global and national totals to personal levels of consumption provides an easily understandable measure of the importance of wood. According to 1992 values, a volume of 3.5 billion m^3 used annually by 5.2 billion people works out at 0.67 m^3/person/yr. That is equivalent to a cube of wood with edges equal to just under 0.9 m (or about 1 in shorter than 1 yd). If we average 0.67 m^3/person over a year, that represents, in volume terms, a daily wood consumption of 1.8 liters (or just under 0.5 gal or just over 3 Imperial pt/person/day). Because it is rare to express wood volumes (which are of solid form) in liters, gallons, or pints (values which are normally associated with liquids), a more appropriate measure is to demonstrate the actual amount.

Take a wooden disk of even thickness cut from a log with an underbark diameter of 60 cm (24 in). That disk is then cut into equal-sized wedges or segments. It can make eight 45° wedges, ten 36° wedges, or twelve 30° wedges. To represent the average daily consumption of wood globally, each wedge should have a volume of 1.8 liters. To have that volume the 60 cm disk would need to have a thickness of: 8.0 cm (3.0 in) when cut into twelve 30° wedges; 6.5 cm (2.5 in) if cut into ten 36°wedges; and 5.0 cm (2.0 in) if cut into eight 45° wedges. For every one of the 5 billion plus global population, that wedge represents the average daily consumption of both fuelwood and industrial wood.

There are of course major variations in consumption levels. The United States has one of, if not the highest per capita wood consumptions in the world. Bowyer (Bowyer 1992) has estimated that the annual consumption of wood for the United States is 2.3 m^3 (80 ft^3)/person/yr. That represents a consumption of 3.4 times the global average. On a daily basis that consumption translates as 6.3 liters (1.65 gal or 1.4 Imperial gal)/person/day.

This all demonstrates how very important wood is as a raw material, especially in developed economies. Demand for wood seems insatiable. Surprisingly, little attention has been devoted globally to wood production and consumption. Especially in the media, other forest values like species and biodiversity protection, concerns over clearcuts, and preserving old growth have received much attention. Yet wood remains one of our most important raw materials, and wood is the most important product of the forest.

Appendix 2

The optimistic outlook for wood

I have recently researched the long-term market prospects for wood. I want to consider a range of options from optimistic to pessimistic. This is a preview of the optimistic outlook for the wood market.

The United States has the world's highest per capita consumption of wood. A significant increase in energy consumption is required if that wood is to be substituted. This is because, relative to wood, all wood substitutes are major energy users. Without an energy breakthrough, large-scale substitution of wood is not possible. The annual per capita consumption of wood in the United States is 2.3 m^3 (Bowyer 1992). Let us reduce that per capita consumption level to 2 m^3/person/yr (to allow for fuel and some other uses). Let us now be optimistic – the whole world achieves a U.S. standard of living with the same industrial wood consumption. In 1995, there were 5.6 billion people on earth now and that would give a global consumption of 11.2 billion m^3 of industrial wood – or just over seven times the present level of global consumption!

But the population of earth will increase. It is expected to be 8.5 billion by 2020 and, maybe, 10 billion by the year 2050. This would mean an eventual global industrial wood demand of 20 billion m^3 – thirteen times the current industrial wood harvest!

Many will see such a vision as totally unrealistic. It probably is, but not for the reason most would consider, namely, that the world could never achieve that level of wood demand. Can we be that certain? Remember, many people today have living standards that were available to only the really wealthy 50 years ago. Few of us deliberately seek lower living standards; the vast majority strive for improvement. The major reason this prediction cannot come true is because we cannot supply that volume of demand. Because of environmental pressures, less wood will come from the remaining natural forests of the world, and the existing planted forest resource is far too small.

Let us assume that there is a potential demand for 20 billion m^3 of industrial wood, and that all of that is supplied from planted forests. Let us also assume an average growth rate equal to the current New Zealand average viz 20 m^3/ha/yr. This means that we would need at least 1 billion ha globally of planted forests to satisfy the demand for industrial wood. That is an area nearly forty times the total area of New Zealand, or just a little more than the total area of the United States. Fuelwood demand is additional! The total cost is huge. A direct investment of U.S.$3 trillion would be needed (assuming an investment of $3,000/ha and no interest charges).

This may be a totally unrealistic vision. But, given that most people want to better their lot (or at least that of their children), there can be little doubt that a huge potential market for wood will exist – probably within one human generation. One thing is for sure, we cannot improve living standards without raising wood consumption levels. We cannot increase the global wood supply without a massive planted forest resource. The effort required is of a scale few have yet considered.

References

Bowyer, J.L. 1992. Responsible Environmentalism – Forests, People, Raw Material Needs and Environmental Protection. Hamilton Roddis Memorial Lecture Series No. 1. Department of Forestry, College of Agricultural and Life Sciences, University of Wisconsin, Madison.

FAO (Food and Agriculture Organization of the United Nations) 1995. FAO Yearbook: Forest Products 1982–1993. FAO, Rome.

Koch, P. 1992. Wood versus non-wood material in USA residential construction: Some energy related global implications. For. Prod. J. 42(5): 31–42.

Sutton, W.R.J. 1993. The world's need for wood, pp. 21–28. In: Proceedings of the Forest Products Conference on the Globalization of Wood: Supply Processes, Products and Markets. Forest Products Society, Portland, OR.

Sutton, W.R.J. 1995a. Plantation forests protect our biodiversity. N. Z. For. 40(3): 2–5.

Sutton, W.R.J. 1995b. The story of the Tasman and forestry accords in New Zealand. Unpublished manuscript.

New Forests **17:** 111–118, 1999.

Four common myths about plantation forestry

REX B. McCULLOUGH
Forestry Research, Weyerhaeuser Company, Tacoma, WA, USA

Key words: Weyerhaeuser, high yield forestry, sustainability

Abstract. Privately owned tree farms and plantations share many characteristics of natural forests, multiple purpose forests, and fiber plantations. Tree farms and plantations are managed primarily for the sustainable production of solid wood and fiber; however, they increasingly play an important role in protecting all forest values. This rapidly changing role poses a challenge to forest managers – to increase productivity to meet spiraling worldwide demand while satisfying public expectations for preserving and enhancing a full range of forest resources.

These heightened expectations are prompting tree farm and plantation managers to modify proven management regimes in order to protect water quality, fish and wildlife habitat, soil productivity, and aesthetics. Proactive measures, such as establishing goals for each stewardship objective, defining measurement criteria, and third-party endorsement or certification, are helping forest managers identify and address improvement opportunities. Each of these stewardship objectives, though, can imply trade-offs in terms of yield and utilization, thus reducing the economic attractiveness of the forestry investment.

In this period of rapid change, one effective strategy is to engage the public in the front end of the forest planning process. By listening and responding to public concerns, establishing common values, and allowing the public to judge us by our actions, we can begin to create awareness and trust. Only then will we attain the stable regulatory platform that is imperative for making long-term forestry investments and implementing science-based mechanisms for meaningful ecosystem enhancement.

Introduction

I'd like to begin with a short story about a fellow who was once introduced as the most gifted businessman in the country. The audience was told he had made a billion dollars in Texas oil. When he rose to speak, the tycoon appeared a bit embarrassed. The facts were essentially correct, he said, but he felt compelled to admit that it wasn't oil ... it was timber ... and it wasn't Texas ... it was Oregon ... and just to keep the record straight, it wasn't a billion ... it was a hundred thousand ... and it wasn't him ... it was his brother ... and he didn't make it ... he lost it! "Matters of fact aside," the fellow said, "I'm grateful for those kind words and this opportunity to talk with you today."

I like this story because it raises a couple of important points. First, I believe very strongly in sharing ideas and opinions, and am grateful for this opportunity. The story about the businessman illustrates a second point as well. It's not always easy to separate fact from fiction – especially in discussions about plantation forests. There are tremendous misperceptions – myths, if you will – surrounding this type of forest. Myths on the part of the public, the media, and the industry itself. Myths that obscure the science and economics of plantation forestry. Myths that you and I must redress through our words and our deeds. I'd like to frame my comments around four of the most common myths about plantation forests.

Myth #1: A plantation is not a forest

What is a plantation forest? If you arranged all forest types in a continuum with short-rotation fiber plantations on one end and unmanaged natural stands on the other, Weyerhaeuser's plantation forests would line up pretty close to the middle. We manage our plantation forests primarily to produce solid wood, with chip production as our secondary objective.

First I want to provide some information about Weyerhaeuser – the company I work for – because the statements presented in this paper draw heavily on our experience. In total, we own over 2.2 million ha of plantation forests in the United States, and these forests are roughly divided between the South and Pacific Northwest. To achieve sustainable productivity, plantation owners follow an intensive management regime. We mainly plant one species on each site: loblolly pine in the South, and Douglas-fir in the Northwest. Where conditions warrant, we also plant other species, such as red cedar, alder, noble fir, western hemlock, and lodgepole and ponderosa pine. Many naturally occurring species such as alder, hemlock, cedar, and cottonwood add diversity to the forest stand.

We practice intensive silviculture, including fertilization, spacing, and pruning, to enhance the volume and value of the stand. We aggressively protect our investment against fire, disease, and pests. We primarily employ an even-aged strategy by clearcutting, although we do use other strategies on a site-specific basis. And we strive to minimize waste. Unfortunately, by improving utilization, we reduce the number of snags and downed logs – important habitat for amphibians and other small critters.

So if "real forests" are defined as "untouched by human hands," then plantations don't qualify. Applying slightly more scientific criteria, an unmanaged stand compares to what we observed in our southwest Washington habitat management planning process as follows:

- Tree species: Unmanaged forests typically have diverse tree species throughout the stand. Plantations usually have less tree species diversity within the stand, but considerable diversity in the riparian management zone.
- Canopy: Unmanaged forests typically feature a multistoried canopy throughout the stand. Plantations usually have multistoried canopies within the riparian management zone.
- Plant diversity: Both unmanaged forests and plantations have considerable plant diversity throughout the stand.
- Snags: Unmanaged forests typically have snags throughout the stand. Plantations usually have snags in the riparian management zones. Snags are ordinarily removed from plantation stands as they pose significant hazards to workers and attract lightning.
- Large woody debris: Unmanaged forests typically contain large amounts of downed logs and woody debris. The amount of large woody debris varies from one plantation to the next, but is typically less than in unmanaged stands.

We recognize that plantations are not old-growth forests – they are not identical to unmanaged natural stands of the same age. However, we *believe* that they are true forests and believe that they are capable of providing most, if not all, of the functions and habitats found in similarly aged, unmanaged natural forests. I admit that we don't have all the answers, but we are committed to collecting data about structure, function, and habitat, and to adjusting our management practices accordingly.

Myth #2: Trees are natural treasures! They should not be cut – period!

Anyone who's spent time in the field knows that forests – especially old-growth forests – evoke a sense of reverence. Unless you are a very special person, you won't get the same feeling when you stand next to a steel mill or a cement mixer that you get when you stand in an old-growth forest. Although we believe that our plantations are true forests capable of providing most, if not all, of the functions and habitats associated with natural forests of similar age, we do recognize that they are not the same as these natural forests. Furthermore, our plantations were established for the purpose of producing wood economically, and, as such, will be harvested at some future date. This is a critical message for several reasons.

To keep up with global demand, we are managing plantations ever-more intensively. Historically, global wood use jumped over 40 percent in the past quarter-century. We can expect the world's appetite for wood to mushroom as the Earth's population increases by 33 percent by the year 2010. Compounding the pressure on the world's forests is an expected 5 percent increase in demand for agricultural land.

Another side effect of population growth is urban sprawl. As suburbs encroach upon America's forests, society demands that more areas be set aside in preserves. At present, we have permanently protected approximately 14 million ha of forest land in the United States. This is land capable of producing at least 1.4 m^3/ha/yr. This, coupled with other set-asides, brings the total area protected up to more than 97 million ha. Recently the U.S.D.A. Forest Service has taken even more land out of timber production in order to protect threatened and endangered species habitat. In addition to these set-asides, recent changes in management policies on our national forests have reduced the harvest to 20 percent of the historic cut.

We believe that our response must be to manage our plantation forests more intensively while continuing to protect important natural resources. Presently, industrial forest lands comprise only about 14 percent of America's total forest land, yet they produce one-third of the annual harvest. Plantation growth rates are already two to five times greater than comparably aged unmanaged stands. The United States is still a net importer of wood and wood products.

Plantations offer society an opportunity to have a win:win situation. It is clearly the right of our society to determine whether or not public lands should be preserved for biological, recreational, and spiritual fulfillment. When this is coupled with increasing worldwide demand for wood products, a void is created in wood production. Much if not all of this void can be replaced with managed plantations. To do so means that we must enable plantation owners to responsibly manage their lands and earn a profit. A consideration of profit brings us to Myth #3.

Myth #3: Profit is all that matters to plantation owners

Profits are as crucial to plantation owners as they are to any businessperson. But if there's one key message from the last 25 years, it's that our industry must temper our profit objectives with an unwavering stewardship commitment. With almost 30 years of experience in intensive forestry, we know that we can make money and, at the same time, improve our care of other forest resources. At issue is how to convince the stakeholders that we are sincere, and that we can and will do it.

Weyerhaeuser's High Yield Forestry pioneered the science and economics of plantation forestry in the 1960s. High Yield Forestry made it feasible to manage forest land in perpetuity, rather than to convert it to other uses.

With High Yield Forestry techniques, plantations can consistently produce real financial returns exceeding the benchmark S&P 500 average – this, according to an analysis of the years 1960–1991 by The Hancock Timber Resource Group. The long-term forest investment of 30–50 years has been attractive because the risks relative to the growth of the investment were known, managed, and accepted.

Yet the economics of long-term forest investments are extremely fragile. Forest landowners and shareholders alike are concerned about the ability to recover their investment in the future. During a rotation, we can invest more than U.S.$740/ha in regeneration, silviculture, and protection – a figure likely to grow as we ratchet up intensity and stewardship activities. Given that lengthy horizon and the effects of compound interest, we must carefully weigh the value of each dollar invested. It is reasonable for people to expect a higher rate of return on investments that they view as risky. Increasing risk without a corresponding increase in market value results in less money for all aspects of forestry, including environmental protection.

Watershed analysis provides an interesting example of a forestry expense that we did not envision 10 years ago. This process is an innovative approach to protecting water quality and fish habitat. The costs of watershed analysis add up quickly. For example, watershed analysis can cost several dollars/ha, and over a 2-year period we have analyzed over 142,000 ha in Washington and Oregon. If we viewed this purely as a silvicultural expense, the economics might be dicey at best. Watershed analysis doesn't make our trees grow any faster, straighter, or closer to the mill!

But watershed analysis does provide cleaner water and better fish habitat, and it establishes a relationship of trust with the public. All of this combines to give us a better chance of recovering our entire investment decades from now. Ten years ago we might have based our decision on cost alone, but that world no longer exists. Weyerhaeuser's view is that sound stewardship and solutions-oriented steps will produce superior returns for our shareholders. Further, they will help foster a more cooperative and stable approach to forest regulation.

Myth #4: We need more laws and regulations to protect our forests

At the heart of this issue rages a debate on how to best manage and protect forest values. The current "species-by-species" approach to habitat protection is perhaps the single greatest obstacle to achieving an equilibrium between environmental and economic objectives. Who can forget the northern spotted owl and the impact it has had on Northwest forestry? Tremendous uncertainty is created every time a new species is listed as threatened or endangered. This uncertainty can make it extremely difficult to justify the long-term investment required for plantations or mills.

Again, our experience shows that there's an alternative to the species-by-species approach – one that makes sense both scientifically and economically. That approach is to manage habitats through landscape management.

Indeed the first ever all-species habitat conservation plan was just approved for Murray Pacific, thus assuring them that they can continue to harvest and manage their lands within the defined parameters, regardless of new species listings. This provides Murray Pacific the confidence to invest in long-term forestry projects.

At Weyerhaeuser, our commitment to forest stewardship is reflected in our own Weyerhaeuser Forestry Resource Goals – uniform goals pertaining to forest products, water quality, wildlife habitat, soil productivity, and aesthetics. By the end of this year, each of our geographic regions will have in place its own set of measurable targets for each goal. The resource goals illustrate how our company is moving toward habitat management. Goals are very important for directing our internal organization. They will take on greater significance with our concerned stakeholders when we can demonstrate that we are making progress.

Resource goals

Forest products: We manage our forests for the sustainable production of wood and other forest products. These days we don't look at our practices just to see if they increase yields. We also evaluate how they can help to create more or better habitat. Two practices that fit this category are pruning and thinning. These practices are helping us to create a higher canopy lift in young forests, which is an essential condition for dispersal habitat for the northern spotted owl.

Water quality: We protect water quality and fish habitat in carrying out forest management. The most pressing issue we face here in the Northwest is that involving native salmon. Along with dams, fishing, and hatcheries, habitat degradation poses a threat to the survival of salmon. Watershed analysis is an example of a scientific process that we use to assess the cumulative effects of forestry operations on forested watersheds.

Watershed analysis brings together stakeholders to assess the condition of a watershed and write corrective prescriptions. We feel that watershed analysis is less costly for private landowners than are blanket regulations. More importantly, we feel that watershed analysis can be more effective at protecting water quality and fish habitat, because we tailor the forest management to the specific conditions and requirements of the watershed.

Wildlife habitat: We provide habitat for wildlife associated with managed forests. Many landowners are taking a landscape approach to managing wildlife habitat in plantation forests. We project that science-based processes can help us to provide certain habitat conditions while achieving target returns on our investment. Earlier this year we signed a Habitat Conservation Plan with the U.S. Fish and Wildlife Service to create spotted owl dispersal habitat in our Coos Bay, Oregon, region. The agreement covered 85,000 ha, and enables us to contribute to the owl's recovery while continuing forest operations.

Soil productivity: We use scientifically based practices to protect soil stability and ensure long-term soil productivity. Because soil is the basic asset we manage, Weyerhaeuser and other plantation owners are turning to science-based practices to protect soil productivity and stability. Because plantation forestry requires substantial on-the-ground activity, decisions about equipment and practices must take into account soil, topography, and weather through defined and measurable, reliable processes.

Aesthetics: We consider aesthetics in our forest practices. At Weyerhaeuser we recognize that much of the public considers clearcuts to be unsightly and offensive. In Washington our foresters receive basic landscape architecture training. We're also using aerial photography, computer modeling, and a rating system to plan harvests that are more compatible with the natural landscape.

The point of all of this is that we want to be part of the solution. Through the American Forestry and Paper Association, the forest products industry has endorsed a similar set of principles, known as The Sustainable Forestry Initiative. Plantation owners are taking progressive steps to address all forest resources. However, this must lead toward regulatory stability and freedom to operate and compete within a defined set of principles. Regulatory stability is built largely upon a foundation of trust. We must prove to people that they can trust us to do the right thingþto take proper care of public resources without layer upon layer of onerous regulations.

Listening to the public. One approach Weyerhaeuser is using to cultivate public trust is listening. We use focus groups, hold town hall meetings, and collect survey information. Listening can be a rather intense and stressful process for the listeners – we find that most folks are not quiet and reserved! Through focus groups we have learned that we share common values about recycling. We also have learned that the public feels we view trees as a commodity. Conversely, most people consider trees to be a precious resource.

Gaps like this pose an obstacle to establishing trust, and, if the public cannot trust us, they won't hesitate to ask the government to step in.

By listening to public input, we can identify where our values overlap. We can grow our business by expanding in the areas where there's strong alignment. And we can use this common ground for establishing a dialogue and explaining what we're doing. Only then will these myths start to fade away.

Conclusions

In a sense, the myths I've described are real. They are real in that they create confusion about plantation forestry in the eyes of the public and our elected officials. Unfortunately, this fog obscures how vital plantations are in meeting worldwide demand for forest products. It also hides the fact that stewardship is a prominent part of our plantation management strategy. I emphasize the following four key thoughts:

- Plantations are forests and are capable of providing most, if not all, of the functions and habitats found in natural stands of the same age.
- Plantation forestry is a long-term investment, and requires a predictable economic platform to stabilize the level of risk. To achieve this stability, we need a regulatory environment that favors landscape approaches over species-by-species mandates.
- Plantations will play an increasing role in meeting world wood demand.
- We need to listen to the public, understand their values, and align our operations and processes with these values.

New Forests **17:** 119–127, 1999.

Multiple purpose forests and the national forest system

KAREN BARNETTE
USDA Forest Service, Sacramento, CA, USA

Key words: planted forests, sustainability, U.S. National Forests, management philosophy

Abstract. The history of the legislative basis for managing National Forests in the USA has led to forests managed for multiple uses for a wide diversity of users and interests. Evolution of emphasis on National Forests from mandates to "preserve and protect" and provide "a continuous supply of timber" to concerns about owls, ecosystem management and contemporary rural development have included changes in on-the-ground forest management. All of these phases include needs for planted forests established and maintained for multiple purposes and benefits.

Introduction

> This land is your land, this land is my land, from California to the New York Island. From the redwood forests to the Gulf Stream waters, this land was made for you and me.
>
> (Woody Guthrie, American songwriter and folksinger, *This Land is Your Land*, 1956).

These lyrics speak volumes when applied to the United States' National Forest System (NFS) lands. Not only are these lands "yours" and "mine", but they belong to the birds and the beasts, and to John Q. Citizen from Hoboken who will never set foot inside a National Forest boundary. And to Native Americans past and present who lived and worshipped, and sometimes still worship, inside those boundaries. In the United States, National Forests serve many masters.

Given the competing demands of the American public as expressed directly and through their representatives in Congress, the requirement for all NFS lands, including planted forests, to fulfill multiple purposes is a given. But we did not come to this realization easily.

Setting the stage

Forested public lands in the United States have gone through an evolution of management philosophies. Throughout most of the country's first century, forests were cleared for homesteading or for exploitation of the natural resources they offered with little concern for the impacts of those activities and no attempt at reforestation. Judged with today's climate toward natural resource management in mind, the lands were managed wastefully and destructively. Given the public attitude toward settling and developing the Western United States at the time, and the seemingly inexhaustible supply of land and resources, a moral judgment of "right" or "wrong" has little value. It happened.

There was a gradual awakening to the impacts and potential losses that would likely result from adherence to such a management philosophy. Legislation and policy shifted from a position of encouraging disposition of the public domain to fostering development of systems of parks and forest reserves. The Forest Reserve Act of 1891 was the foundation of the National Forest System. However, the 1891 legislation provided little insight into how such reserves should be managed, or even the philosophy behind the Act, since it appeared to encourage both preservation of the proposed reserves for watershed protection and development for the benefit of all citizens. The potential conflicts between these two laudable goals were not discussed in the Act.

In 1897 Congress passed a second Forest Reserve Act, this time defining the purposes of the reserves as

- to "preserve and protect the forest within the reservation";
- "for the purpose of securing favorable conditions of water flows"; and
- "to furnish a continuous supply of timber for the use and necessities of the people of the United States".

This is the familiar "Organic Act" for the National Forest System. The Organic Act and the National Forest Management Act (NFMA) passed in 1976 are the primary statutory authorities for management of NFS lands by the United States Department of Agriculture's Forest Service.

Beginning with passage of the 1897 Forest Reserve Act, Congress, Federal forest managers, and the public have continually clarified, refined, and modified Federal policy for National Forest management. Since the 1960's this has been a matter of responding to a developing body of environmental law as well as to changes in legislation specific to National Forest management.

Chasing Multiple Use

One piece of legislation which attempted to clarify, refine, *and* modify forest management practices, all at the same time, was the Multiple Use Sustained Yield Act of 1960. The concept of multiple use has been discussed, attacked, and defended *ad nauseam*. As a management philosophy it has inspired devotion in legions of National Forest employees, been embraced belatedly by commodity development interests, and always viewed negatively by the Sierra Club.

The Multiple Use Act passed Congress at a time when numerous single interest groups were claiming rights to National Forest resources. The Forest Service crafted the legislation as a means of responding to those outside pressure groups, most notably the growing numbers of people desiring greater recreation opportunities within National Forests. At the same time the Forest Service sought to position multiple use as simply another way of describing traditional practices for managing NFS lands.

The Multiple Use Act did a rather poor job of setting standards for forest management. The definition of multiple use in Section 4(a) of the bill follows:

> Multiple Use means the management of all the various renewable surface resources in the combination that will best meet the needs of the American people; making the most judicious use of the land for some or all of these resources or related services over areas large enough to provide sufficient latitude for periodic adjustments in use to conform to changing needs and conditions; that some land will be used for less than all of the resources; and harmonious and coordinated management of the various resources, each with the other, without impairment of the productivity of the land, with consideration being given to the relative values of the various resources, and not necessarily the combination of use that will give the greatest dollar return or the greatest unit output.

This broad, somewhat vague definition side-stepped important questions about competing uses of the National Forests and left the Forest Service ample room to develop its own management standards. Unfortunately, at the time and for decades afterward, Forest Service leaders did not recognize the depth of the growing public desire for Forest uses other than commodity development and campgrounds. Nor did they fully appreciate the growing concern over the biological condition of the National Forests, though the Sierra Club's negative stance on the Multiple Use Act was an early indicator of that concern.

Even passage of the Wilderness Act in 1964 and the National Environmental Policy Act (NEPA) in 1969 failed to engender significant changes in overall management emphasis. Sincere and significant efforts were made to change the processes involved. Environmental analysis became standard and mitigation of impacts from management activities was required. Still, managers were judged primarily by timber volume sold. Over large portions of the National Forests, using some land for "... less than all of the resources ..." was equated to single resource use or use dominated heavily by a single purpose, timber management.

In 1976 Congress passed the National Forest Management Act (NFMA), legislation which changed the face of National Forest planning. NFMA mandated preparation of Land and Resource Management Plans for all National Forests and required that Plans be developed using an interdisciplinary process. The use of interdisciplinary teams began with NEPA, but involvement in the strategic planning process as required by NFMA gave non-foresters new stature and ultimately has been a major influence toward diversification of management in the Forest Service. That diversification, along with another piece of legislation, the Endangered Species Act (1973), and critical forest health problems over large portions of the western National Forests, brought the Forest Service to its current effort to implement a new management philosophy, ecosystem management, on NFS lands.

Ecosystem management and multiple purpose forests

In October 1994, Forest Service Chief Jack Ward Thomas laid out a "Course to the Future" (10/25/94) which focused attention on the Agency's stated land and service ethics as follows.

> Forest Service land ethic is to: "Promote the sustainability of ecosystems by ensuring their health, diversity, and productivity."
>
> Forest Service service ethic is to: "Tell the truth, obey the law, work collaboratively, and use appropriate scientific information in caring forthe land and serving people."
>
> The Chief further committed the Forest Service to demonstrating its land and service ethics through the "... development and practice of ecosystem management."

The "Course to the Future" defines ecosystem management as "... the integration of ecological, economic, and social factors in order to maintain and enhance the quality of the environment to best meet current and future needs." It goes on to set priorities for providing sustainable benefits as follows:

1. Protect Ecosystems
2. Restore Deteriorated Ecosystems

3. Provide Multiple Benefits for People Within the Capabilities of Ecosystems
4. Ensure Organizational Effectiveness

The dominance of multiple purpose forests in the National Forest System is directly supported by (1) and (3), which are further defined in the "Course to the Future" as (1) protecting ecosystems while meeting people's needs and (3) providing multiple benefits for people within limitations of maintaining ecosystem health and diversity, with forests and rangelands meeting people's needs for uses, values, products, and services. As with the original definition of multiple use, this is a large order for National Forests to fill. A major question to be answered on a case-by-case basis is "At what level can the National Forests meet people's needs while maintaining ecosystem health and diversity?"

Owls and their pals: the influence of rural development legislation and policy

The listing of the Northern Spotted Owl as an Endangered Species has had tremendous impacts on disposition and management of National Forest lands. This is true directly in the geographic area of the owl's habitat, the Pacific Northwest and Northern California. But the impacts are much wider ranging, because the reduction in timber harvest associated in part with listing the spotted owl and managing its habitat was the primary contributor to the rural economic development provisions of yet another piece of legislation, the 1990 Farm Bill.

The 1990 Farm Bill (P.L. 101-624) provided several incentives to rural economic diversification. First, it encouraged Forest Service employees to get involved with efforts to diversify natural resource-dependent economies. It supported the use of public/private community development teams to analyze the existing economic situations of rural communities and develop plans for diversifying resource-dependent communities. Finally, the Farm Bill authorized appropriation of grant monies, to be administered through the Forest Service, to assist in funding community planning efforts and give a jump-start to new businesses seeking to become established in these communities.

Congress actually followed through with limited funds for Farm Bill activities, and has done so each year since the original bill's passage. And although the bill was conceived in the devastated situation of timber-dependent communities affected by reductions in Forest Service timber sale volumes, its provisions were applied widely across the United States where rural communities were suffering due to natural resource-related economic depression.

In implementing the 1990 Farm Bill, some National Forests and their Ranger Districts have experienced an unexpected side effect. As they become increasingly involved in community economic diversification/development planning, Forest Service employees have found themselves taking advantage of partnerships and cooperative relationships to accomplish projects within the Forest boundaries which directly benefit Forest users and indirectly support the communities.

As a result we are increasingly recognizing what we have known in our hearts all along – the relationship is a symbiotic one. A Ranger District has a place in the ranks of a community's economic structure, side-by-side with other local business and service providers. The National Forest cannot occupy a position on the outskirts of town, influencing but never being influenced by local communities. This is not to say that National Forests can or should ignore the man from Hoboken or the needs of the birds and beasts. But it does illustrate that the National Forests have their metaphorical "feet" firmly planted in a number of diverse communities of users. Diversification of forest management, and increasing emphasis on multiple purpose forests of all types, is supported by this situation.

Characteristics of multiple purpose planted forests

Understanding the reasons that planted forests in National Forests are increasingly multi-purpose in nature makes describing them fairly straightforward. Multi-purpose planted forests on National Forest System lands typically:

- Have a history of use by contemporary humans that includes but is not necessarily limited to timber harvesting and tree planting.
- Have variable rotation lengths or no specified rotation length; i.e. are managed as uneven-aged forests.
- Are managed to allow trees to become much older than in conventional National Forest plantations or in tree farms – 80–200 years is, and will become more, common, given the land allocations and management prescriptions in existing Forest Land and Resource Management Plans.
- Often are mixed species forests with a variety of age classes and complex stand structures.
- Incorporate management activities and sometimes silvicultural practices designed specifically to enhance, sustain, or restore characteristics which support multiple uses.

Sweet Home Ranger District: Embracing the concept

Multiple purpose forests should be the dominant component of National Forest lands such as the Sweet Home Ranger District. In the early 1990's opportunity visited the Sweet Home District, a part of the Willamette National Forest, disguised as a massive reduction in its timber harvest program, a new mandate to implement ecosystem management, and a rural community, Sweet Home, Oregon, trying to re-awaken an economy that had been almost entirely dependent on timber from National Forest lands for more than half a century. We at the Sweet Home District (I was District Ranger there) embarked on a sort of "diversification of management philosophy" that was followed by actual diversification of management activities. In many ways, the course was parallel to that facing the community of Sweet Home in its attempt to economically diversify. The result is a roughly 80,900 ha Ranger District that should eventually be predominantly multi-purpose forest over its entire area, planted forests included.

Sweet Home Ranger District is located in west-central Oregon approximately 160 km southeast of Portland. It is one of seven Districts that make up the Willamette National Forest. Sweet Home has traditionally been a heavy contributor to the Willamette's timber harvest, with sales totaling over 500,000 m^3 as recently as 1989–1990. Also traditionally, the District had been widely viewed as largely a single purpose unit, with little to offer its users other than its timber output. This is, of course, an oversimplification. Many other purposes were served by the District throughout its history. However, those other purposes paled in comparison to the dominant timber management program, to the extent that even identifying significant secondary purposes was difficult.

Fortunately, over the years Sweet Home Ranger District's timber program was managed with foresight and uncommon sensitivity, and the land now lends itself well to a more balanced management philosophy. This District and others are in the forefront of what John Gordon (1994) has described as a "paradigm shift" away from management which relies on land allocations, sustained wood yield, and multiple use and toward ecosystem management, or some similar philosophy. In the early 1990's Sweet Home Ranger District began its transition by: (1) dramatically increasing its emphasis on targeted amenity uses; (2) devoting effort to developing a viable special (or miscellaneous) forest products program; and (3) seeking a sustainable timber harvest program planned and implemented on a landscape level, with ecosystem management concepts in mind.

The path embarked on is far from smooth. The President's Plan for Late-Successional and Old-Growth Related Species, for example, resulted in large blocks of District land being devoted largely to habitat and habitat restoration. But the start toward making the District multi-purpose is there. Given the scale of previous timber harvest on the District and the large acreage in second-growth status, the three-pronged attack to create a better balanced management program has included a variety of projects within, and new uses/purposes for, the District's planted forests.

These projects are not always new ideas for the National Forest system. But their combination and implementation as part of a philosophy of management diversification is significant. They include:

- the Longbow Institute, a cooperative project between the District and Oregon State University's College of Forestry, which was based on doing conservation education for teachers in an on-site, resident student program.
- hosting an Elder Hostel program in partnership with Linn-Benton Community College to introduce participants to a *forest* (as opposed to timber) based economy.
- leading a pilot "Jobs-In-the-Woods" program, linking ecosystem management and restoration projects with providing family-wage jobs and training for local community dislocated timber workers; – partners included Bureau of Land Management, Oregon Department of Forestry, Oregon Economic Development Department, U.S. Department of Labor, Oregon State University, and the University of Oregon.
- a variety of Special Forest Products projects, including one that combined thinning plantations with Christmas tree sales.
- enhancing elk habitat in old plantations, as well as in association with new timber sales.
- development of a highly accessible recreation area along the South Santiam River and in adjacent environs, with accessible fishing platforms, wildlife viewing areas, interpretive trails (through a variety of forested conditions, including both old growth forests and current/former plantations) – this targeted a recreation user population known to be largely senior citizens and families.
- beginning to manage forest resources at the landscape level, focusing on identifying an appropriate management mix that satisfies people's needs while preserving ecosystem health – based on the Forest's Land and Resource Management Plan, in areas such as the Santiam Pass.

In 1991 Sweet Home Ranger District was recognized by the Forest Service's Pacific Northwest Region as the "District of the Year", primarily because of its commitment to and actions taken toward making the myth of multiple purpose forests a reality.

Conclusion

Payoff or paying the piper?

Managing U.S. National Forest lands for multiple purposes has many advantages and potentially beneficial outcomes. In general, multiple purpose forests help the Forest Service to serve its many masters by fostering a wide variety of forested conditions that lend themselves to a variety of purposes. The fact that managers can do this while causing minimal or relatively modest disturbances of natural ecosystems, and even capitalize on some uses to restore deteriorated ecosystems, makes multiple purpose forests the natural dominant type for the Forest Service's "Course to the Future." Added to this, multiple purpose forests do a much better job of preserving future management options than the abandoned "timber-dominant" paradigm.

In today's socio-economic climate, a management philosophy which makes extensive use of multiple purpose forests is the only one with a chance of being considered politically and socially acceptable by all of those publics of National Forests who maintain that "... this land is your land, this land is my land ...". At the same time, there are major uncertainties and risks involved in such a management regime. First, the more we learn about ecosystem management, the more we need to know. This includes information about silvicultural practices as well as public satisfaction. There will no doubt be false starts and course corrections for decades to come. Second, managing planted forests for multiple benefits entails a very complex forest management system. Whether we can actually manage large tracts for multiple benefits over the long term is unknown. Finally, our system of democratic government, which serves us well most of the time, could be the falldown of forest management regimes based on multiple purpose forests. The constantly changing political interests coupled with the natural tendency of the majority of our publics to focus on special interest, primarily short-term issues will ultimately shape the success or failure of a National Forest System dominated by multiple purpose forests.

Reference

Gordon, J.C. 1994. From Vision to Policy: A Role for Foresters. J. For. 92: 16–19.

New Forests **17:** 129–143, 1999.

Sustainable plantations of high-yield *Eucalyptus* trees for production of fiber: the Aracruz case

EDGARD CAMPINHOS, JR.
Research and Technology Center, Aracruz Celulose SA, Aracruz, Brazil

Key words: planted forests, short-rotation species, ecological impacts, economic viability

Abstract. The importance of planted forests for human communities is evident today because of the increase in world population year after year. In order to supply the constantly increasing requirements for wood for fiber, energy, civil construction, resin, latex, and other needs, new forests have been established by adopting new concepts of technology. These forests, especially those with fast-growing and short-rotation species, incorporate the objectives of quality, productivity, and environmental protection.

The *Eucalyptus* plantations established by Aracruz Celulose S. A. to produce pulpwood have strongly contributed to the evolution of forest science, especially in the fields of silviculture, propagation, improvement, genetics, biotechnology, nutrition, environment, and disease control related to eucalypt species. Significant investments in research and development were required, as well as technological adaptation to the local conditions. Results of these investments in productivity and quality defined the technical feasibility to build a pulp industry and its expansion to reach 1,025,000 t/yr of bleached pulp.

As a result of these advances, the mean increment of pulp productivity has been improved significantly – from 5.9 air-dried t/ha/yr to 10.9 t/ha/yr (Fenton and Romero 1995; Bertolucci et al. 1995). This became possible by selecting trees more adapted to each site, with high yields of high quality fiber, and with eucalypt clonal plantations with an average rotation age of 7 years. Additional studies of Aracruz have promoted advances in industrial processes, as well as definition of fiber characteristics of mother trees to be in attendance with requirements of markets, thus providing an important competitive advantage (Campinhos and Cláudio-da-Silva 1990).

The care for the environment – both forest and industrial – has been always present and has been the key point for the good performance of the project. This fact has created a world emulation that strongly contributes to decreasing the pressures on native forests, as a result of the consequent utilization of smaller land areas to produce a desired wood volume. This total project is an example of domestication (Eldridge et al. 1994) of the forest genus *Eucalyptus* with exceptional qualities and represented by more than 500 species and hybrids (Campinhos et al. 1992).

Introduction

The global increase in the demand for wood is causing rapid change in the concept of tree cultivation, and is promoting the utilization and development of new technologies. Better quality woods are being produced. Further, more productive and lower cost planting is taking place through methods that protect the environment, and plants are being selected for sites that previously were considered to be marginal (FAO 1988). It is a new and attractive business opportunity for forestry operations that allows the use of modern growing techniques, stimulates the formation of wood cooperatives, and induces governments to provide incentives for the cultivation of trees alongside corn, soya, rice, wheat, coffee, and other crops.

Fast-growing and short-rotation species, without a doubt, offer significant competitive advantages and versatility. They are the species of choice, especially those belonging to the genus *Eucalyptus*, for setting up high-yield projects for a number of reasons: (1) they adapt well to different ecosystems; (2) naturally occurring native populations already exist; (3) they mature precociously and are well shaped; (4) species may be crossed to produce vigorous hybrids and wood with new characteristics; (5) it is now possible to clone various species and hybrids; and (6) they produce wood with good properties for the purposes of fiber, sawtimber, chipboard, charcoal, posts, and civil construction (FAO 1988).

Demand for eucalypt market pulp has been growing at 11.2 percent/yr since 1980 (Wilson et al. 1995). Consequently there has been an increase in plantings, pulp mill expansions, and new mill construction. The traditional center of pulp production has moved from the temperate and subtemperate to tropical and subtropical countries, where there is plenty of available land, more light, and year-long growing seasons.

This paper focuses on the genus *Eucalyptus*. Other fast-growing forest genera such as *Acacia*, *Gmelina*, and *Pinus* are also used for pulp production, although plantings in tropical and subtropical regions are dominated by *Eucalyptus* and *Pinus* (Fenton and Romero 1995).

Eucalyptus around the world

Natural occurrence

The plant kingdom produces few species as numerous or versatile as the *Eucalyptus* – there are some 500 species. Only two of these species are not found in Australia: *E. urophylla* in Timor and nearby islands in Indonesia, and *E. deglupta* in Papua New Guinea, Irian Jaya, and Moluccas in Indonesia, and on Mindanao in the Philippines. The limit to natural occurrence is between latitudes 9° N and 44° S (7).

The genus *Eucalyptus* occurs from sea level up to 1,800 m in altitude. There are some species that are bushlike and some of great size, whose crowns soar to 90 m and whose trunks can grow to 6 m in diameter (Kelly 1993).

Planted areas and products

Eucalyptus is the genus of trees that is most planted around the world because of its large number of species, the ability to adapt to many different sites, and the possibilities of producing wood for many different uses. It can be planted in most regions of tropical and temperate climates between latitudes 45° S and 40° N, with a high degree of tolerance of extremes in latitudes and altitudes. *Eucalyptus globulus*, for example, occurs naturally between latitudes 38.5° S and 43.5° S below 1,000 m. Nevertheless, it has adapted well in Ethiopia (latitude 12° N and 2,500 m altitude) and in Peru (latitude 5°–10° S and 3,000 m altitude) (FAO 1988).

Large tracts of eucalypt forest were first planted in Brazil with the introduction of many species in 1904 by Edmundo Navarro de Andrade. Today there are an estimated 3,000,000 ha of planted areas, principally of *E. grandis*. India is in second place, with 550,000 ha, mostly *E. tereticornis*. It was estimated in 1973 that Spain had 390,000 ha, primarily *E. globulus* and *E. camaldulensis*, whereas Portugal had 300,000 ha of *E. globulus* in 1983. These two countries significantly increased their eucalypt plantings during the 1980s. In 1989, South Africa had about 470,000 ha planted in eucalypt, of which some 300,000 ha were *E. grandis*. Angola also appears on the list, with about 390,000 ha of *E. tereticornis* in 1985 (Eldridge et al. 1994) (Table 1).

China has been rapidly expanding its eucalypt plantations; in 1982, it was estimated to have more than 400,000 ha in, primarily, *E. citriodora*, *E. exserta,* and *E. globulus*. In the south, China recently introduced the *E. grandis* × *E. urophylla* hybrid with good results. In Ethiopia, it is estimated that there are over 250,000 ha planted with *E. globulus* and *E. camaldulensis* (Eldridge et al. 1994).

In Chile, eucalypt plantations also have recently undergone rapid expansion. There are now an estimated 250,000 ha under cultivation, half in *E. globulus*. In 1981, Uruguay had 110,000 ha of *E. globulus* and more recently began plantations with *E. grandis*. Argentina has excellent environmental conditions for growing *E. grandis,* and in 1985 there were some 240,000 ha in eucalypt forests. It is estimated that eucalypt-planted forests exist, in smaller sizes, in over 100 other nations (FAO 1988; Turnbull). Because of the need

Table 1. Estimates of approximate area of eucalypt plantations in the countries with the largest plantation areas (*Source*: Eldridge 1984).

Country	Area (1,000 ha)	Principal species	Source
Angola	390	*E. tereticornis*	FAO 1985[a]
Argentina	240	*E. grandis*	Golfari 1985
Brazil	2,500	*E. grandis*	Galvão and Couto 1984
Chile	60	*E. globulus*	Prado and Barros 1989
China	300	*E. exserta*	Turnbull 1981
Ethiopia	250	*E. globulus*	Davidson 1989
India	550	*E. tereticornis*	FAO 1985[a]
Morocco	180	*E. camaldulensis*	Jacobs 1981
Peru	90	*E. globulus*	Jacobs 1981
Portugal	430	*E. globulus*	Pereira and Pereira 1988
South Africa	470	*E. grandis*	Anonymous 1990
Spain	390	*E. globulus*	San Miguel 1988
Uruguay	110	*E. globulus*	Jacobs 1981
More than 60 other countries	500[b]		
Approximate world total in 1985	6,000		

[a] "FAO" is from 1985 summaries of Inventory and Global Environmental Monitoring Projects by the Food and Agriculture Organization of the United Nations, Rome.
[b] Estimate is less reliable than those for the thirteen countries listed.

to preserve native forests, plantations of fast-growing trees are on the rise. It is estimated that around the world some 6,000,000 ha of eucalypt-planted forests currently exist.

The uses and by-products of eucalypts are the most varied imaginable, e.g., for firewood, charcoal, windbreaks, shade, pulp, mine props, fencing, volatile oils, honey, civil construction, railroad sleepers, furniture, fiberboard, plywood, chipboard, cosmetics, and cleaning products. The greatest industrial use for eucalypt wood is for the production of pulp and paper because of its qualities, availability, and price. The demand for eucalypt pulp is growing rapidly and, thus, new mills are coming on line, others are being expanded, and new forests are being planted. World production of bleached eucalypt kraft pulp in 1994 was 4,000,000 t, of which 2,200,000 t came from Brazilian mills. The second most important use of eucalypts is as wood for fuel – firewood for domestic use and in light industry, and as charcoal for the iron and steel industries in Brazil and elsewhere (Eldridge et al. 1994).

Table 2. Rotation time and productivity (mean annual increment) of some forest species for pulp.

Species	Country	Rotation (yr)	MAI (m^3/ha/yr)
		Short fiber	
Eucalyptus spp.	Brazil (Aracruz)	7	40.0
Eucalyptus grandis	South Africa	8–10	20.0
Eucalyptus globulus	Chile	10–12	20.0
Eucalyptus globulus	Portugal	12–15	12.0
Eucalyptus globulus	Spain	12–15	10.0
Betula spp.	Sweden	35–40	5.5
Betula spp.	Finland	35–40	4.0
		Long fiber	
Pinus radiata	Chile	25	22.0
Pinus radiata	New Zealand	25	22.0
Pinus spp.	Brazil	15–20	16.0
Pinus elliottii and *P. taeda*	United States	25	10.0
Pseudotsuga menziesii	Canada (coast)	45	6.0
Picea abies	Sweden	70–80	4.5
Picea abies	Finland	70–80	3.6
Picea glauca	Canada (interior)	55	2.5
Picea mariana	Canada (east)	90	2.0

Eucalypt plantations: Prospects and challenges

The increases in volume yields of eucalypt forests around the world are quite varied, and depend on genetic potential, the species, its interaction with the ecosystem, and the type of cultivation utilized. Mean Annual Increments (MAI) of from 10 m^3/ha/yr (Table 2) to as much as 90 m^3/ha/yr on small sample plots (Eldridge et al. 1994), with rotation periods of between 5 and 25 years, are recorded; values depend on the final use of the wood and whether clearcut or thinning is used. One of the great advantages of eucalypt is its ability to coppice, thus generating new wood production from the same plantation. With these potentially high increments and an ability to grow in and improve marginal or poor soils, eucalypt culture has become a great ally to mankind, thus helping to preserve natural forests and the use of good soils for food production, as well as generating jobs.

A frequent question about eucalypts regards their better growth performance as exotics compared to natural growth in regions of origin. Scientists point out a number of reasons for this. In the natural habitat, regeneration suffers from the attack of insects and diseases that are absent in other environments. Moreover, there is competition for light, water, and nutrients on the part of other plants that grow in the same environment and dispute space. The physical condition of the soil in many cases is inadequate for the development of a good root system. And there also is the danger of fire and other negative factors (Eldridge et al. 1994).

Industrial plantations of eucalypt must be economically viable and, therefore, seeds of good quality are selected for use, seedlings are carefully selected, the soil is well prepared and fertilized, there is control of potentially damaging competing plants, there is sufficient space between trees, and protection is provided against erosion and fire. Eucalypt plantations on a short-rotation basis are true cultivation. Because plantation trees have been genetically improved for each site, they adapt and interact well with the local environment, coexisting with other organisms and using the soil's available nutrients to grow in accordance with the existing rainfall (Eldridge et al. 1994).

In various parts of the world eucalypt plantations have generated much debate regarding their ecological impacts. The biggest argument against extensive eucalypt plantations is that they reduce the availability of water in a given region. On the other hand, soil that is well covered with trees, dead leaves, and branches promotes greater availability of water in rivers during periods of drought, and diminishes the possibility of erosion. If the trees are removed there will be periodic cycles of flooding alternating with little usable river water (FAO 1988). The characteristics of a region where a eucalypt or other fast-growing tree forest is planted must be carefully studied to assess feasibility in relation to local hydrology (FAO 1988).

Another argument against eucalypt forests is that the plantations consist of gigantic silent environments, a green desert without biodiversity, monotonous, and not aesthetically pleasing. It is necessary to understand that a eucalypt plantation is an ecologically simplified crop similar to sugar cane, coffee, or corn, and not an exuberant tropical rain forest with its hundreds of plant species and many animals. In fact, there is an additional advantage. Even in its shortest rotation periods (6–8 years for pulp), the understory of a eucalypt forest contains considerable biodiversity, something that does not occur in wheat, soy bean, or corn fields, for example (Eldridge et al. 1994). Industrial plantations of eucalypts are monocultures, normally of short rotation and high productivity, that take up relatively small amounts of land area.

These sites are planted using the highest levels of tree farming techniques in order to be economically sustainable.

One of the great indirect advantages of high-yield eucalypt forests is their biomass fixation of excess atmospheric carbon. A large part of this carbon remains fixed for long times, transformed into paper, furniture, buildings, and other wood by-products (FBDS 1994).

In order to set up large, homogeneous tree plantations, especially those of exotics, planning for the maintenance of large tracts of native vegetation should be done, particularly near rivers, lakes, and steep areas, to avoid the total disappearance of local biodiversity. This measure diminishes the possibility of attack by pests, especially if the forest is made up of material genetically adapted to and improved for the region. Consequently, there will be satisfactory interaction between the trees and the environment.

Regarding questions raised about plantations of the genus *Eucalyptus*, solutions are being developed as a result of the accumulated work of dozens of scientists around the world who are studying all aspects of this tree, and adding to the existing extensive base of experience with it. Fortunately there is international recognition of the great importance of the genus *Eucalyptus*, both as a source of renewable products and as a participant in its native ecosystems. Australia has developed a fantastic conservation program to protect its native eucalypt forests, through national parks, forest reserves, and state forest systems, that should guarantee the perpetuity of the genetic variability of the species in natural habitats (Eldridge et al. 1994).

Each year the CSIRO and state forestry services in Australia organize the collection of thousands of eucalypt seeds from many different species and provenances, to supply scientists and the special assistance programs that Australia, the UNDP, and the Food and Agriculture Organization of the United Nations (FAO) carry out in developing nations. Also encouraged are research projects conducted by private companies and forestry services in other countries. Large amounts of seeds are collected by authorized professionals to be used both at home and abroad.

World trends in the production of industrialized wood for pulp making

The production of industrialized wood around the world grew by nearly 16 percent over the past 15 years, with a number of countries being standouts, especially Brazil, Chile, and Uruguay (9) (Table 3). A large portion of the eucalypt plantations that have come into being over the past 15–20 years in the tropics and subtropics are destined for producing wood for pulp. According to the FAO, the total amount of pulp exports around the world has increased about 40 percent during the last 15 years, and the greatest increases

Table 3. Industrial wood production in 1977 and 1992 (*Source*: adapted from Fenton and Romero 1995).

Country	Production (1,000,000 m^3)		Change
	1977	1992	(%)
Brazil[a]	37.32	77.71	108
Chile[a]	6.02	17.77	195
Uruguay[a]	0.35	1.04	201
New Zealand[a]	9.49	14.99	58
South Africa[a]	9.18	12.60	37
Spain[a]	10.60	15.11	43
Portugal[a]	7.42	10.31	39
Sweden	44.26	49.15	11
United States	320.44	402.50	26
Canada	143.37	179.22	25
World	1,381.20	1,603.38	16

[a] These countries have relied extensively on industrial forest plantations mainly with exotic species.

have been in those countries that have been most active in establishing plantations utilizing fast-growing species, e.g., Brazil, New Zealand, Chile, and Spain. In South America, pulp exports have increased from 433,000 t in 1977 to 2,665,000 t in 1992. Brazil and Chile are the principal exporting nations (Fenton and Romero 1995). In Brazil, the export of pulp refers to bleached eucalypt, which increased from 927,000 t in 1987 to 1,897,000 t in 1994, with 52 percent produced by Aracruz Celulose S. A. (Aracruz 1995). Brazilian pulp exports in 1994 accounted for revenues of U.S.$788 million (ANFPC 1993). In North America, the United States has nearly tripled pulp exports because of productions from its forests (mainly pines) in the Southeast. Tree improvement and intensive silviculture have increased, and will continue to increase, the productivity in all of the world's plantations. As indicated in Table 4, industrial woods derived from fast-growing forests have shown significant increase over the past two decades, thus becoming an important part of the world pulpwood supply (Fenton and Romero 1995).

Eucalyptus in Aracruz Celulose S. A.

Aracruz Celulose S. A. is one of the world's leading suppliers of bleached market eucalypt pulp, for tissue, writing, and printing papers. The Aracruz

Table 4. Pulp exports in 1977 and 1992 (*Source*: adapted from Fenton and Romero 1995).

Country	Exports (1,000 t)		Change
	1977	1992	(%)
Brazil	95	1,630	1,616
Chile	339	939	177
New Zealand	477	592	24
Spain	83	569	586
United States	2,415	6,552	171
Canada	6,090	8,860	45
Fast-growing plantations[a]	1,312	17,070	1,201
World	17,070	25,500	43

[a] These figures represent 7.7 and 66.9 percent, respectively, of the world total for 1977 and 1992.

Celulose mill is located in the municipality of Aracruz in the state of Espírito Santo (500 km north of Rio de Janeiro), surrounded by 38,170 ha of eucalypt plantations (Figure 1). The company owns another two areas planted in *Eucalyptus*: one in the north of Espírito Santo containing 44,764 ha, and another in the south of the state of Bahia with 49,213 ha, totaling 132,147 ha. Moreover, the company owns 56,667 ha of land containing native vegetation and 14,137 ha of land with roads, firebreaks, reservoirs, and buildings, for total landownings of 202,951 ha (Aracruz 1995).

The industrial site consists of two plants with a nominal annual production capacity of 1,025,000 t; facilities for recovering chemical products, for storing and treating water and steam, and for generating electricity; electrochemical plants to supply chemicals for the pulping process; and the only port in Brazil specialized in shipping pulp. The company exports about 90 percent of its production, principally to clients in the United States (31 percent), Europe (37 percent, of which 7 percent is to Germany) and Asia (23 percent). Currently Aracruz accounts for 3 percent of the total world short-fiber market pulp production and 22 percent of the world's market eucalypt pulp.

All of the company's activities are conducted in accordance with the highest principles of technology, quality, and sustainable development, which means the promotion of social and economic development in harmony with nature. Aracruz is an active member of the Business Council for Sustainable Development, and was one of the original signatories of the ICC's Business Charter for Sustainable Development.

The company is self-sufficient in high-quality eucalypt wood and conducts all of the operations necessary to provide its two plants with the 4,000,000 m^3

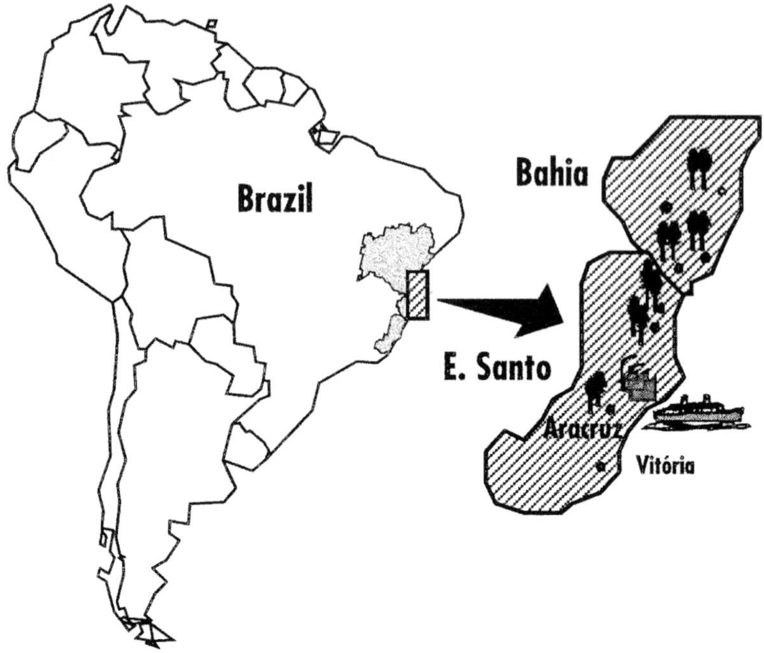

Figure 1. Integrated project location: forests, mills, and Port of the Aracruz Celulose Mill, located in the municipality of Aracruz in the State of Espírito Santo (500 km north of Rio de Janeiro).

of wood with bark required each year. These include a nursery for the production of seedlings and clones, planting, weed and pest control, fire protection, tree felling, timber transport, road and firebreak maintenance, forest inventory taking, and research and development.

Aracruz began its activities in 1967 in Espírito Santo state, where it purchased flat, bare, and unworked land near the coast, low-lying (under 60 m in altitude), with water supply sufficient for its industrial needs and with climatic conditions favorable for the cultivation of fast-growing eucalypt for the production of bleached pulp. Because of the lack of local experience with eucalypt plantations and the lack of improved seeds for the local conditions around Aracruz, commercial Brazilian *E. grandis*, *E. saligna*, and *E. urophylla* seeds were introduced, because they are the traditional pulp-producing eucalypt species in Brazil. However, after 2 years, some problems appeared, including the following:

- Occurrence of a large amount of phenotypic problems in individual trees, thus demonstrating that the seeds were of low genetic quality as a result of interspecific crosses.

- A heavy mortality rate and suppressed growth affecting 40 percent of trees in the seventh year.
- A strong attack on tree trunks of canker disease caused by the *Criphonectria cubensis* fungus, which particularly affected *E. saligna* (90 percent). *Eucalyptus grandis* was affected to some extent (35–40 percent), whereas *E. urophylla* suffered very little (5 percent).
- Death of trees attributed to defective root systems as a result of the seedling production system in use at the time.
- The system used to control leaf-cutting ants was inefficient and, consequently, large areas, reaching as much as 2 percent of the total, were killed off.

In the face of these facts and because Aracruz was a pioneer to the region, the need to put together a technical team to develop and adapt the necessary technologies to the local conditions soon became evident. This began in 1971, with the execution of a well-planned research program. There was never any doubt what should be the priority for developing adequate genetic material for the local ecosystem and final use of the wood. After studies regarding the potential eucalypt species for the region were carried out, a number of *E. grandis* mother seed collection trips were made to different North Australian provenances, and *E. urophylla* was collected on Indonesian islands. With this material, genetic work was conducted to improve the two species. Another 52 eucalypt species also were introduced, with the idea that hybrids for pulp and other future uses – such as wood for sawmills and energy generation – would be developed from them (Campinhos and Cláudio-da-Silva 1990).

With relation to the problems found in seedling production, a new technology adapted in Hawaii by G. Walters of the U.S.D.A. Forest Service (Walters 1981) was employed. It consisted of using conic-shaped polyethylene tubes, 14.0 cm in height and 2.5 cm in diameter, with a light substrate vermiculite and eucalypt bark compost and trays to support the tubes. Later, the seedlings were fertilized with a slow-release fertilizer that supplies nutrients over a 90-day period, the time required for the seedlings to develop.

Ant outbreaks were monitored and systematic control was instituted, thus avoiding new problems with them. Today, damage caused by ants is insignificant, and they are controlled through an efficient, low-cost program.

Despite the fact that eucalypts are fast-growing trees, it took 12 years to start up the program of producing seeds in the first stage of improvement. Thus it was in 1974 that the first studies for adapting vegetative propagation methods to local conditions through rooting cuttings were undertaken. The objective was to take advantage of the spontaneously generated hybrids from the initial plantings that were behaving well and that showed good characteristics for pulp making. Cloning would rapidly create the proper conditions for better and more productive forests.

A search was begun to discover the best trees out of a population of 60 million individuals. Some 6,000 were found with the satisfactory characteristics: disease and pest free, high volume, well-shaped, self-pruning, thin branches, and the ability to coppice and to produce rooted cuttings (Campinhos et al. 1992). Later, full studies were conducted on the wood characteristics of the clones developed for paper and pulp purposes, and those that did not qualify were eliminated. In parallel, pure *E. grandis* and *E. urophylla* trees were being chosen and hybrids were being developed in seed orchards.

Advances came about rapidly once advantage was taken of the clones and their ability to produce a specialized eucalypt that maintained the characteristics desirable for both tree culture and industrial purposes. Further advances were recorded when the existence of clones that adapted well to various sites was discovered along with others that were site specific. The soils were mapped throughout the company's 132,000 ha of potential forest lands, and tests were conducted to evaluate the degree of genetic interaction in a given environment. Based on the results, clones that were well adapted to each type of site were planted.

Risks for mass propagation of clones are significant and are associated with the planting of one genotype over a wide area. In this case, the genetic base may become so restricted that it limits selections and later gains. This can lead to a low adaptation to fast-changing pest and pathogen mutations that then attack the trees. On the other hand, the potential for gains with clonal plantings cannot be obtained because of the low clonal capacity for adaptation to different environmental conditions (Campinhos et al. 1992). Presently, thanks to technological advances, these risks have been reduced through better planting management.

Forests should be established with a large genetic base of clones adapted to different environmental conditions. To do this, it is indispensable that competitive tests with clones be carried out in different sites or on predominant soils. In addition, the study of physiological parameters of water relations and their relationship with primary productivity and the partitioning of photosynthates takes on great importance in the selection of clones that are tolerant of water deficits, are productive, and have a diminished environmental impact. Through these strong environmental and genotype tests, selection can be made to obtain clones that are well-adapted to specific conditions and others that are well-adapted to a range of environmental conditions (Campinhos et al. 1992). A mosaic distribution of clones at each site is recommended in order to reduce the risk of low local genetic variability. Although many works discuss the size of a site and the ideal number of clones per block, there is as yet no scientific conclusion defining the best model.

Aracruz-cloned forests, begun in 1989, have shown good environmental stability. This seems to be mostly associated with the presence of areas preserving native forests, and with environmental adaptation, because most of the cloned mother trees have coexisted in their ecosystems for a number of years before they were selected.

The biodiversity, and consequently the stability offered by this system, can be seen because only 0.02 percent of the pest attacks that have been monitored over the past 12 years required any type of outside intervention. Thus, the large majority of pest attacks in the eucalypts was naturally controlled by the local fauna, principally avifauna. It has, therefore, been possible to maximize wood production, to adopt the best management practices for each soil type, and to create an adequately balanced fertilization program for each clone at optimized cost, thus giving the proper conditions for maintaining sustainable production of wood.

The MAI of the initial forests was 28 m^3/ha/yr, rising gradually to the 1995 level of 45 m^3/ha/yr, with a 7-year rotation. Specific consumption of wood to produce 1 t of air-dried bleached pulp decreased from 4.9 m^3 wood/t pulp to 4.1 m^3wood/t pulp (outside bark measure). The most important result has been the increase in productivity of bleached pulp, which went from 5.9 air-dried t/ha/yr to 10.9 t/ha/yr (Ikemori et al. 1994).

To assure the sustainable production of wood, the company has invested in research and development to improve its forests, its pulp-making processes, and its products. Each year it has been possible to improve and select genetic material for the forests, this resulting in better quality. To achieve this level of quality, Aracruz had to develop and adapt technologies, especially those related to eucalypt culture and soil management. The key point was the creation of good channels of communication and cooperation with researchers, forestry and industrial professionals from both the public and private sectors, in order to exchange experiences.

The world short fiber pulp market is especially competitive, as was proven during the 1991–1994 period, when prices fell drastically and many mills had to close because of the high cost of wood and production. In such a competitive scenario it is necessary to be constantly improving forestry operations, as well as developing new technologies associated with the processes of pulp production and new product development.

Conclusion

Wood is the principal raw material for the production of paper and pulp, whose consumption has increased significantly and, as a consequence, generated an increase in the plantations of short-rotation, high-increment forests in tropical and subtropical regions. *Eucalyptus* and *Pinus* are the tree genera preferred for these forests because of the diverse number of species and available technologies for growing them. Productivity achieved by improved *Eucalyptus* forests makes them highly competitive, as well as capable of producing fibers with excellent characteristics. Through concerted efforts the negative points raised about *Eucalyptus* planted forests are being technically answered. The positive results of the Aracruz project are merely one example of well-managed forestry based on technology developed and adapted to a local ecosystem, and on managers who recognize that care for the environment is decisive for success.

References

ANFPC (Associação Nacional dos Fabricantes de Papel e Celulose) 1993. Relat. Estat. Florest. São Paulo.

Aracruz Celulose, S.A. 1995. *Fatos & Números*. Rio de Janeiro.

Bertolucci, F.L.G., Demuner, B.J., Garcia, S.L.R. and Ikemori, Y.K. 1995. Increasing fiber yield and quality at Aracruz. In: Potts, B.M., Borralho, N.M.G., Reid, J.B., Cromer, R.N., Tibbitts, W.N. and Raymond, C.A. (Eds) Eucalypt Plantations: Improving Fibre Yield and Quality. Proceedings of the Cooperative Centre for Temperate Hardwood Forestry-International Union of Forestry Research Organizations Conference. The Cooperative Centre for Temperate Hardwood Forestry, Hobart, Tasmania, Australia.

Campinhos, E. 1991. Plantation of Fast Growing Species for Tropical Areas. Special Paper 10° World Forestry Congress, Paris, September.

Campinhos, E. and Cláudio-da Silva, E. 1990. Development of the Eucalyptus Tree of the Future. ESPRA Spring Conference, Seville, May.

Campinhos, E., La Torraca, S.M., Laranjeiro, A.J. and Penchel, R.M. 1992. The Use/Place of Clonal Multiplication in Tree Breeding and Propagation Programs – Prerequisite for Success and Reason for Failures. Aracruz Celulose, Rio de Janeiro.

Eldridge, K.G., Davidson, J., Harwood, C. and van Wyk, G. 1994. Eucalypt Domestication and Breeding. Clarendon Press, Oxford.

FAO (Food and Agriculture Organization of the United Nations) 1988. The Eucalypt Dilemma. FAO, Rome.

FBDS (Fundação Brasileira para o Desenvolvimento Sustentável) 1994. O Seqüestro de CO_2 e o Custo de Reforestamento, com *Eucalyptus* spp e *Pinus* spp, no Brasil. Rio de Janeiro.

Fenton, R. and Romero, J.L. 1995. An Overview of Fast Growing Plantations. In: Zobel, B.J. (Ed).

Ikemori, Y.K., Penchel, R.M. and Bertolucci, F.L.G. 1994. Integrating biotechnology into eucalypt breeding. In: International Symposium of Wood Biotechnology, Tokyo, Aug. 31–Sep. 1.

Kelly, S. 1993. Eucalypts (Volume 1). Van Nostrand Reinhold Company, New York.

Walters, G.A. 1981. Why Hawaii is changing to the dibble-tube system of reforestation. J. For. 79(11): 743–745.

Wilson, R.A., Astorga, R., Gomes, C. and Gonzales-Rio, F. 1995. Papermaking with DNA. "Intelligent Fibre." In: Potts, B.M., Borralho, N.M.G., Reid, J.B., Cromer, R.N., Tibbitts, W.N. and Raymond, C.A. (Eds) Eucalypt Plantations: Improving Fibre Yield and Quality. Proceedings of the Cooperative Centre for Temperate Hardwood Forestry-International Union of Forestry Research Organizations Conference. The Cooperative Centre for Temperate Hardwood Forestry, Hobart Tasmania, Australia.

New Forests **17:** 145–174, 1999.

Trees outside forests: agro-, community, and urban forestry

ALAN J. LONG[1] and P.K. RAMACHANDRAN NAIR[2]
[1] *School of Forest Resources and Conservation, University of Florida, Gainesville, FL, USA;*
[2] *University of Florida, Gainesville, FL, USA*

Key words: planted forest, sustainability, agroforestry, community forestry, social forestry

Abstract. Planted forests are often considered to consist of tree plantings at a scale large enough to satisfy such objectives as commercial production of timber and fiber, protection of watersheds, and preservation of natural habitats. However, trees are planted also at greatly reduced scales in agroforestry systems or as community woodlots to provide a mixture of products and services to resident households, local communities, and regional cultures.

Agroforestry systems represent a major form of small-scale tree planting, where trees are grown in purposeful combinations with agricultural crops and/or livestock in order to take advantage of tree-crop interactions, and thereby enhance crop production, diversify farm output, stabilize or improve soils, or ameliorate harsh environmental conditions. Some important examples of these systems in tropical countries include homegardens, alley cropping, improved fallows, intercropped trees for shade and fodder production, and trees planted in hedgerows and along fence lines. Throughout the tropics, there is a large variety of indigenous practices and species mixtures that represent adaptations of these systems to meet localized needs and opportunities. Research and development programs have supported the expansion and refinement of many of these systems during the last 20 years, but substantial constraints on tree planting still exist in the form of land-tenure practices, population pressures that relegate agroforestry practices to degraded lands, subsistence needs that prevent extended periods of tree growth, and insufficient technical information or technology dissemination.

Agroforestry systems in temperate, industrialized countries include combinations of trees, pasture, and livestock; fruit or nut trees interplanted with vegetable or grain crops; windbreaks and shelterbelts; multispecies riparian buffer strips; and forest farming systems for specialty crops. Compared to the tropics, however, temperate-zone systems tend to focus on one or two high-value crops, often involve some level of mechanization, and frequently represent an opportunistic approach to improving the economic profitability of farms rather than meeting subsistence needs. In both tropical and temperate regions, agroforestry systems and community woodlots will be an important component of new sustainable agriculture and environmental protection programs.

Although species diversity is an essential feature of all agroforestry systems, community forests generally involve planting only a few species in small woodlots near farms, around villages, along roads, and as riparian buffers. Provincial or state governments and the local populace are often involved in landownership and plantation establishment. Major objectives of these forests are production of fuelwood for local consumption and of other tree products for market; soil stabilization, reclamation, or improvement; and protection of water quality. As with many other planted forests, the number of species widely used in community forests has been relatively small, with the genera *Eucalyptus, Pinus*, and *Acacia* providing the bulk of the species. Major issues with these "planted forests" focus on rights for use of the products, tending responsibilities once trees are established, protection until trees are large enough for their designated use, increasing interest in using "native" species, and greater community involvement in planning and management.

Trees planted along streets and waterways, or as woodlots in parks and other public places, represent a major group of planted forests in many urban and periurban landscapes. In addition to providing many of the same environmental services that agroforests and community forests do, these urban plantings have unique aesthetic and recreational value. For much of the world's ever-increasing urban population, these may be the only tangible reference points for understanding planted forests.

These relatively little-recognized forms of planted forests – planted trees, to be more appropriate – are now receiving much greater attention. There are, however, some serious technical and sociopolitico-institutional constraints to their development as more widely adopted systems in both tropical and temperate regions.

Introduction[1]

Contrary to popular perception, trees and forests are not synonymous; trees alone do not make up forests, and trees are not found only in forests. The emphasis of the term "Planted Forests," the theme of this symposium, is undoubtedly on trees, much less on forests, but it implies that discussion is centered around large-scale tree plantings, as in forests. In this context, trees that are planted outside the domain of the generally implied large-scale planting are also important. Although several of such tree-planting activities are relatively small operations individually at a nonforestry scale, collectively they constitute substantial tree planting efforts. During the past couple of decades, we have witnessed a dramatic increase in interest in such "trees outside forests" and, as a consequence, the emergence of a myriad of terms and ideas to represent the concept. Whereas some of these terms and ideas denote new efforts in integrating trees on landscapes for recreational and aesthetic purposes, others represent age-old practices of combining trees with crops. All are a consequence of the perceived role of trees in addressing some of the recently recognized land-management problems and ecological concerns. In this paper, we discuss developments in this general area of trees outside forests.

Agroforestry, community forestry, farm forestry, social forestry, and urban forestry are the most important among the terms that have become prominent among those representing small-scale tree planting efforts. Urban forestry, as the name indicates, refers to tree planting in urban areas, whereas all of the other terms have a rural orientation. Agroforestry emphasizes the interaction between woody perennials and agricultural crops or animals for multiple products and services, and the other terms refer specifically to tree planting, often in small woodlots. Social forestry is considered to represent tree planting specifically for social objectives (usually betterment of the poor) through delivery of ensuing benefits to the local people. Community forestry refers to tree planting undertaken by a community on common lands, and is based on peoples' direct participation – either by growing trees themselves or by processing tree products locally. Farm forestry, a term used mainly in South and Southeast Asia, indicates tree planting on farms, either alone or in combination with crops. However, these distinctions are mostly of an academic nature; in practice, all of these designations refer directly or indirectly to growing and using trees to provide a variety of products and services. Little wonder, then, that, in common parlance, these terms of rural tree planting are often used synonymously and sometimes out of place.

We will first present a descriptive overview of these important planted-tree, land-use systems. The last part of the paper will evaluate the larger policy, social, and administrative issues that influence the development of these land-use opportunities.

Agroforestry

Agroforestry systems represent the most widespread form of small-scale tree planting. Although the practice in one form or another can be traced back a long time in the history of land use in different parts of the world, it was only during the past two decades that the concept attained prominence as a sustainable land-use option. Today, the term encompasses a wide range of land-use systems and practices in a variety of situations, whereby trees and crops and/or animals are grown in interacting combinations for a number of objectives (Nair 1983, 1984). As an interface between agriculture and forestry, it is considered to be a promising approach to land use, particularly in small landholdings and as an important part of agricultural development and ecosystem management.

Agroforestry systems of wide variety are found throughout the world, but more predominantly in the tropical and subtropical regions (hereafter referred to as "the tropics") than in the temperate, industrialized world. There are ecological as well as socioeconomic reasons for this tropical predominance. Ecologically the tropics are home to a wide variety of plants, and the socioeconomic conditions of the developing countries in this region favor and necessitate integrated land-use systems much more than in industrialized nations. A distinct feature of the multitude of these agroforestry systems is their extremely site-specific nature. Each system is a specific local example of

Table 1. Major agroforestry practices in the tropics (*Source*: Nair 1993).

Alley cropping (Hedgerow intercropping): Fast-growing, preferably leguminous woody species grown in crop production fields; the woody species are periodically pruned at low height (<1.0 m) to reduce shading of crops; the prunings are applied as mulch into the alleys as a source of organic matter and nutrients, or removed from the field to be used for other purposes, such as animal fodder.

Homegardens: Intimate, multistory combinations of a large number of various trees and crops in homesteads; livestock may or may not be present.

Improved fallow: Fast-growing, preferably leguminous woody species planted and left to grow during the fallow phase of shifting cultivation; the woody species cause site improvement and may yield economic products.

Multilayer tree-gardens: Multispecies, multilayer combinations of different woody species of various forms and growth habits.

Multipurpose trees (MPTs) on farms and rangelands: Fruit trees and other MPTs scattered haphazardly or according to some systematic planting arrangement in crop or animal production fields; trees provide fruit, fuelwood, fodder, timber, and so forth.

Pasture under plantations: Cattle grazing on pasture under widely spaced rows of plantation species.

Plantation-crop combinations (Shaded perennial-crop systems): Integrated multistory mixtures of tree crops, such as coconut, cacao, coffee, and rubber, with other tree crops, shade trees, and/or herbaceous crops.

Protein banks: Production of protein-rich tree fodder on farms/rangelands for cut-and-carry fodder production.

Shelterbelts and windbreaks: Use of trees to protect fields from damage from such sources as wind, sea encroachment, and flooding.

Taungya: Growing agricultural crops during the early stages of establishment of forestry (timber) plantations.

the association or combination of the components, characterized by environment, plant species, and their arrangement, management, and socioeconomic functioning. Thus, several hundred agroforestry systems have been identified and documented from different parts of the tropics (Nair 1989; Zhaohua et al. 1991; Montagnini et al. 1992). However, common to many of these systems are some distinct spatial and temporal patterns of component arrangement, or agroforestry practices. In spite of the multitude of localized agroforestry systems, the number of such distinct agroforestry practices is small (Table 1; Nair 1993).

Tropical agroforestry practices

Shifting cultivation and improved fallows

Farmers in the tropics have traditionally used certain species to restore fertility and suppress weeds in shifting agriculture (Budowski 1987; Unruh 1990) and have encouraged their presence (Raintree and Warner 1986; Padoch and deJong 1987; Unruh 1990). Following the recognition of the scientific merit of this practice, improved fallows were proposed as a management alternative to shifting cultivation (Nye and Stephens 1962). A common feature of most improved fallows is that certain valuable trees are left standing when the forest is cut and burned and/or valuable species are planted during the cropping phase to enrich the subsequent fallow.

Two types of improved fallows are visualized: economically enriched fallows, which increase the economic utility of the fallow vegetation by enrichment with trees valued for cash or subsistence purposes, and biologically enriched fallows, which enhance and accelerate the vegetative regeneration of soil fertility and control of weeds. The validity of the economic enrichment approach has been amply proven by such indigenous examples as the fallow enrichment planting of rattan by the Luangan Dayaks of Borneo (Weinstock 1983), of cedar and bamboo by the Lingnan Yao of China, of *Casuarina* by the Siane of Papua New Guinea (Salisbury 1962), of gum arabic in the Sahel (von Maydell 1986), and of multipurpose fallow woodlots with species diversities that exceed that of the natural forest by the *Ifugao* terrace-builders of the Philippines (Conklin 1980). Biologically enriched fallow practices using *Dactyladenia barteri* Engl., *Anthonotha macrophylla* P. Beauv., *Alchornea cordifolia* (Schum. & Thonn.) Mull. Arg., *Gliricidia sepium* (Jacq.) Nalp., and *Leucaena leucocephala* (Lam.) de Wit have been reported by Dijkman (1950); Benneh (1972); Okigbo and Lal (1979); Agboola et al. (1982); Getahun et al. (1982); Olofson (1983); FAO (1984); and Raintree and Warner (1986).

Attempts to replace spontaneous fallow vegetation with selected soil-improving woody species have been made in the Peruvian Amazon with a range of herbaceous and woody acid-tolerant legumes during a 4.5-year fallow period (Szott et al. 1991). Experiments with a mixed *Inga* and *Desmodium* fallow in a maize-cassava-plantain cropping sequence showed great potential for suppressing weeds and reducing the length of required fallow. However, the enriched fallow species required more time to plant, competed with other trees and shrubs, and required greater management skill to avoid competition with crops (Staver 1989).

Taungya

Taungya, which originated from the Burmese word "taung" (hill) and "ya" (cultivation), refers to the practice of growing agricultural crops along with forest species during the early phases, usually the first 2–3 years, of forest plantation establishment. The practice, initially developed in British colonial India in the 1850s as a means to involve farmers in forest plantation establishment with a view toward reducing forest encroachment by these farmers, has since spread to most parts of the tropics. Many tropical forest plantations have been established with the taungya approach; forest species include *Tectona grandis* L. f., *Eucalyptus camaldulensis* Dehnh., *Gmelina arborea* Roxb., *Shorea robusta* Gaertn. f., *Terminalia superba* Engl. & Diels, *Pinus* spp., and *Cupressus* spp. (Lamprecht 1989; MacDicken 1990).

In many places, there is no clear-cut distinction between shifting cultivation and taungya, and both systems are known by the same local terms. Similar to the improved fallows, improvements to taungya have also been attempted in a number of places (Jordan et al. 1993). Perhaps the "forest village" schemes in Thailand (Boonkird et al. 1984) come closest to the ideal. In some variants of this approach, participation in forestry is made more attractive to traditional shifting cultivators, not only by encouraging them to grow long-term perennial cash crops by widening the between-row spacing of the commercial forest species, but also by allocating permanent agricultural plots for them to use as they see fit.

Alley cropping

In alley cropping, arable crops are grown between hedgerows of planted shrubs and trees, preferably leguminous species, which are periodically pruned to prevent shading of companion crop(s) (Kang et al. 1981, 1984). The shrubs and trees grown in the hedgerows retain the same functions of recycling nutrients, suppressing weeds, and controlling erosion on sloping land as do those in the bush fallow. Prunings from the trees and shrubs are a source of mulch and green manure. Leguminous woody species also add fixed nitrogen to the system.

The alley cropping concept is currently being evaluated in many parts of the tropics under different names. Integrating small-ruminant production with alley cropping is referred to as alley-farming (Sumberg et al. 1987) in which prunings from the hedgerows provide high-quality supplementary fodder. The International Centre for Research in Agroforestry uses the term "hedgerow intercropping" (Torres 1983), whereas in Sri Lanka the term "avenue cropping" is used (Wijewardene and Waidyanatha 1984). The criterion most widely used to assess the desirability of alley cropping is the effect of this practice on crop yields. Indeed, most alley cropping trials produce little data other than crop yield data, and these are usually derived from trials conducted over relatively short periods of time. Much of the research on alley cropping during the past decade has addressed the ability of the practice to replace or reduce the use of chemical fertilizer, and to maintain soil and crop productivity. Most results show that alley cropping with fast-growing, repeatedly pruned shrub and tree species is a biologically sustainable technology in the humid and subhumid tropics (Kang et al. 1990; Szott and Kass 1993).

Although many research results suggest that alley cropping offers considerable potential in the humid tropics, it may not be a suitable crop production technology for the semiarid tropics (Rao et al. 1990; Ong et al. 1991; Ong 1994; Jama et al. 1995). Thus, a key issue concerning alley cropping is its ecological adaptability. The provision of nutrients through decomposing mulch, a basic feature of alley cropping, depends on the quantity of the mulch, as well as on its quality and time of application. If the ecological conditions do not favor the production of sufficient quantities of mulch, then there is no perceptible advantage in using alley cropping on this score.

Homegardens

"Homegarden" has been used rather loosely to denote diverse practices, from growing vegetables behind houses to complex multistoried systems. Numerous terms have been used by various authors to denote these practices. Analyzing the structure and function of several types of tropical homegardens, Fernandes and Nair (1986) define homegardens as "land use practices involving deliberate management of multipurpose trees and shrubs in intimate association with annual and perennial agricultural crops and invariably livestock within the compounds of individual houses, the whole crop-tree-animal unit being intensively managed by family labor."

Homegardens can be found in almost all tropical and subtropical ecozones where subsistence land-use systems predominate (Landauer and Brazil 1990). Although the choice of plant species in homegardens is determined to a large extent by environmental and socioeconomic factors, dietary habits of people, and market demands of the locality, there is a remarkable similarity with respect to species composition among different homegardens in various places. A conspicuous character of the tree crop components of homegardens is the predominance of fruit and other food-producing trees. Apart from providing a steady supply of various types of edible products, these fruit and food trees also combine well – biologically and environmentally – with other components of the system (Nair 1984). Fruit trees such as guava (*Psidium guajava* L.), rambutan (*Nephelium lappaceum* L.), mango (*Mangifera indica* L.), and mangosteen (*Garcinia mangostana* L.), along with such other food-producing trees as *Moringa* sp. and *Sesbania grandiflora* (L.) Poir., dominate the Asian homegarden. West African compound farms, on the other hand, are dominated by indigenous trees that produce leafy vegetables (*Pterocarpus* spp.), fruit for cooking (*Dacroydes edulis* [G. Don] Lam.), condiment (*Pentaclethra macrophylla* Benth.), and so on.

Shaded perennial-crop systems and tree gardens

Tropical perennial plantation crops, such as oil-palm, rubber, coconut, coffee, cacao, tea, cashew, and black pepper, occupy only about 8 percent of the total arable area in developing countries, but play a major role in many national economies because of their high-value economic products. Traditionally, most of the major plantation crops were developed as monocultural production enterprises which required high labor input during harvesting and, in some cases, during processing. As a result, modern commercial plantations of crops such as rubber, oil-palm, coffee, and tea are well-managed, profitable, land-use enterprises in the tropics, and are supported by excellent research. However, contrary to popular belief, substantial proportions of these crops are also grown by small landholders (Ruthenberg 1980; Nair 1983, 1989; Watson 1983).

The perennial crop/shade-tree systems, especially of cacao (*Theobroma cacao* L.) and coffee (*Coffea* sp.), under the shade of such species as *Erythrina* spp. and *Inga* spp. are often cited as good examples of sustainable agroforestry (Nair 1990; Fassbender et al. 1991). One of the major biophysical factors that contribute to sustainability of these systems is the nutrient cycling facilitated through the leaf litter from the overstory trees, either through natural leaf fall or periodic leaf pruning. Other advantages arise from the relatively low amounts of plant nutrients removed during the harvest of these high-value crops, higher soil organic matter from litterfall, and the reduction of soil erosion made possible by the presence of continuous plant cover.

Among other plantation crop combinations that have been described are crops grown with cashew (*Anacardium occidentale* L.) and coconut on the Kenyan coast (Warui 1980); crops grown with plantation crops in northeastern Brazil (Johnson and Nair 1984) and in Bahia, Brazil (Alvim and Nair 1986); crops grown with babassu palm (*Orbignya oleifera* Burret) in Brazil (May et al. 1985); crop associations with arecanut palm (*Areca catechu* L.) in India (Bavappa et al. 1982); and crop associations with oil palm (*Elaeis guineensis* Jacq.) and rubber (*Hevea brasiliensis* [H. B. K.] Muell. Arg.) in West Africa (Watson 1983).

In addition to these shaded plantation crop systems, there are some other multistory tree gardens that are mixed tree plantations consisting of forest species and other commercial tree crops, thus forming a forestlike system. As opposed to homegardens, which surround individual houses, these tree gardens are usually located away from houses, and are mostly on communally owned lands surrounding villages with dense clusters of houses as in Indonesia (Java and Sumatra). Depending on the characteristics and conditions of the places where the systems are practiced, various forms of tree garden systems can be found. The most important among these include those in Southeast Asia, especially Indonesia, humid West Africa, and the Amazon Basin. The role of these tree gardens in food production will depend upon their species composition. In general, it is not as significant as that of homegardens. However, in most tree gardens, fruit trees are important components, and these are exploited for their fruits. An important value of tree gardens is their contribution to the general cash economy of the farmers through the sale of various commercial products.

Windbreaks and shelterbelts

Farmers throughout Africa use windbreaks to protect crops, water sources, soils, and settlements. Hedgerows of *Euphorbia tirucalli* L. protect maize fields and settlements in the dry savannas of Tanzania and Kenya. Tall rows of *Casuarina* line thousands of canals and irrigated fields in Egypt. In Chad and Niger, multispecies shelterbelts protect wide expanses of cropland from desertification. These practices are not new, but the design of multipurpose windbreaks for small landholdings will require new agroforestry skills.

The most effective windbreaks provide a semipermeable barrier to wind over their full height. Because their shapes change as they grow, it is usually necessary to mix several species of different growth rates, shapes, and sizes in three or more rows. Some fast-growing species, such as *Eucalyptus, Cassia, Prosopis, Leucaena*, and *Casuarina*, should be used to establish the desired effect as rapidly as possible. However, no tree will grow if it is not well adapted to the environmental conditions of the site. In addition, some of these species are not as long-lived as slower growing trees. Fast- and slow-growing species should be mixed to extend the useful life of the windbreak. Mixing species also provides protection against attack from diseases or pests that can easily destroy single-species stands.

Cashew trees used in a windbreak in Senegal are yielding a fair amount of fruit and nuts. Although not in sufficient quantity and quality to be commercially viable on a large scale, these by-products are important additions to local diets. *Acacia scorpioides* (L.) W. F. Wight trees planted in windbreaks in Niger are now producing seed pods used for traditional leather tanning. As there is a steady market for this product, the windbreaks make a modest but important contribution to the local economy. In other cases, where *Prosopis* species are used in windbreaks, seed pods are collected daily to supplement livestock feed and for sale in local markets. In northwestern China, shelterbelts of *Paulownia* have been planted to stop desert encroachment. A 21–55 percent decrease in wind speed was measured in the protected area, together with increases of 13 percent in air humidity and 19 percent in soil moisture. Maximum summer and winter temperatures were reduced, crop yields were increased, and wood was produced.

Agroforestry for soil conservation

A substantial amount of information is now available on the role and use of agroforestry in soil conservation (Young 1989; Nair et al. 1995). The use of agroforestry in soil erosion can involve a number of practices, depending on the supplementary and/or direct use of trees and shrubs for the purpose. In supplementary use, the trees and shrubs are not the primary means of checking erosion. They fulfill the function of supporting other soil conservation structures through stabilizing effects of tree root systems, and at the same time make some productive use of land occupied by such conservation structures. Direct use refers to the use of trees and shrubs themselves as a method of reducing erosion, effected through various means and factors, such as increased soil cover by trees and shrubs, permeable hedgerow barriers, natural terrace formation by soil accumulation upslope of hedgerows, and increased soil resistance to erosion by maintenance of organic matter. Various agroforestry practices discussed earlier can be employed for soil conservation purposes based on these direct and indirect uses of trees, and depending on site-specific situations.

Other agroforestry practices in the tropics

The other agroforestry practices listed in Table 1 are not discussed in detail here, mainly because there is, as yet, little scientific information to draw upon. This in no way belittles their importance or potential for improvement. Practices on which some information – albeit qualitative in nature – is available include silvopastoral practices (Baumer 1987; Le Houérou 1987; von Maydell 1987; Rocheleau et al. 1988), role of indigenous trees as food producers in agroforestry systems (Nair 1989, 1990), and potential of agroforestry in combating desertification (Baumer 1987).

Agroecological analysis of tropical agroforestry systems

The type of agroforestry system found in a particular area is determined to some extent by agroecological factors. However, several socioeconomic factors, such as human population pressure, availability of labor, and proximity to markets, also come into play, and result in considerable variations among systems operating under similar or identical agroclimatic conditions. Sometimes socioeconomic factors take precedence over ecological considerations. Even in the case of systems that are found in most ecological and geographical regions, systems such as shifting cultivation and taungya, there are numerous variants that are specific to certain socioeconomic contexts. As a general rule, it can be said that, although ecological factors determine the major type of agroforestry system in a given area, the complexity of the system and the intensity with which it is managed increase in direct proportion to the population density and land productivity of the area.

The multispecies, multistoried homegarden systems serve to illustrate some of these points. Although these systems are found mainly in humid lowlands, they are also common in pockets of high population density in other ecological regions. In their analysis of the structural and functional aspects of ten homegarden systems in different ecological regions, Fernandes and Nair (1986) found that, although the average size of a homegarden unit is less than 0.5 ha, the units generally consist of a large number of woody and herbaceous species. Further, the units are carefully structured so that the species form three-five canopies at varying heights, with each component having a specific place and function within the overall design.

Agroecological factors also have a considerable bearing on the functional emphasis of agroforestry practices. For example, the primary function of agroforestry practices in sloping lands is erosion control and soil conservation; in wind-prone areas the emphasis is on windbreaks and shelterbelts; and in areas with a fuelwood shortage the emphasis is on fuelwood production. There are also specific agroforestry approaches for the reclamation of degraded lands or wasteland (for example, land that has been badly eroded or overgrazed, or is highly saline or alkaline). The preponderance of homegardens and other multispecies systems in fertile lowlands and areas with high agricultural potential at one end of the ecological scale, and extensive silvopastoral practices at the other end, with various systems in between, indicates that the ecological potential of an area is the prime factor that determines the distribution and extent of adoption of specific agroforestry systems. A significant feature that emerges from this analysis is that, irrespective of the sociocultural differences in different geographical regions, the major types of agroforestry systems are structurally similar in areas with similar or identical ecological conditions.

Agroforestry practices in industrialized countries

As in the tropics, there is a long tradition in industrialized (used synonymously as temperate) countries of the purposeful growing or deliberate retention of trees with crops and/or animals for multiple products or benefits from the same management unit. Whereas combinations of trees and agriculture are essential for many subsistence-level farmers and production systems in the tropics, agroforestry in the temperate zone usually represents economic opportunities to enhance the profitability of current farming or forestry operations and to diversify crops and harvested products as an insurance against difficult markets, high tree establishment and fuel costs, or excess production of staple crops. Another major benefit from agroforestry in industrialized countries arises from its incorporation into land-use plans that emphasize long-term sustainable farming or forestry practices. On marginal cropland or erodible slopes, many landowners combine trees and crops for soil conservation benefits along with crop or timber revenue. This benefit will be increasingly derived as agroforestry systems are integrated into urban-rural interface lands, forest stewardship plans, and buffer zones around public lands (Long and Nair 1991).

Unlike the great variety of agroforestry systems in the tropics, the breadth of different tree-plant-animal combinations is considerably less than in the tropics. The number of species used is also much smaller than in the tropics, although it is increasing as new combinations are tested and new markets are opened. Farming systems are usually much less labor intensive than in the tropics, with mechanization and large landholdings an important characteristic of agroforestry operations in many industrialized countries.

Most agroforestry strategies in industrialized countries are grouped under three major practices, each with different forms in different situations. These practices are intercropping under hardwood species, silvopastoral, and windbreaks or shelterbelts. However, significant expansion in the scope of temperate-zone agroforestry is occurring, with similar expectations that this meshing of agriculture and forestry will generate new solutions to current land-use problems.

Intercropping under hardwood species

Some of the earliest records of agroforestry in Europe and the Middle East include biblical descriptions of tree-based agriculture (olives and figs) and Renaissance paintings that show crop cultivation among trees (Harris 1977). Similarly, the Chinese intercropped soybeans between rows of chestnut trees in the 1600s, and perhaps used other species at even earlier dates. Native Americans frequently grew various grain, melon, and vegetable crops under existing pine or hardwood forest trees (Byington 1990). In the United States in the early 1900s, J. Russell Smith first promoted the idea of agricultural systems that used nut trees, oaks, persimmons, and other hardwoods as an overstory for regular agricultural crops.

Today, two major types of hardwood intercropping systems can be differentiated: those with fruit- and nut-producing trees, and those with high-value timber species such as poplar (*Populus* spp.). Generally, multicropping offsets plantation establishment costs, allows for more intensive use of both forest and agricultural land (especially close to processing facilities, decreasing the likelihood of conversion to other land uses), and reduces cultivation costs of individual crops because cultural operations and fertilization can benefit, and be allocated jointly, to all crops (Gold and Hanover 1987).

A wide variety of orchard tree-crop combinations has been adopted. For example, leeks, corn, and strawberries are grown in peach orchards in Ontario; oats are grown in some New York apple orchards; potatoes, grains, soybeans, squash, and peaches have been planted in pecan orchards in the southern United States (Williams and Gordon 1991); and citrus and other fruit trees are intercropped with winter vegetables in India (Tejwani 1987, 1994). When orchard trees are large enough to shade out the ground crops, forages and livestock are often introduced to the orchards.

More prevalent currently than the combinations of fruit trees and crops in North America are multicropping and grazing in black walnut (*Juglans nigra* L.), pecan (*Carya illinoensis* [Wangenh.] K. Koch), and other nut orchards. Trees in these orchards may attain larger sizes than fruit trees and, therefore, require wider spacings with longer intercropping periods. Also, unlike the fruit orchards, walnut and pecan trees will likely be harvested for high-value wood products after 50–80 years.

Black walnut orchards have been more widely studied than any other set of orchard agroforestry practices in North America (Noweg and Kurtz 1987; Garrett et al. 1989; Campbell et al. 1991). A current multicropping strategy for these systems includes intercropping black walnut orchards with wheat, milo, or soybeans during the first 10–12 years; forage and limited cattle grazing for another 10 years; nut collection for 30–50 years; and a final harvest of the walnut timber. Economic analyses of similar strategies suggest that, even with the high value of walnut lumber, the long rotation plantations could not be justified without the early returns from other crops (Kurtz et al. 1984). Similar, but simpler, strategies have been tested with pecan in both plantations and natural forest stands.

In Europe, Australia, and China, poplar (and *Paulownia* sp. in China) has been widely planted for short rotation fiber, fuel or lumber production. Interplanting with various crops during the first 3–6 years provides greater returns to landowners than do tree monocultures. Despite these economic benefits, the system has fallen into disfavor in some countries because of low wood prices, marketing difficulties, and unfavorable cultural interactions between the trees and the crops. Such problems underscore the need to understand owner objectives, infrastructure requirements, and joint cultural problems before proceeding with widespread adoption of agroforestry systems.

Livestock grazing in managed plantations

Natural hardwood and conifer forests have been freely grazed by cattle, hogs, and sheep for hundreds of years. Today, this practice has largely been replaced by managed livestock grazing in plantations, especially conifer plantations. The simplest approach allows livestock to graze freely in plantations established essentially for timber production. Examples include sheep grazing in Douglas-fir (*Pseudotsuga menziesii* [Mirb.] Franco) in the northwestern United States and cattle grazing in open pine plantations in the southern United States. Grazing in these stands affords a reasonable level of vegetation control, but requires herd or flock management during the first few years after establishment to avoid damage to new seedlings.

More complex silvopastoral systems involve deliberate management of both pasture and trees, primarily with *Pinus radiata* D. Don in New Zealand and the southern pines in the United States. Wide tree spacing, pasture and forage improvement, and rotational livestock grazing have been demonstrated repeatedly to provide a substantially greater financial return to the landowner than does managing for pine or livestock alone (Dangerfield and Harwell 1990; Knowles 1991; Mead 1995). An important aspect of this financial return is the early revenue from livestock production that offsets costs of plantation establishment.

Windbreaks and riparian buffers

Wind erosion and water quality in streams are serious problems in many parts of the temperate zone, especially near ongoing agricultural and farming operations. Windbreaks have been planted in Europe and China for hundreds of years, and in North America since the late 1800s, to protect fields and dwellings from winter snow and hot, dry, windy summers. Additionally, they have been used for rehabilitation of wasteland, development of vegetation for the control of sandstorms, and control of soil and water erosion.

The benefits of windbreaks for agriculture have long been recognized and have been the subject of much research. Despite these efforts, significant problems remain. Windbreak establishment continues to be difficult; there is a very limited choice of medium-to-tall species that are well adapted and long-lived; better methods are needed for silvicultural manipulations; and windbreak design for hilly country is currently inadequate.

Riparian buffers are increasingly being established along streams as barriers and filters between agricultural operations and waterways. They reduce both overland water flow and movement of agricultural chemicals, and thereby help to improve water quality in streams. Other benefits include expanded wildlife habitat and shade for grazing animals and adjacent crops (Schultz et al. 1995). Various configurations have been proposed for these systems, and we should expect to see them implemented on much more land in the future in response to initiatives to improve the sustainability of agricultural lands and water resources.

Community forestry

As described earlier, social forestry and community forests generally seek to meet many human and environmental needs from naturally regenerated or planted trees on common access or other nonprivate lands. The potential recipients of benefits from these forests are often rural and landless poor who are unable to procure subsistence products such as fuelwood from their own farms. However, there are many examples from countries as diverse as China (Bruce et al. 1995), Italy (Merlo 1995), and Mali (Thomson and Coulibaly 1995) which demonstrate that community forestry also represents local management of common resources to meet the needs of all local citizens.

The technical skills needed to establish and manage community forests are similar to those used in most other planted forests, although there is still a great deal to learn about local preferences for species selection and about methods for growing mixed-species stands (FAO 1992). The emphasis in this section is on the social, economic, and political issues that have been the motivating force behind community forests. They are the dominating factors that must be unraveled, understood, and incorporated in planning if these forests are to meet their full potential for enhancing natural resource sustainability. We will first present an overview of general trends in the occurrence, use, and management of community forests. A case study of India will provide more detailed information for a set of principles and recommendations for future community forest programs.

Historical perspective

As with many of the agroforestry systems described previously, community forests actually have roots that extend much deeper into the past than is often implied when they are maintained or established under rural development programs. In many regions of the world, rural villages, communities, and tribes historically managed local forests for the production and harvest of fuelwood, fodder, fruit, nuts, and a variety of other services and products for local consumption. Over the last 100 years, these common property forests have disappeared or changed for a number of reasons: technological and economic change (McKean and Ostrom 1995); individual privatization of forest and agricultural land, sometimes excluding common property systems in attempts to formalize and codify property rights; government appropriation and management of common property natural resources; and conversion to agricultural production to meet the needs of increasing populations.

The loss of local forests and wood sources, coupled with increasing populations of rural landless and poor, has forced villagers to travel farther to collect wood on common properties or government lands while also placing intense pressure on remaining community forests close to villages. To counter this cycle of degradation and unsustainable resource use, national governments and development agencies have initiated programs in many countries for tree planting on off-forest lands and for control of other previously common-access forests. Although diverse in design, these social forestry programs are similar in their general goal to increase local forests to meet the needs of the rural poor. Some of these programs have been successful; many, however, have fallen far short of their goals, often because of a failure to account for sociological issues and problems. For example, imposing planting plans on local communities without participation of the community, and privatization of previously open-access land are two of the more common reasons for weak programs. Collective management of local resources is being reevaluated in many regions, with an increasing focus on the restoration or strengthening of common property resource management systems.

A case study from India

National policies initiated in the late 1800s gradually removed opportunities for many of India's rural communities to manage or control forest or common land use, despite long traditions of local community management and collection of forest products (Poffenberger 1990). However, many rural communities were still able to maintain indigenous management systems and some level of traditional access control on public (common) land through the first half of the 1900s. In the 1950s, increased forest nationalization and land reform led to the deterioration of most of the community systems, and the loss of 50 percent or more of the common property resources (Jodha 1995). The pressure on the remaining common property resources increased dramatically as a result of the reduced area of community property and a burgeoning population with needs for fuelwood and other forest products.

In attempts to counter these trends, national and state governments have undertaken measures since the 1950s to increase productivity of community and other local forest lands. Under the broad banner of social forestry, a variety of community forest programs have been initiated, with equally variable results. The overall social forestry goals were to provide adequate quantities of fuel, fodder, timber, and other forest products; conserve existing forests and avert ecological disasters; and directly involve the public in planting, care, harvest, and use of forests and their associated products (Khosla and Kohli 1987). Although the programs vary from state to state, most of them include one or more of the following four components (Kedharnath 1987):

- Strip plantations along roads, rails, and canals (government lands) planted by government forest departments.
- Village woodlots or fodder banks on government and community land planted by forest departments and/or by community self-help.
- Rehabilitation of degraded forest or agricultural land by forest departments or private landowners.
- Farm forestry on private land with seedlings provided by forest departments.

The diversity of projects is demonstrated by the following specific examples: plantations of fodder trees for rearing milk cattle by landless people; mulberry trees planted along canals to provide the raw material to engage a large number of people in the silk industry; neem trees for various solid wood products and seed oil; and *Eucalyptus* in strips along roads for local fuelwood or for sale for paper or rayon manufacturing. Various authors in Khosla and Kohli (1987) list over 100 tree species, and their uses, that have been recommended for social forestry programs in India. Choice of which species to use, singly or in combination, should depend on the local needs of the people, potential markets, and soil and climatic conditions. Projects such as these were a marked contrast to conventional forestry projects which received much of the earlier attention, and funding, of state forestry departments.

Whereas conventional forestry seeks to meet the long-term demands of society for certain forest products and environmental protection, trees were planted in the community systems to meet immediate requirements of local societies. The large landless population in many areas was the intended major beneficiary as well as participant in managing the lands. To the extent that community forest programs have been properly implemented (Kumar and Kohli 1987):

- They provide employment to people in rural areas as the people sell the products harvested from community lands or through various small-scale industries based on the variety of forest and tree products.
- Subsistence products such as fuel, fodder, fruit, and timber are available for use by local communities.
- Land unfit for agriculture is brought into production of tree crops with potential improvements in environmental protection.
- Population pressures on remaining natural forests are reduced.

As successful as these programs may have been in some regions, there have also been numerous criticisms, centering on the choice of tree species, the wisdom of monoculture plantations, and the nature of actual beneficiaries. Whereas the emphasis has been on nursery production and plantations of certain *Eucalyptus* species for local wood-using industries, reforestation efforts have focused on large landholdings, often on previous agricultural cropland. Tree growing is less labor intensive and provides more revenue for landowners. Consequently, both agricultural production and availability of jobs for the rural poor or landless people are reduced, in direct opposition to the underlying intent of the social forestry program. Similarly, tree planting on village common lands has displaced grazing in some areas, again to the detriment of local citizens. Another point of criticism has been the emphasis on certain exotic tree species in some programs, rather than on understanding the preferences of local people for the species they use and that might provide multiple benefits.

In addition to these criticisms, other deep-rooted problems and issues affect the success and continuation of many social and community forest programs. Some of the most important issues center on questions of land tenure and usufruct. Many social forestry programs in the 1950s and 1960s emphasized woodlots on common lands around villages to meet the rural people's primary needs for fuelwood, fodder, and small timber. Although government projects were successful in the establishment of plantations, subsequent management and allocation of products by community leadership often bypassed the rural poor for whom the benefits were intended. Continued livestock grazing and cutting trees prematurely also contributed to the failure of these community forests to meet local fuelwood needs. On the other hand, planting of farm forests on private land was much more successful than anticipated, and several states expanded this success by initiating a "tree tenure scheme" (Hegde 1987). Under this scheme, landless and marginal farmers were leased community land on which they planted trees, and then became owners of those trees. It was anticipated that the farmers' vested interest in the trees would encourage participation by low-income families in the community forest programs.

Along with increased planting on government-owned nonforest land, the government also hoped to increase the production of fuelwood to start to close the gap between fuelwood consumption and supply. Unfortunately, fuelwood from planted trees in many of these areas has seldom provided as much profit to owners as have products such as poles and pulpwood. As long as fuelwood can be collected free from other government lands, or has lower market value, community forests under the "tree tenure system" might just add to the commercial timber market, rather than to the fuelwood supply. A critical step for making fuelwood production more attractive financially to poor farmers is improvement in the spectrum of reforestation activities, from types of seedlings to restrictions on how seedlings are planted to cultural treatments after planting. Many government agencies and nongovernmental organizations recognize the necessity of providing poor farmers with low-cost seedlings of species that will survive well, require minimal planting effort, and grow quickly into useful product sizes. Some of these features have been noticeably lacking in many community forestry programs.

Another deep-rooted problem has been the inability to bring poor, landless people into social forestry programs. Communications are essential for securing the active consent and participation of community members for tree planting and cultivation, and focused efforts on such communications have been lacking in many programs.

As the people and government agencies of India, and many other developing countries, look forward to the challenges of sustainable forest resources, expansion of community forestry programs will require resolution of the criticisms and issues that have characterized the last 40 years of social forestry. Perhaps the most important major advancement, already recognized and addressed in many programs, will be the understanding of the attitudes, aspirations, and needs of rural populations. Most foresters are well versed in technical matters related to forest establishment and management, but their technical opportunities will be mute unless they work with social scientists to incorporate the local populations in participatory programs. Historical, social, cultural, economic, and political issues must all be factored into the design of community forestry projects if outside agencies or organizations are involved in project design and implementation. Increasingly, these organizations and agencies are realizing that local development and control of community resources may be the most certain pathway to sustainability.

Urban forests

One additional type of planted forest – the urban forest – has not been discussed elsewhere in these proceedings, perhaps because in terms of scale, and to some extent design, it fits more closely with farm and community forestry than with any of the other major topics. Urban forests in the United States alone have been estimated to cover 28,350,000 ha of municipal land, and a much greater area if suburbs and urban-rural interface areas are included (Moll 1989). Street tree valuation for the United States may be at least $30 billion (Schoeneman and Ries 1994). The abundance of planted trees in urban areas is less well documented in many other countries, but, to whatever extent they occur, for much of the world's ever-increasing urban population, urban forests may be people's only tangible reference point for understanding "planted forests."

Tremendous opportunities exist for proponents of tree planting to greatly increase the size and quality of urban forests throughout the world. In areas with high poverty levels, urban trees could potentially impact the local supply of necessities such as fruit and fuelwood. Elsewhere, as city dwellers learn to appreciate the fact that trees are planted to provide particular services and benefits, they will hopefully also learn to accept that planted trees serve a major function and purpose in rural landscapes (Dwyer and Schroeder 1994). Their support of tree-planting activities in rural areas may be crucial in future resource policy formation.

As with agroforestry systems, urban forests assume myriad forms and functions. They are planted in widely spaced rows along streets or property boundaries, just as fruit trees may be spaced along a farm boundary in Jamaica. The former provide beauty and shade, the latter shade and food. They are planted as tightly spaced rows or barriers to filter sound or sight, just as hillside hedgerows in Haiti filter erosive soil materials and check overland water flow. Urban trees may be planted in parks, interspersed with recreation facilities and vegetation, whereas overstory fruit, nut, or fodder trees are interplanted with short-term crops on a Missouri agroforestry farm. They are planted in more dense blocks to cover degraded lots or protect hillsides from erosion, whereas the same type of solid stands of trees in a community forest in India may similarly protect a degraded site while providing a supply of fuelwood. Urban forests also include the variety of trees planted around individual homes and yards to provide shade, fruit, aesthetic value, and recreational activities (including aerobics of leaf-raking); similarly, homegardens in the tropics return to the landowner a plethora of foods and meet other essential needs. Thus, there are many similarities in the values and functions that planted trees contribute to urban and rural landscapes. Urban forest managers, in concert with local citizens and organizations, must decide which functions are most important and the species and vegetation structures that will provide these functions (Nowak 1994).

Just as social forestry programs were initiated to provide benefits for rural people and families, urban forestry projects ought to be promoted to provide the same types of benefits for the huge low-income populations which live in and around all major cities. Careful species selection and cultural practices could increase the numbers of trees in urban settings that provide food and fuelwood for burgeoning populations. Of course, the availability of such trees and crops would probably result in much more difficult tenure, usufruct, and protection problems than have hampered many community forest programs.

A thorough evaluation of opportunities and constraints for urban forests must await a focused consideration like this one on planted forests, but devoted to the topic of urban forests. Until that time, we who tend to view planted forests through rural-colored glasses cannot forget that our goals and programs might receive much greater support, and protection, with the presence of healthy multipurpose urban forests.

Social and policy considerations

In this last section, the term "agroforestry" is used in a generic sense to encompass all the different forms of "trees outside forests" discussed in this paper.

Socioeconomic factors

One of the basic premises of agroforestry is the inseparable mixture of biophysical principles and social objectives that determine land use. Agroforestry entails a special emphasis on tree-people relationships. It is expected to be especially suitable and adoptable by small-scale land users with low capital and energy requirements, and to yield products and benefits directed to immediate human needs, rather than commercial advantages. Thus, agroforestry development has, from the beginning, involved substantial inputs from social scientists. Nevertheless, the development of methods and procedures to quantify the socioeconomic benefits of agroforestry has been slow. This is perhaps because social and economic parameters are far more difficult to quantify than are biophysical factors.

Economic studies of agroforestry have been ad hoc, and are generally *ex ante* analyses based on assumptions, rather than *ex post* analyses based on field data. This is undoubtedly because of the dearth of actual field data. Moreover, there are serious difficulties in evaluating agroforestry systems by applying economic and financial analyses, for which well-established and widely used procedures are available. Inadequate attention has been given to the economic value of agroforestry outputs, including fodder, green manure, fuelwood, and timber, as well as to the significant, but harder-to-quantify environmental effects, such as enhanced soil fertility and watershed protection. The problem is compounded by the fact that different social groups can have a variety of criteria for evaluating agroforestry.

Valuation is relatively simple when agroforestry inputs or products are marketed through commercial channels, but it is much more difficult when most or all of production is either bartered or consumed on-farm. Compounded with inadequate estimates made in market-oriented systems, the failure to include on-farm consumption can result in grossly underestimated actual returns to agroforestry investments.

Assessment of economic sustainability of agroforestry enterprises should consider their indirect effects, such as erosion control and watershed maintenance, on the economic and social welfare of the people inside and outside a project boundary. This inclusion is a critical issue from society's viewpoint, because these environmental benefits can be key factors in the decision to promote agroforestry (Mercer 1992). Unfortunately, these effects are often neither obvious nor easy to quantify, especially in the short term. Nevertheless, markets can provide considerable information about the demand for similar marketed goods and their valuation (Randall 1987; Prinsley 1990). For example, in rural Nepal, Gregersen et al. (1989) calculated the economic value of fuelwood by comparing it to the opportunity cost of the decreased soil fertility and maize production incurred when cattle dung was used as a fuel source. In the case of soil conservation, benefits can be estimated from the market value of the sustained or increased crop production made possible by agroforestry.

Efforts are now underway to undertake detailed economic analyses of agroforestry systems based on actual field data. For example, Sullivan et al. (1992) provide some indication of the economic advantages and sustainability of some agroforestry systems. An example is the study of agricultural sustainability by dissemination of improved agroforestry practices during 1985–1990 on the Coca Agroforestry Project in Amazonian Ecuador. The study showed that the Net Present Values and Internal Rates of Return calculated for the improved systems were substantially higher than those for the traditional practices. More importantly, the Internal Rates of Return of improved systems were higher than the 8 percent opportunity cost of capital which prevailed at the time of the evaluation (Ramirez et al. 1992).

Institutional issues

A serious deficiency that limits the realization of the benefits of agroforestry is the lack of adequate policy and institutional infrastructure in most developing countries (Follis and Nair 1994). It is well known that the success of every agricultural project or program will be affected by the prevailing governmental policies (Gittinger 1982). Experience from past forestry and agricultural development programs indicates that institutional support, exerted through equitable land tenure legislation, research and extension, market support and development, and credit availability, is as important to farmer response as the biological performance of the promoted technologies (Gregersen and McGaughey 1985; Chew 1989). The experience to date in agroforestry dissemination confirms the importance of institutional commitment and public policy support, as related in Gregersen et al. (1992) and Mercer (1992).

In most developing countries, secure property rights for small farmers are more often the exception than the rule. If an individual cannot enforce claims on a particular parcel to the exclusion of others, there is little incentive for his or her improvement of its agricultural productivity. In such situations, the tendency is to disregard soil conservation and fertility enhancement practices in which the majority of benefits are realizable only in the long term (Gregersen and McGaughey 1985; Southgate 1988). The use of trees and shrubs further highlights the importance of secure tenure. As discussed in Vergara and MacDicken (1990), in many societies, planting of perennials on a farm or property is predicated on the right to extended possession of the property. This has been corroborated by studies in Central America and the Caribbean that have clearly demonstrated a small-farmer preference for tree-crop production on more securely held land and, conversely, for devoting less securely held parcels to short-term crops (Ehrlich et al. 1987; Tschinkel 1987). Tenure insecurity and the lack of land-based collateral can also reduce access to the credit that may be necessary to finance agroforestry implementation; farmers without land title may be restricted to informal credit sources with higher interest charges and shorter terms.

Market uncertainty can be a major barrier to investments in sustainable agriculture and forestry technologies, and agroforestry is no exception. Experience has shown that adequate markets for many products of agroforestry cannot be taken for granted (Hoskins 1987; Chew 1989). Conversely, agroforestry projects in Central America have succeeded where farmers were able to meet specific demands, or correctly anticipated increased demands for fuelwood and food crops (Tschinkel 1987).

Subsidies and credit facilities are two other major factors that determine the ultimate realization of the sustainability attributes of agroforestry technologies. Faced with a choice between maximizing present income and investing in sustainable agricultural technologies with a significant delay in benefit realization, the often overwhelming survival considerations of small farmers dictate quick returns, even if the consequence is the destruction of agricultural productivity. Consequently, as with traditional agriculture, some degree of subsidy and financial support will be required for most agroforestry projects, particularly during the first few years (Follis and Nair 1994). Unfortunately, credit facilities, which have featured prominently in many state forestry and agricultural programs, have been notably absent from most agroforestry projects (Arnold 1992). Small borrowers may thus be relegated to higher-cost, shorter-term credit sources which will depress the economic viability of long-term agroforestry.

Formidable constraints and obstacles stand in the path of expanded application of agroforestry land-use systems in industrialized countries, just as they do in the tropics. One of the major constraints is the lack of an institutional infrastructure that specifically recognizes, and supports, agroforestry as a land-use system for sustainable development (Lassoie and Buck 1991). Although there is an increasing number of agency programs that support agroforestry research and application, there are, as of yet, no specific agencies or departments in any government body in the United States that have the specific objectives of advocacy and planning for agroforestry. All institutional lines are still basically drawn between forestry and agriculture, although there are some signs that these may begin to overlap.

Linked to the difficulty of institution-building in agroforestry in industrialized countries is the inherent lack of interest in agroforestry by many landowners. Even if the ecological advantages of agroforestry are scientifically established and well understood by the farming community, it will remain less appealing to landowners and professional land managers in industrialized countries so long as its economic advantages remain unconvincing. The lack of an adequate research base does not help this situation, but there are again signs that this is changing. We now see more universities and agency research groups developing expertise in agroforestry disciplines. However, even as the base develops, there remains the challenge of training and recruiting the extension and other specialists on whom will fall the responsibility of "carrying the news" to landowners.

Conclusion

India's experiences with social and community forestry have been repeated in many other countries over the last four decades. With few exceptions, the results have been similar. Many trees have been planted, but with far-from-successful impacts on the distribution of goods and services to intended populations. Of equal or more importance has been the inability of most of the programs to stem resource degradation. Future success of all programs involving trees outside forests will depend on the application of principles that have been learned with great difficulty. First and foremost of these principles is the involvement of local communities in the planning, implementation, and reaping of benefits. This local involvement is crucial for social forestry and agroforestry programs in poor rural areas, as well as for urban forests in large cities in North America (Schoeneman and Ries 1994). Planning and implementation include the selection of what will be planted or managed, sharing of responsibilities, and often some form of trade-off with other activities and needs. Carrying out these steps will, in many countries, require new organizational arrangements that are effective in enabling local management to function without interference from higher levels (Gregersen et al. 1993). The potential for successful local management will be improved if species and product choices focus on local interests, such as grazing and fuelwood, rather than on outputs, such as timber, which tend to encourage privatization or government control. The local community must also maintain control over a set of use and protection rules which affect how products and benefits will be allocated. These should be based on local tenure customs and not on government or development agency directives. On the other hand, these same agencies need to play stronger roles in extension and communication as they work with local communities to understand and adapt appropriate technologies. Successful implementation of these principles will be a major step in achieving the "hopes for farm and community forestry [which] are framed within the broad context of combating food shortages, energy scarcities, environmental degradation, and the continuing scourge of material poverty" (Laarman and Sedjo 1992).

Note

1. This paper is journal series number R-05442 of the Florida Agricultural Experiment Station.

References

Agboola, A.A., Wilson, G.F. and Yamoah Getahun, A. 1982. *Gliricidia sepium*: A possible means to sustained cropping, pp. 141–143. In: MacDonald, L.H. (Ed) Agroforestry in the African Humid Tropics. United Nations University, Tokyo.

Alvim, R. and Nair, P.K.R. 1986. Combination of cacao with other plantation crops: An agroforestry system in southeastern Bahia, Brazil. Agrofor. Syst. 4: 3–15.

Arnold, J.E.M. 1992. Policy issues related to the role of trees in rural income and welfare security, pp. 15–32. In: Gregersen, H., Oram, P. and Spears, J. (Eds) Priorities for Forestry and Agroforestry Policy Research: Report of an International Workshop. International Food Policy Research Institute, Washington, DC.

Baumer, M. 1987. Agroforesterie et désertificatión. Centre Technique de Coopération Agricole et Rurale, Wageningen, The Netherlands.

Bavappa, K.V.A., Nair, M.K. and Kumar, R.P. 1982. The Arecanut Palm (*Areca catechu* Linn). Central Plantation Crops Research Institute, Kasaragod, India.

Benneh, G. 1972. Systems of agriculture in tropical Africa. Econ. Geogr. 48(3): 244–257.

Boonkird, S.A., Fernandez, E.C.M. and Nair, P.K.R. 1984. Forest villages: An agroforestry approach to rehabilitating forest land degraded by shifting cultivation in Thailand. Agrofor. Syst. 2: 87–108.

Bruce, J.W., Rudrappa, S. and Zongmin, L. 1995. Experimenting with approaches to common property forestry in China. Unasylva 180 46: 44–49.

Budowski, G. 1987. The development of agroforestry in Central America, pp. 69–88. In: Steppler, H.A. and Nair, P.K.R. (Eds) Agroforestry: A Decade of Development. International Centre for Research in Agroforestry, Nairobi, Kenya.

Byington, E.K. 1990. Agroforestry in the temperate zone, pp. 228–289. In: MacDicken, K.G. and Vergara, N.T. (Eds) Agroforestry Classification and Management. John Wiley and Sons, New York.

Campbell, G.W., Lottes, G.J. and Dawson, J.O. 1991. Design and development of agroforestry systems for Illinois, USA: Silvicultural and economic considerations. Agrofor. Syst. 13: 203–224.

Chew, S. 1989. Agroforestry Projects for Small Farmers: A Project Manager's Reference. Special Study No. 59. United States Agency for International Development, Washington, DC.

Conklin, H.C. 1980. Ethnographic Atlas of Ifuagao: A Study of Environment, Culture and Society in Northern Luzon. Yale University Press, New Haven, CT.

Dangerfield, C.W., Jr. and Harwell, R.L. 1990. An analysis of a silvopastoral system for marginal land in the southeast United States. Agrofor. Syst. 10: 187–197.

Dijkman, M.J. 1950. *Leucaena* – a promising soil erosion control plant. Econ. Bot. 4: 337–347.

Dwyer, J.F. and Schroeder, H.W. 1994. The human dimensions of urban forestry. J. For. 92(10): 12–15.

Ehrlich, M., Conway, F., Adrien, N., LeBeau, F., Lewis, L., Lauwerysen, H., Lowenthal, I., Mayda, Y., Paryski, P., Smucker, G., Talbot, J. and Wilcox, E. 1987. Haiti: Country Environmental Profile. United States Agency for International Development, Washington, DC.

FAO (Food and Agriculture Organization of the United Nations). 1984. Changes in Shifting Cultivation in Africa. FAO Forestry Paper No. 50. FAO, Rome.

FAO (Food and Agriculture Organization of the United Nations). 1992. Mixed and Pure Forest Plantations in the Tropics and Subtropics. FAO Forestry Paper No. 103. FAO, Rome.

Fassbender, H.W., Beer, J., Heuveldop, J., Imbach, A., Enrique, G. and Bonnemann, A. 1991. Ten-year balances of organic matter and nutrients in agroforestry systems of CATIE, Costa Rica, pp. 173–183. In: Jarvis, P.G. (Ed) Agroforestry: Principles and Practice. Elsevier, Amsterdam, The Netherlands.

Fernandes, E.C.M. and Nair, P.K.R. 1986. An evaluation of the structure and function of tropical homegardens. Agri. Syst. 21: 279–310.

Follis, M.B. and Nair, P.K.R. 1994. Policy and institutional support for agroforestry: An analysis of two Ecuadorian case studies. Agrofor. Syst. 27: 223–240.

Garrett, H.E., Jones, J.E. and Slusher, J.P. 1989. Integrated forestry-farming (agroforestry) with eastern black walnut – a case study, pp. 248–259. In: Proceedings of the Fourth Black Walnut Symposium. Walnut Council, Indianapolis, IN.

Getahun, A., Wilson, G.F. and Kang, B.T. 1982. The role of trees in farming systems in the humid tropics, pp. 28–35. In: McDonald, L.H. (Ed) Agroforestry in the African Humid Tropics. United Nations University, Tokyo.

Gittinger, J. 1982. Economic Analysis of Agricultural Projects. Johns Hopkins University Press, Baltimore, MD.

Gold, M.A. and Hanover, J.W. 1987. Agroforestry systems for the temperate zone. Agrofor. Syst. 5: 109–121.

Gregersen, H.M. and McGaughey, S.E. 1985. Improving Policies and Financing Mechanisms for Forestry Development. Economic and Social Development Department, Inter-American Development Bank, Washington, DC.

Gregersen, H., Draper, S. and Elz, D. 1989. People and Trees: The Role of Social Forestry in Sustainable Development. The World Bank, Washington, DC.

Gregersen, H.M., Oram, P. and Spears, J. 1992. Priorities for Forestry and Agroforestry Policy Research. International Food Policy Research Institute, Washington, DC.

Gregersen, H., Arnold, J.E.M., Lundgren, A., Contreras, A., de Montalembert, M.R. and Gow, D. 1993. Assessing Forestry Project Impacts: Issues and Strategies. Food and Agriculture Organization of the United Nations (FAO) Forestry Paper No. 114. FAO, Rome.

Harris, J.W. 1977. Publius Vergilius Maro on trees plus crops plus cattle. Farm For. 19(3): 83–86.

Hegde, N.G. 1987. Scope for increasing the profitability of social forestry programme, pp. 68–85. In: Khosla, P.K. and Kohli, R.K. (Eds) Social Forestry for Rural Development. Indian Society of Tree Scientists, Solan, H. P., India.

Hoskins, M.W. 1987. Agroforestry and the social milieu, pp. 191–203. In: Steppler, H.A. (Eds) Agroforestry: A Decade of Development. International Centre for Research in Agroforestry, Nairobi, Kenya.

Jama, B.A., Nair, P.K.R. and Rao, M.R. 1995. Productivity of hedgerow shrubs and maize under alley cropping systems in semiarid Kenya. Agrofor. Syst. 31: 257–274.

Jodha, N.S. 1995. Common property resources and the dynamics of rural poverty in India's dry regions. Unasylva 180 46: 23–29.

Johnson, D.V. and Nair, P.K.R. 1984. Perennial-crop based agroforestry systems in northeast Brazil. Agrofor. Syst. 2: 281–292.

Jordan, C.F., Gajaseni, J. and Watanabe, H. 1993. Taungya: Forest Plantations with Agriculture in Southeast Asia. CAB International, Wallingford, United Kingdom.

Kang, B.T., Wilson, G.F. and Sipkens, L. 1981. Alley cropping maize (Zea mays) and Leucaena leucocephala (Lam.) in southern Nigeria. Plant Soil 63: 165–179.

Kang, B.T., Wilson, G.F. and Lawson, T.L. 1984. Alley Cropping: A Stable Alternative to Shifting Cultivation. International Institute of Tropical Agriculture, Ibadan, Nigeria.

Kang, B.T., Reynolds, L. and Atta-Krah, A.N. 1990. Alley farming. Adv. Agron. 43: 315–359.

Kedharnath, S. 1987. Social forestry in India – some research issues, pp. 1–6. In: Khosla, P.K. and Kohli, R. K. (Eds) Social Forestry for Rural Development. Indian Society of Tree Scientists, Solan, H. P., India.

Khosla, P.K. and Kohli, R.K. 1987. Social forestry for Rural Development. Indian Society of Tree Scientists, Solan, H. P., India.

Knowles, R.L. 1991. New Zealand experience with silvopastoral systems: A review, pp. 251–267. In: Jarvis, P.G. (Ed) Agroforestry: Principles and Practice. Elsevier, Amsterdam, The Netherlands.

Kumar, R. and Kohli, R.K. 1987. Social forestry – a hope for rural uplift, pp. 7–13. In: Khosla, P. K. and Kohli, R.K. (Eds) Social Forestry for Rural Development. Indian Society of Tree Scientists, Solan, H. P., India.

Kurtz, W.B., Garrett, H.E. and Kincaid, W.H., Jr. 1984. Investment alternatives for black walnut plantation management. J. For. 82: 604–608.

Laarman, J. and Sedjo, R. 1992. Global Forests: Issues for Six Billion People. McGraw-Hill, New York.

Lamprecht, H. 1989. Silviculture in the Tropics: Tropical Forest Ecosystems and Their Tree Species: Possibilities and Methods for Their Long-Term Utilization. Gesellschaft für Technische Zusammenarbeit, Eschborn, Germany.

Landauer K. and Brazil, M. (Eds) 1990. Tropical Homegardens. United Nations University Press, Tokyo.

Lassoie, J.P. and Buck, L.E. 1991. Agroforestry in North America: New challenges and opportunities for integrated resource management, pp. 1–19. In: Garrett, H. (Ed) Proceedings of the Second Conference on Agroforestry in North America. University of Missouri, Columbia.

Le Houérou, H.N. 1987. Indigenous shrubs and trees in the silvopastoral systems of Africa, pp. 139–151. In: Steppler, H.A. and Nair, P.K.R. (Eds) Agroforestry: A Decade of Development. International Centre for Research in Agroforestry, Nairobi, Kenya.

Long, A.J. and Nair, P.K. 1991. Agroforestry system design for the temperate zones: Lessons from the tropics, pp. 133–139. In: Garrett, H. (Ed) Proceedings of the Second Conference on Agroforestry in North America. University of Missouri, Columbia.

MacDicken, K.G. 1990. Agroforestry management in the humid tropics, pp. 98–149. In: MacDicken, K.G. and Vergara, N.T. (Eds) Agroforestry: Classification and Management. John Wiley and Sons, New York.

May, P.H., Anderson, A.B., Frazao, J.M.F. and Balik, J.M. 1985. Babassu palm in the agroforestry systems in Brazil's mid-north region. Agrofor. Syst. 3: 275–295.

McKean, M. and Ostrom, E. 1995. Common property regimes in the forest: Just a relic from the past? Unasylva 180 46: 3–15.

Mead, D.J. 1995. The role of agroforestry in industrialized nations: The southern hemisphere perspective with special emphasis on Australia and New Zealand. Agrofor. Syst. 31: 143–156.

Mercer, D. E. 1992. The economics of agroforestry, pp. 111–143. In: Burch, W.R. and Parker, J.K. (Eds) Social Science Applications in Asian Agroforestry. Winrock International, USA and South Asia Books, Arlington, VA.

Merlo, M. 1995. Common property forest management in northern Italy: A historical and socio-economic profile. Unasylva 180 46: 58–63.

Moll, G. 1989. The state of our urban forest. Am. For. 95(11/12): 61–64.

Montagnini, F. and 18 collaborators 1992. Sistemas Agroforestales: Principios y Applicaciones en los tropicos (second edition). Organization for Tropical Studies, Inc., San José, Costa Rica.

Nair, P.K.R. 1983. Agroforestry with coconuts and other tropical plantation crops, pp. 79–102. In: Huxley, P.A. (Ed) Plant Research and Agroforestry. International Centre for Research in Agroforestry, Nairobi, Kenya.

Nair, P.K.R. 1984. Soil Productivity Aspects of Agroforestry. International Centre for Research in Agroforestry, Nairobi, Kenya.

Nair, P.K.R. 1990. The Prospects for Agroforestry in the Tropics. Technical Paper No. 131. The World Bank, Washington, DC.

Nair, P.K.R. 1993. Agroforestry, pp. 9–11. In: McGraw-Hill Yearbook of Science and Technology. McGraw-Hill, New York.

Nair, P.K.R (Ed). 1989. Agroforestry Systems in the Tropics. Kluwer Academic Publishers, Dordrecht, The Netherlands.

Nair, P.K.R., Kang, B.T. and Kass, D.C.L. 1995. Nutrient cycling and soil-erosion control in agroforestry systems, pp. 117–138. In: Juo, A.S.R. and Freed, J.D. (Eds) Agriculture and the Environment: Bridging Food Production and the Environmental Protection in Developing Countries. American Society of Agronomy, Madison, WI.

Nowak, D.J. 1994. Understanding the structure of urban forests. J. For. 92(10): 42–46.

Noweg, T.A. and Kurtz, W.B. 1987. Eastern black walnut plantations: An economically viable option for conservation reserve lands within the corn belt. North. J. Appl. For. 4(3): 158–160.

Nye, P.H. and Stephens, D. 1962. Soil fertility, pp. 127–143. In: Wills, J.B. (Ed) Agriculture and Land Use in Ghana. Oxford University Press, London.

Okigbo, B.N. and Lal, R. 1979. Soil fertility maintenance and conservation for improved agroforestry systems in the lowland humid tropics, pp. 41–77. In: Mongi, H.O. and Huxley, P.A. (Eds) Soils Research in Agroforestry. International Centre for Research in Agroforestry, Nairobi, Kenya.

Olofson, H. 1983. Indigenous agroforestry systems. Philipp. Q. Cult. Soc. 2: 149–174.

Ong, C.K. 1994. Alley cropping – ecological pie in the sky? Agrofor. Today 6(3): 8–10.

Ong, C.K., Corlett, J.E., Singh, R.P. and Black, C.R. 1991. Above- and below-ground interactions in agroforestry systems. For. Ecol. Manage. 45: 45–57.

Padoch, C. and deJong, W. 1987. Traditional agroforestry practices of native and Ribereno farmers in the lowland Peruvian Amazon, pp. 179–194. In: Gholz, H.L. (Ed) Agroforestry: Realities, Possibilities and Potentials. Martinus Nijhoff, Dordrecht, The Netherlands.

Poffenberger, M. 1990. Joint Management for Forest Lands: Experiences from South Asia. Ford Foundation, New Delhi, India.

Prinsley, R. 1990. Agroforestry for Sustainable Production: Economic Implications. Commonwealth Science Council, London.

Raintree, J.B. and Warner, K. 1986. Agroforestry pathways for the intensification of shifting cultivation. Agrofor. Syst. 4: 39–54.

Ramirez, A., Sere, C. and Uquillas, J. 1992. An economic analysis of improved agroforestry practices in the Amazonian lowlands of Ecuador. Agrofor. Syst. 17: 65–86.

Randall, A. 1987. Resource Economics. John Wiley and Sons, New York.

Rao, M.R., Sharma, M.M., and Ong, C.K. 1990. A study of the potential of hedgerow intercropping in semiarid India using a two-way systematic design. Agrofor. Syst. 11: 243–258.

Rocheleau, D., Weber, F. and Field-Juma, A. 1988. Agroforestry in Dryland Africa. International Centre for Research in Agroforestry, Nairobi, Kenya.

Ruthenberg, H. 1980. Farming Systems in the Tropics (third edition). Oxford University Press, Oxford.

Salisbury, R.F. 1962. From Stone to Steel. Cambridge University Press London.

Schoeneman, R.S. and Ries, P.D. 1994. Urban forestry: Managing the forests where we live. J. For. 92(10): 6–10.

Schultz, R.C., Colletti, J.P. and Faltonson, R.R. (Eds) 1995. Agroforestry opportunities for the United States of America. Agrofor. Syst. 29(3): 181–331.

Southgate, D. 1988. The Economics of Land Degradation in the Third World. Environmental Department Working Paper No. 2. The World Bank, Washington, DC.

Staver, C. 1989. Shortened bush fallow rotations with relay-cropped *Inga edulis* and *Desmodium ovalifolium* in wet central Amazonian Peru. Agrofor. Syst. 8: 173–196.

Sullivan, G.M., Huke, S.M. and Fox, J.M. (Eds) 1992. Financial and Economic Analyses of Agroforestry Systems: Proceedings of a Workshop. Nitrogen-Fixing Tree Association, Paia, Hawaii.

Sumberg, J.E., MacIntire, J., Okali, C. and Atta-Krah, A.N. 1987. Economic analysis of alley farming with small ruminants. Int. Livest. Cent. Afr. Bull. 28: 2–6.

Szott, L.T. and Kass, D.C.L. 1993. Fertilizers in agroforestry systems. Agrofor. Syst. 23: 177–194.

Szott L.T., Fernandes, E.C.M. and Sanchez, P.A. 1991. Soil-plant interactions in agroforestry systems. For. Ecol. Manage. 45: 127–152.

Tejwani, K.G. 1987. Agroforestry practices and research in India, pp. 109–136. In: Gholz, H.L. (Ed) Agroforestry: Realities, Possibilities and Potentials. Dordrecht, Martinus Nijhoff, The Netherlands.

Tejwani, K.G. 1994. Agroforestry in India. Oxford and IBH Publishing, New Delhi.

Thomson, J.T. and Coulibaly, C. 1995. Common property forest management systems in Mali: Resistance and vitality under pressure. Unasylva 180 46: 16–22.

Torres, F. 1983. Potential contribution of leucaena hedgerows intercropped with maize to the production of organic nitrogen and fuelwood in the lowland tropics. Agrofor. Syst. 1: 323–345.

Tschinkel, H. 1987. Tree planting by small farmers in upland watersheds: Experience in Central America. Int. Tree Crops J. 4(4): 249–266.

Unruh, J.D. 1990. Iterative increase of economic tree species in managed swidden-fallows of the Amazon. Agrofor. Syst. 11: 175–197.

Vergara, N.T. and MacDicken, K.G. 1990. Extension and agroforestry technology delivery to farmers, pp. 354–373. In: MacDicken, K.G. and Vergara, N.T. (Eds) Agroforestry: Classification and Management. John Wiley and Sons, New York.

von Maydell, H.J. 1986. Trees and Shrubs of the Sahel: Their Characteristics and Uses. Gesellschaft fur Technische Zusammenarbeit, Eschborn, Germany.

von Maydell, H.J. 1987. Agroforestry in the dry zones of Africa: Past, present and future, pp. 89–116. In: Steppler, H.A. and Nair, P.K.R. (Eds) Agroforestry: A Decade of Development. International Centre for Research in Agroforestry, Nairobi, Kenya.

Warui, C.M. 1980. Research on agroforestry at the coast, Kenya, pp. 361–366. In: Buck, L. (Ed) Proceedings of the Kenya National Seminar on Agroforestry. International Centre for Research in Agroforestry, Nairobi, Kenya.

Watson, G.A. 1983. Development of mixed tree and food crop systems in the humid tropics: A response to population pressure and deforestation. Exp. Agri. 19: 311–332.

Weinstock, J.A. 1983. Rattan: A complement to swidden agriculture. Econ. Bot. 37(1): 58–68.

Wijewardene, R. and Waidyanatha, P. 1984. Conservation Farming. Department of Agriculture, Peredeniya, Sri Lanka; Commonwealth Secretariat, London.

Williams, P.A. and Gordon, A.M. 1991. The potential of intercropping as an alternative land use system, pp. 166–175. In: Garrett, H.E. (Ed) Proceedings of the Second Conference on Agroforestry in North America. University of Missouri, Columbia.

Young, A. 1989. Agroforestry for Soil Conservation. CAB International, Wallingford, United Kingdom.

Zhaohua, Z., Mantang, C., Shiji, W. and Youxu, J (Eds) 1991. Agroforestry Systems in China. Chinese Academy of Forestry, Beijing, China; International Development Research Centre, Ottawa, Canada.

New Forests **17:** 175–190, 1999.

Forests planted for ecosystem restoration or conservation

CONSTANCE A. HARRINGTON
Resource Management and Productivity Research Program, USDA Forest Service, Pacific Northwest Research Station, Olympia, WA, USA

Key words: planted forests, sustainability, site rehabilitation, forest restoration

Abstract. Although the phrase, "planting for ecosystem restoration," is of recent origin, many of the earliest large-scale tree plantings were made for what we now refer to as "restoration" or "conservation" goals. Forest restoration activities may be needed when ecosystems are disturbed by either natural or anthropogenic forces. Disturbances can impact (1) basic components of the system (e.g., plant and animal composition, soil pools, and atmospheric pools), (2) ecosystem processes, i.e., interactions among basic components, or (3) both components and processes. Early efforts at restoration or site rehabilitation focused primarily on reducing off-site impacts, such as sediment introduced into streams from ecosystems that had been severely disturbed. More recent restoration programs include ecosystems in which only some of the components are missing or some of the processes have been impacted. Restoration activities can begin immediately after the disturbance has ended. Although forest restoration projects can include many activities, planting is almost always a key component.

When planning an ecosystem restoration project, land managers need to be aware that commonly used plant establishment and management procedures may need to be altered to meet project objectives. Some systems may have been so severely impacted that ameliorative activities, e.g., fertilization, liming, land contouring, and microsite preparation, will be necessary prior to planting. Managers may also need to take special measures to reduce herbivory, control competing vegetation, or reduce physical damage from wind or sun. Choice of species needs careful consideration. Desired species may not grow well on degraded sites, may need a nurse species to become established, or may not provide an opportunity to harvest a short-term crop to reduce restoration costs. New methods may need to be developed for projects that require underplanting or interplanting. The end result of restoration should be an ecosystem with the same level of heterogeneity inherent in an undisturbed system; thus, managers should consider how pre- and postplanting activities will affect system variability.

As our understanding of ecosystems has increased, so has our expectation that restored ecosystems have the same components and function in the same manner as do undisturbed systems. These expectations require that land managers have more sophisticated information than was considered necessary previously. In the absence of more pertinent information, we can prescribe restoration activities based on results from related ecosystems or on theoretical considerations. Additional research, careful monitoring, and adaptive management are critical to our long-term success.

Introduction

Forests can be planted for many objectives. These include producing fiber, wood, or other forest products; providing services, such as recreational opportunities or watershed protection; and achieving ecological objectives, such as ecosystem restoration. Other papers in this proceedings deal with production of various products or a mix of products and services from planted forests. In this paper I deal specifically with forests planted for conservation objectives or for ecosystem restoration. Because these terms can mean different things to different people, I will start with some definitions. Next I will indicate which forces can trigger the need for conservation or restoration, stress the importance of having specific project objectives and the need to consider the constraints, provide examples of practices that might be used as part of conservation or restoration plantings, and, finally, conclude with a short discussion of future needs.

The fields of restoration ecology, habitat management or re-creation, watershed management, and mine spoil rehabilitation are not new ones; however, knowledge in all of these fields has expanded tremendously in recent years and several symposia or synthesis texts provide good overviews of the current state of knowledge (e.g., Jordan et al. 1987; Buckley 1989; Luken 1990; Chambers and Wade 1992; Naiman 1992; and Ferris-Kaan 1995). My purpose is not to summarize these diverse fields, but to provide a brief overview of some of the concepts that should be of interest to resource managers, and to provide some references for the reader interested in pursuing the subject.

Definitions

There are many possible definitions of most of the terms that are used in this field. These range from terms having general meanings to those having very specific connotations, and terms defined by statute or other type of formal mechanism. A general definition of conservation is: "A careful preservation and protection of something: esp. ... planned management of a natural resource to prevent exploitation, destruction, or neglect" (Woolf 1977). Common conservation objectives include protecting soil and water resources by preventing erosion or altering water yields, or preserving wildlife diversity by preserving or creating suitable habitat conditions. The term "conservation" has been in use for so many years that more restrictive definitions are not usually used.

Defining restoration and associated terms is more complicated, because there are many alternative definitions, and, even within the field of restoration ecology, they are not always used consistently. A general dictionary definition of restoration (Woolf 1977) is: "A bringing back to a former position or condition." People understand this definition when someone says that they are restoring a 1955 Chevrolet; that is, they are trying to re-create the vehicle condition as it would have been when it left the factory. But what factory did the ecosystem come from? And what model year are we trying to create?

I think it is helpful in this context to discuss ecosystems and their developmental pathways with the terms proposed by Magnuson et al. (1980) and the graphical approach as modified by Bradshaw (1987). In this approach ecosystems are described by two axes, one measuring species diversity and complexity and the other measuring biomass accumulation and nutrient cycling. In normal ecosystem development, systems aggrade along both axes (Figure 1A). If either structure or function has been reduced, then the ecosystem has been degraded. In this example I have assumed that an ecosystem has been degraded in both structure and function. It is possible, however, to have ecosystems that are degraded primarily in only one of the dimensions. Thus, it's important for foresters who are familiar with the term "site degradation" to realize that "ecosystem degradation" can occur without reducing the basic productivity potential of the site if the system is simplified (i.e., reduced in species and complexity).

We can discuss the alternative pathways that a degraded ecosystem can follow by seeing how the end result compares to the original or target system.[1] If no human activities occur, that is, if the system is neglected (Figure 1B), it could continue to degrade for some period of time, or it could begin to aggrade (i.e., increase in biomass and complexity). Another term that can be used to describe intentional neglect is "preservation." I think that it is important to recognize that neglect or preservation does not always result in an aggrading system (at least within human time frames). Neglect can result in a system that continues to degrade in productivity or a system that loses complexity. For example, it is clear that in a system where active erosion is occurring, the erosion will continue (and the productive capacity of the system will continue to decrease) until the land surface stabilizes. If a forest stand is just moving into the competitive exclusion phase of stand development (i.e., the canopy is closing and mortality is occurring), the stand is likely to lose additional understory and midstory plant species, and the vertebrates and invertebrates that depend on them. On the other hand, a neglected ecosystem could begin aggrading with respect to either or both axes as soon as a disturbance event ends. The pattern of ecosystem development in a neglected system could be either quite similar to the path of development the system had followed prior to being degraded, or very different, depending

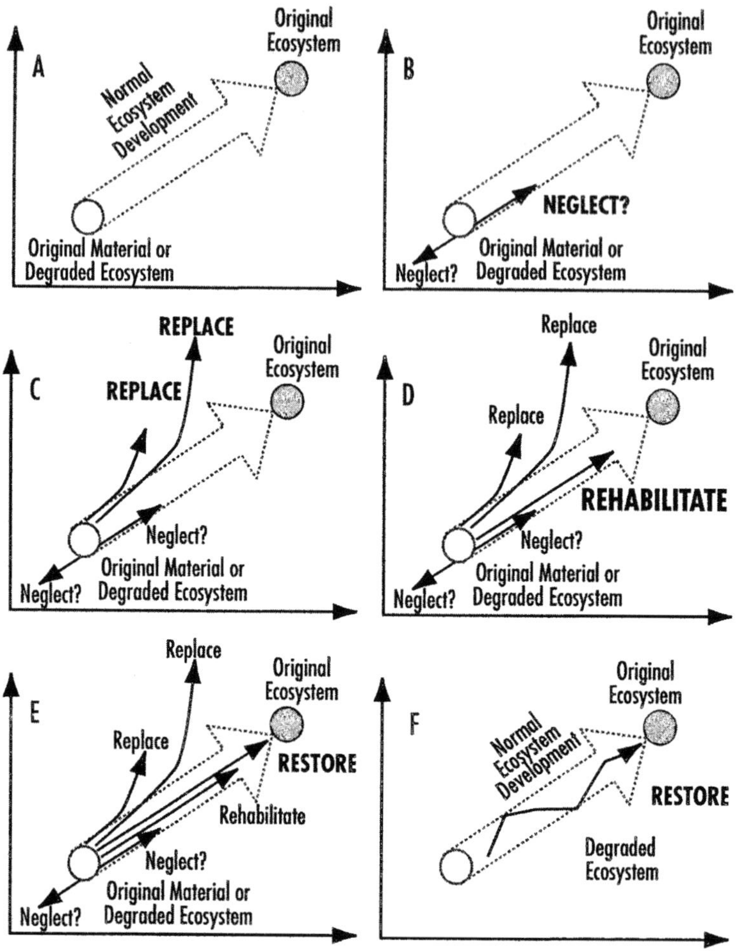

Figure 1. Ecosystems Can Be Characterized by their Function (biomass and nutrient content) and their Structure (species composition and complexity). (A) Systems are degraded if their function or structure is reduced. Normal ecosystem development may return the system to its previous status. (B) A neglected system may continue to degrade, or begin to recover. (C) A replacement system is one with less structure, and may be more or less productive than the original system. (D) A rehabilitated system is one with less structure and function than the original. (E) A restored system has the same structure and function as the original system. It may not follow the same path as the original system (*Modified from*: Bradshaw 1987).

on the condition of the system and the availability of colonizing plants and animals.

Instead of neglecting an ecosystem, one could intentionally plan to replace the original ecosystem with one that is structurally simpler (Figure 1C). The replacement system may be more or less productive than the original ecosystem. For example, if a forested ecosystem had been severely degraded after years of row cropping, one could establish a grassland ecosystem which would have less structural complexity and be less productive than the original (i.e., nondegraded system). Alternatively, if a bottomland hardwood system had been degraded and then planted with *Populus* clones, the replacement system would have a simplified structure compared to the original system, but it might be more productive. It's important to remember that all of these comparisons are in relation to the original system.

Based on our graphical representation, if our goal is to establish an ecosystem that is simpler and less productive than the original, but similar in concept, we would be rehabilitating the ecosystem (Figure 1D). Our intent would be less ambitious than if our goal were complete restoration. The goal of ecological rehabilitation (a term more specific than just rehabilitation) is to use site-indigenous species to reestablish a stable ecosystem capable of replacing the ecological functions of the original (Newton 1992). In many situations, rehabilitation or partial restoration is much more feasible than complete ecosystem restoration.

Looking at our graph again, if ecosystem structure and function were brought back to the same level as had existed previously (or exists in an undisturbed system), we would say the system had been restored (Figure 1E). Ecosystem restoration has been defined by the Society for Ecological Restoration as, "The process of intentionally altering a site to establish a defined, indigenous, historic ecosystem. The goal is to emulate the structure, function, diversity and dynamics of the specified ecosystem" (Newton 1992). The emphasis in definitions of restoration is on re-creating or bringing back to a previous condition, or to a condition that might have occurred without disturbance. Although some may quibble about how one determines exactly what system was present on a particular site or what would have occurred, the important aspect of restoration is the intent to create an ecosystem based on a biological model. Depending on the degree of degradation and the resources available, restoration may be a multistep process. The path followed in each step may differ somewhat from the original path of ecosystem development (Figure 1F).

One other term that is worth mentioning is "reclamation." This term has been used for years in association with reclamation of mined lands or in the sense of "reclaiming lands from the sea" to describe activities in the Netherlands. There are many definitions of reclamation (Table 1); some are legal definitions provided in statutes, whereas others are focused on how the end result can be used. Although there is not complete consensus on this, I believe that most connotations of reclamation involve human use or impose a human value system, rather than describe the project's activities in ecological terms.

Table 1. Alternative definitions of reclamation.

Definition	Source
Combined processes that minimize adverse environmental effects of surface mining and return land to a beneficial end use.	California Surface Mining and Reclamation Act 1975
The act or process of making available for human use by changing natural conditions. Restoration to use; recovery.	Woolf 1977
An attempt to create an ecosystem similar to the original or predisturbance ecosystem. It can include introduced species that respond similarly to the native species which they replace.	National Academy of Sciences 1974 (in Wade and Chambers 1992)

To avoid shifting between biologically based and human-oriented terms, I will not use the term "reclamation" in this paper. However, readers should be aware that a great deal of pertinent work available in the literature is associated with this term.

Causes of degradation

Forest ecosystems can be degraded as a result of many different forces. Some of these forces are the result of human activities; others may be the result of natural phenomena. Some degrading forces can result from either natural or anthropogenic forces; for example, fire can result from natural ignition sources, can be prescribed, or can escape from a human source.

Activities which can result in degradation include mining; land clearing, with subsequent use for crops or pastures; traffic (mechanical or biological); chemical deposition; off-site nutrient or biomass export; mass soil movement off site (slope failure or erosion); flooding; volcanism; fire; and alteration of species, genotype, or stand structure. Some of these impacts are obvious, such as the degradation that occurs after years of cropping on steep slopes where, not only was the original vegetation completely destroyed, but also the surface soil layers were removed and what remained was often compacted and gullied. Other effects are more subtle, such as the loss of animals that may result from a simplification of stand structure.

The effect of a particular force depends on its nature and magnitude, and on the condition of the predisturbance system. For example, repeated light fires may maintain productivity and complexity, whereas high-intensity fires reduce site pools of nitrogen and organic matter (which could reduce productivity) and may reset succession (probably reducing species complexity). Although many restoration projects may have been triggered by human activities, it is important to keep in mind that even natural forces such as wildfires can result in ecosystem degradation. The issue is not whether the force acting on a system is a natural part of succession, but whether within some reasonable time period it results in loss of structure or productivity.

Objectives

In an idealized world, it might be possible to restore all ecosystems and also meet all other objectives. However, in most situations, there are priorities for setting objectives and for funding, and it is not reasonable to expect that meeting one objective will necessarily meet other objectives. Thus, it is important to clearly identify the priority objectives. Increasing diversity of bird species may require an entirely different strategy than preserving a particular rare species. Similarly, restoring an ecosystem may not be the most effective way to achieve a particular watershed objective. Allen and Hoekstra (1987) provide examples of how differences in the scale of a project may influence its objectives, and how an objective at one scale may preclude achieving an objective at a different spatial or temporal scale.

I heard a recent interview about the television series "Star Trek," in which the actors were laughing about the portions of the script that call for them to use technical jargon that has been invented by the script writer to discuss a particular problem or a proposed solution. The actors call the resulting dialogue "techno-babble." If you don't listen carefully or aren't familiar with the technical details, techno-babble appears to make sense. Unfortunately some of the dialogue associated with restoration and conservation plantings appears to me to be the ecological equivalent, or "eco-babble." Although "eco-babble" may sound OK, it almost always results in projects with "fuzzy" objectives. Without clear objectives it is almost impossible to prioritize projects or to judge whether or not a project was successful. In addition, because many conservation or restoration projects are very long term, changes in project personnel will probably be inevitable; thus, clear project objectives will help keep future efforts focused on the highest priority objectives.

If the factor that caused the system degradation has not ceased, e.g., if atmospheric deposition continues, then the best that can be hoped for is system rehabilitation, rather than complete restoration (Lenz and Haber 1992). In addition, the degree of degradation sets practical limits on what types of restoration goals are reasonable (Werner 1987). For example, if soil layers as well as vegetation have been removed, the likely goal is to restore basic ecosystem functions; if the soil layer is intact but the vegetation has been severely impacted, then the goal may be to restore specific growth forms of vegetation; and if the composition or structure of the vegetation has merely been simplified, then the goal may be community or species restoration. Based on the definitions discussed previously, if the goal is not a complete ecosystem restoration, then it is usually better termed "replacement" or "rehabilitation."

Determine the constraints

Once the biological objectives have been clearly determined, then it is important to clearly understand the legal and social considerations involved in implementation. In reality, these steps (objective setting and recognition of constraints) often proceed jointly. Thus, if there is a legal mandate to return the land to a specified condition and to do so within a specified time period, it is clear that meeting these legal requirements will be at least part of the project objectives. Other considerations in defining project objectives include current attitudes of opinion makers, the need for short- versus long-term inputs, and costs. In addition, if the system was degraded by human activities, are there alternative areas readily available for those activities? If not, the project may be doomed from the beginning. It is clear to most people that, if an area has traditionally been used for agriculture or livestock grazing, it is not reasonable to expect the users to react positively to new fences and a planting project which restrict their access (Michaelson 1994). These considerations do not just apply to developing countries. Closing off an area popular with drivers of off-road vehicles or trying to reduce public access for harvesting of special forest products (e.g., berries, floral greens, and mushrooms) may also require substantial preproject planning to ensure that users accept the need for reduced access and are aware of alternative sites for their activities.

In addition to clearly specifying objectives, resource managers also need to identify reasonable goals or short-term targets by which to judge project success. For example, we have known for many years that planting eroded land can help achieve soil and water conservation objectives (c.f., Duffy and Ursic 1991). The success of a project with soil or water conservation objectives should not be judged primarily on a forestry measure, such as the number of trees planted, but rather on the actual benefits associated with that particular project. An example may make this clearer. It has been shown that older teak plantations can actually result in greater erosion than if the area had never been planted (Michaelson 1994). This occurs because the large leaves of the genus collect rain, and this results in a canopy drip with much larger drop size than exists above the canopy. As the tree crowns lift over time, the force at which the drops impact the ground increases. Teak leaf litter decomposes rapidly, and the ground is not protected during much of the year and substantial erosion can result. Thus, if the project objective is primarily one of soil conservation, then the benefits of the project should be judged by measuring change in soil parameters rather than by counting numbers of trees planted.

Practices associated with conservation and restoration Plantings

Once project objectives have been set, the next step is to determine which activities need to be accomplished, and in what order. It is clearly unreasonable to be considering fine points like how to increase stand structure if the topography is not stable. For severely degraded sites, the first priorities are usually site-related; that is, how can we stabilize soil, reduce erosion, reduce compaction, improve macroporosity, increase nitrogen and organic matter, and reduce toxicity (Table 2)? Most of the answers are based on years of experience in rehabilitating mine spoils or other special sites. Forest managers should be aware of the voluminous literature on this topic, and not assume that their experiences in revegetating cut-over forest land are comparable.

Some people may question the need to plant degraded areas, and suggest that neglect will result in normal ecosystem development. However, planting for conservation or restoration objectives is done for the same reasons as planting for timber production or other objectives – it allows greater control of species composition and spacing, and can result in a forested area more rapidly than the no-action alternative (Bradshaw 1987; Harmer and Kerr 1995). The presence of many invasive weed species worldwide and the lack of seed sources for some desirable species may also mean that historical patterns of ecosystem development will not occur without intervention.

Once the decision has been made to plant, there are several considerations. First, can the desired species be directly planted, or must other species be planted first to ameliorate the site conditions or the exposure? Some plants can alter soil conditions in very specific ways; thus, a manager could slowly increase soil nitrogen levels by planting species capable of nitrogen fixation, could increase soil organic matter by planting species capable of producing large amounts of biomass, or could utilize species that act as cation "pumps" (Alban 1982; Aber 1987). (Cation pumps take up high levels of cations from lower soil levels, and this increases pH and cation availability in surface soil.) Species might also be selected for their berry or mast production, their ability to provide cattle forage or a cash crop, their ability to penetrate compacted soils or tolerate sodic conditions, the type of shade they provide, the lack of root competition, or merely their availability at a particular point in time.

Table 2. Project activities and methods used in restoration and conservation plantings.[a]

Goal of activity	Possible methods
Stabilize soil surface	Contour. Construct debris dam. Mulch. Drain.
Reduce compaction	Use ripping (mechanical disruption).
Improve macroporosity	Incorporate wood or shale. Plant deep/strong "rooters." Tolerate burrowing animals.
Reduce toxicity (pH, metals, and salts)	Have patience (natural weathering). Replace soil. Use irrigation. Lime if pH is too low. Provide other amendments. Plant resistant species or cultivars.
Increase organic matter	Mulch with organic matter. Establish vegetation.
Increase nitrogen	Fertilize. Establish nitrogen-fixing plants.
Increase other nutrients	Fertilize or add other soil amendments. Establish selected plants.
Improve microsite conditions	Scalp. Do microsite preparation. Add shade or shelter. Use irrigation. Use snow fence, clumped plantings.
Reduce herbivory or physical damage by animals	Control access. Use repellents, protectors. Provide alternative browse. Use direct methods such as trapping.
Alter species composition of plants	Have patience. Do planting, seeding. Attract plant disseminators (e.g., with artificial perches).
Alter composition of vertebrate and invertebrate species	Have patience (suitable habitat). Use inoculation. Relocate.

[a] See Bradshaw 1987; MacMahon 1987; Gupta 1990; Luken 1990; and Moffat and Buckley 1995 for greater detail.

For example, species planting decisions after the 1980 eruption of Mount St. Helens (state of Washington, United States) were partially based on what species were already growing in the nurseries or for which seed had already been collected and could be sown in a greenhouse or nursery.

Genetic considerations are also important in selecting plant materials. For some situations it may be necessary to select adapted populations or even specific cultivars to grow on toxic surfaces (McNeilly 1987). If the goal is ecosystem restoration, rather than replacement or rehabilitation, then the manager will want to select native species and local sources. When ecological variants occur that differ in form or tolerances, managers must also take care to select the appropriate variants. Even with the most diligent care, it is almost inevitable that the resulting gene pools will be different from what would have been present without the degrading and rehabilitation activities. This may not always be a serious problem, though, as selection pressures may act on the new gene pool to result in similar selection pressure, and thus the population may evolve in a similar way (McNeilly 1987).

When dealing with severely degraded systems, managers should consider that there may be several alternative approaches to meet their long-term objectives. Lugo et al. (1993) have suggested that with suitable adjustments in management activities, monoculture tree plantations in the tropics can represent a pathway to restore biodiversity. With proper attention to species-site relationships, a tree species can be selected which can tolerate the stressful conditions that exist on sites abandoned for agricultural production, and begin the process of soil rebuilding. Once a plantation is well established, it can begin to function as a nurse crop, shading the ground, modifying the microclimate, and adding organic matter to the soil. If the species are chosen properly and later management practices are evaluated for their compatibility with long-term objectives, then the number of understory species should begin to increase. In the scenario proposed by Lugo et al. (1993), the overstory would probably be managed for wood yields to recover the initial investments in plantation establishment. However, the incorporation of management practices that facilitate natural successional processes (or avoidance of those which would impede them) would help ensure meeting the long-term conservation or restoration goals. This concept is supported by data on understory development from several monoculture tree plantations in the tropics (Lugo et al. 1993; Figure 2). These plantations were not managed for goals other than production of wood or other forest products; presumably, we could do at least as well in encouraging understory development if we were trying to do so deliberately. Some people will have philosophical disagreements with this approach (especially if it involves the use of exotic species in the overstory), because they might think that restoration or rehabilitation would occur more rapidly under an alternative approach (especially if one did not need to wait until products were harvested). The pragmatic approach suggested by Lugo et al. (1993), however, might allow many more hectares to be rehabilitated than would be possible in a scenario that required large front-end costs without any likelihood of a future return.

Similarly, some non-native, nontree species can be beneficial in the restoration process, and the timing of their eradication should be carefully evaluated. For example, recovery of species diversity in a forest in Argentina that had been subject to fire, overgrazing, and overtrampling was improved by the nurse crop behavior of a non-native shrub (Pietri 1992). The native species had not evolved with the current disturbances and had not developed strategies for dealing with them.

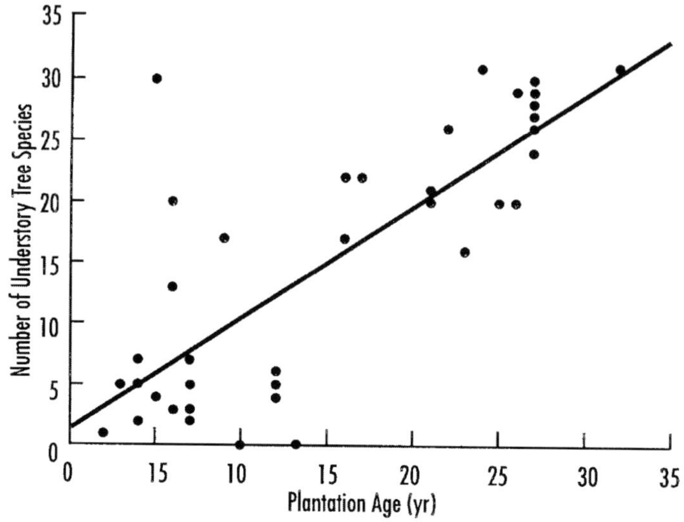

Figure 2. Understories of native tree species developed inside tree plantations in Puerto Rico and India, and the number of species present generally increased over time (*Source*: Lugo et al. 1993).

If there are native species available that are suited to the current soil and site conditions, the regeneration methods have been worked out for the desired species, and the resources are available to cover the project's costs, then clearly the reestablishment of native species rather than the use of exotics would be preferred for a project with restoration goals. Recent work (e.g., Butterfield 1995) has increased our knowledge of the growth rates and reforestation technology for several tropical tree species; however, there are many species for which our knowledge base is still extremely limited.

Once the plant materials have been chosen, managers must decide on the appropriate spacing and spatial pattern. Most undisturbed forest ecosystems are more heterogeneous than those that have been planted or managed. Natural systems (and, thus, the targets of our restoration activities) generally have greater species diversity, structural features, and spatial heterogeneity than planted stands. Ignoring the desirability of this variation can result in stands that are less functional than might otherwise be the case. For example, Maehr and Marion (1984) and Marion et al. (1981) report less species diversity of small mammals and birds on mined lands planted with trees than on lands unplanted. They attribute these results to the fact that the preplanting (site preparation) and planting activities reduced microsite variation. Planting in clumps rather than in a uniform distribution (Figure 3) may be more effective in re-creating desired forest conditions. Initial establishment of clumps can alter microclimate, can trap blowing seeds of other species, and may serve

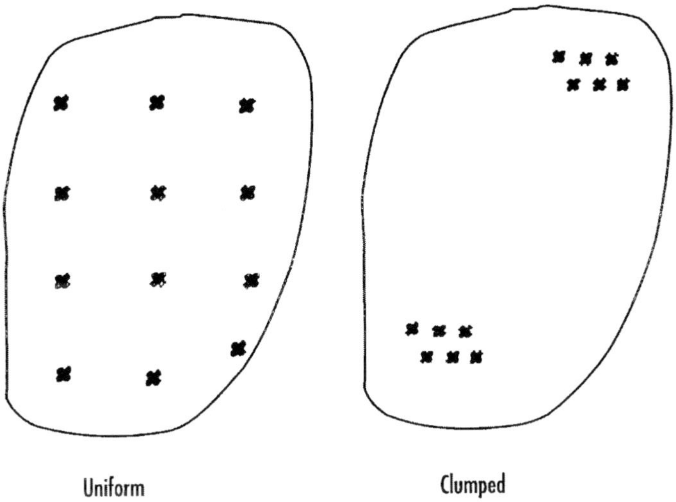

Figure 3. Planting can be planned for a uniform or clumped pattern. Although uniform plantings are simpler to administer, clumped plantings may result in a more natural pattern of variation in both plant distribution and microclimate.

as perches or cover for birds or mammals important in seed dissemination (Luken 1990). Future management activities, such as underplanting or thinning, should also be conducted with some knowledge of the spatial pattern that we are trying to emulate.

Those involved in planning restoration or conservation plantings may also want to consider incorporating wildlife enhancement activities such as adding artificial perches or nest boxes, or creating cavities. Such practices can be quite effective in shortening the time necessary for an area to serve as habitat or increasing the carrying capacity for desired species of birds (Caine and Marion 1991) and arboreal rodents (Carey and Gill 1983; Carey et al. 1996).

Future needs

Conservation and restoration projects are usually very long term. It is important to have continuity through time; however, the methods for achieving this continuity are not clear. Ashby (1987) suggests that preestablished institutional arrangements may be a necessity and that biologists (researchers or managers) should be aware of these long-term needs before establishing a project. Legislation to encourage forest restoration projects would be helpful (as suggested by Ashby 1987), but, if such legislation exists, I am unaware of it. For example, last year I began working on a project on the Olympic Peninsula that was started with congressional direction to "begin a restoration demonstration in young forests." The funding for the project was for 1 year. We have received funding for the second year, but the lack of an alternative funding mechanism to appropriations that must be spent in 1 fiscal year makes long-term project planning a very frustrating exercise. Although we have accumulated a substantial body of information on rehabilitating mine spoils in some parts of the world, our knowledge of the best way to proceed in many site-specific and objective-specific conservation or restoration projects is quite limited. Werner (1987) suggests that two types of research are needed in the future. First, we need manipulative studies – studies in which we make a change, step back, and see what happens. These studies are relatively easy to carry out and can result in rules-of-thumb for future activities. These types of "black-box" studies (i.e., you know what went in and what goes out, but don't understand what happens inside the box) are hard to extrapolate to other situations. These black-box studies are needed to provide answers to urgent questions as quickly as possible (Werner 1987). On the other hand, mechanistic studies are designed to provide information about why a particular result occurs under a given set of conditions. This type of "microscope" study provides a lot of detail, but with a limited field of vision (Werner 1987). Mechanistic studies should ultimately allow better prediction of results under a range of conditions. Thus, both types of studies[2] need to be included in planning research for restoration or conservation objectives. In addition to research activities, substantial information on effective and ineffective practices could also be gained by monitoring operational activities.

Notes

1. Substantial differences may exist in the desired ecosystem depending on whether one is targeting the re-creation of a previous system or the system as it would have developed without the degrading activities. See Bonnicksen 1988 and Cairns 1989 for discussions of this issue. For our purposes, however, the important issue is that one must determine the basis for system comparisons.
2. It would also be desirable to include studies which meld the two approaches. That is, mechanistic approaches can be used in manipulative studies.

References

Aber, J.D. 1987. Restored forests and the identification of critical factors in species-site interactions, pp. 241–250. In: Jordan III, W.R., Gilpin, M.E. and Aber, J.D. (Eds) Restoration Ecology: A Synthetic Approach to Ecological Research. Cambridge University Press, Cambridge.

Alban, D.H. 1982. Effects of nutrient accumulation by aspen, spruce and pine on soil properties. Soil Sci. Soc. Am. J. 46: 853–861.

Allen, T.F.H. and Hoekstra, T.W. 1987. Problems of scaling in restoration ecology: A practical application, pp. 289–305. In: Jordan III, W.R., Gilpin, M.E. and Aber, J.D. (Eds) Restoration Ecology: A Synthetic Approach to Ecological Research. Cambridge University Press, Cambridge.

Ashby, W.C. 1987. Forests, pp. 89–108. In: Jordan III, W.R., Gilpin, M.E. and Aber, J.D. (Eds) Restoration Ecology: A Synthetic Approach to Ecological Research. Cambridge University Press, Cambridge.

Bonnicksen, T.M. 1988. Restoration ecology: Philosophy, goals, and ethics. Envirn. Prof. 10: 25–35.

Bradshaw, A.D. 1987. The reclamation of derelict land and the ecology of ecosystems, pp. 53–74. In: Jordan III, W.R., Gilpin, M.E. and Aber, J.D. (Eds) Restoration Ecology: A Synthetic Approach to Ecological Research. Cambridge University Press, Cambridge.

Buckley, G.P. (Ed). 1989. Biological Habitat Reconstruction. Belhaven Press, London.

Butterfield, R.P. 1995. Promoting biodiversity: Advances in evaluating native species for reforestation. For. Ecol. Manage. 75: 111–121.

Caine, L.A. and Marion, W.R. 1991. Artificial addition of snags and nest boxes to slash pine plantations. J. Field Ornithol. 62: 97–106.

Cairns, J., Jr. 1989. Restoring damaged ecosystems: Is predisturbance condition a viable option? Envirn. Prof. 11: 152–159.

Carey, A.B. and Gill, J.D. 1983. Direct habitat improvements – some recent advances, pp. 80–87. In: Snag Habitat Management: Proceedings of a Symposium. General Technical Report RM-99. Rocky Mountain Forest and Range Experiment Station, Fort Collins, CO.

Carey, A.B., Thysell, D.R., Villa, L.J., Wilson, T.M., Wilson, S.M., Trappe, J.M., Colgan III, W., Ingham, E.R. and Holmes, M. 1996. Foundations of biodiversity in managed Douglas-fir forests, pp. 62–82. In: Pearson, D.L. and Klimas, C.V. (Eds) The Role of Restoration in Ecosystem Management. Society for Ecological Restoration, Madison, WI.

Chambers, J.C. and Wade, G.L. (Eds). 1992. Evaluating Reclamation Success: The Ecological Consideration – Proceedings of a Symposium. Northeastern Forest Experiment Station GTR-NE-164. Northeast Forest Experiment Station, Radnor, PA.

Duffy, P.D. and Ursic, S.J. 1991. Land rehabilitation success in the Yazoo basin, USA. Land Use Policy 8: 196–205.

Ferris-Kaan, R. (Ed). 1995. The Ecology of Woodland Creation. John Wiley and Sons, London.

Gupta, G.N. 1990. Integration of moisture and fertilizer management practices for early growth and establishment of *Eucalyptus* on a skeletally degraded dry land. Int. Tree Crops J. 6: 123–142.

Harmer, R. and Kerr, G. 1995. Creating woodlands: To plant trees or not, pp. 113–128? In: Ferris-Kaan, R. (Ed) The Ecology of Woodland Creation. John Wiley and Sons, New York.

Jordan III, W.R., Gilpin, M.E. and Aber, J.D. (Eds) 1987. Restoration Ecology: A Synthetic Approach to Ecological Research. Cambridge University Press, Cambridge.

Lenz, R. and Haber, W. 1992. Approaches for the restoration of forest ecosystems in northeastern Bavaria. Ecol. Model. 63: 299–317.

Lugo, A.E., Parrotta, J.A. and Brown, S. 1993. Loss of species caused by tropical deforestation and their recovery through management. Ambio 22(2–3): 106–109.

Luken, J.O. 1990. Directing Ecological Succession. Chapman and Hall, London.

MacMahon, J.A. 1987. Disturbed lands and ecological theory: An essay about a mutualistic association. In: Jordan III, W.R., Gilpin, M.E. and Aber, J.D. (Eds) Restoration Ecology: A Synthetic Approach to Ecological research. Cambridge University Press, Cambridge.

Maehr, D.S. and Marion, W.R. 1984. Bird abundance and distribution in a north Florida phosphate mine. Proc. Ann. Conf. Southeast. Assoc. Fish Wildl. Agencies 38: 111–120.

Magnuson, J.J., Regier, H.A., Christie, W.J. and Sonzogni, W.C. 1980. To rehabilitate and restore Great Lakes ecosystems, pp. 95–112. In: Cairns, J. (Ed) The Recovery Process in Damaged Ecosystems. Ann Arbor Science, Ann Arbor, MI.

Marion, W.R., Maehr, D.S. and Frohlich, R.K. 1981. Phosphate mine reclamation and habitats for wildlife, pp. 501–505. In: Symposium on Surface Mining Hydrology, Sedimentology and Reclamation. University of Kentucky, Lexington.

McNeilly, T. 1987. Evolutionary lessons from degraded ecosystems, pp. 271–286. In: Jordan III, W.R., Gilpin, M.E. and Aber, J.D. (Eds) Restoration Ecology: A Synthetic Approach to Ecological Research. Cambridge University Press, Cambridge.

Michaelson, T. 1994. Management for soil and water conservation, pp. 67–80. In: Readings in Sustainable Forest Management. Food and Agriculture Organization of the United Nations (FAO) Forestry Paper No. 122. FAO, Rome.

Moffat, A.J. and Buckley, G.P. 1995. Soils and restoration ecology, pp. 75–90. In: Ferris-Kaan, R. (Ed) The Ecology of Woodland Creation. John Wiley and Sons, New York.

Naiman, R.J. (Ed) 1992. Watershed Management: Balancing Sustainability and Environmental Change. Springer-Verlag, New York.

Newton, G.A. 1992. Assessing the rehabilitation potential of disturbed lands, pp. 27–30. In: Proceedings, Western Forest Nursery Association. Rocky Mountain Forest and Range Experiment Station GTR-RM-221. Rocky Mountain Forest and Range Experiment Station Ft. Collins, CO.

Pietri, D.E.D. 1992. Alien shrubs in a national park: Can they help in the recovery of natural degraded forest? Biol. Conserv. 62: 127–130.

Wade, G.L. and Chambers, J.C. 1992. Introduction, pp. 1–2. In: Chambers, J.C. and Wade, G.L. (Eds) Evaluating Reclamation Success: The Ecological Consideration – Proceedings of a Symposium. Northeastern Forest Experiment Station GTR-NE-164. Northeast Forest Experiment Station, Radnor, PA.

Werner, P. 1987. Reflections on "mechanistic" experiments in ecological restoration, pp. 321–328. In: Jordan III, W.R., Gilpin, M.E. and Aber, J.D. (Eds) Restoration Ecology: A Synthetic Approach to Ecological Research. Cambridge University Press, Cambridge.

Woolf, H.B. (Ed). 1977. Webster's New Collegiate Dictionary (7th edition). G. & C. Merrian Co, Springfield, MA.

Technical and Social Considerations for Planted Forests

New Forests **17:** 193–212, 1999.

How can we feign sustainability with an increasing population?

DAVID B. SOUTH
School of Forestry and Alabama Agricultural Experiment Station, Auburn University, Auburn, AL, USA

Key words: wood supply, land base, limited resources, consumption, plantation

Abstract. Attitudes of human societies toward tree plantations can be a critical factor in determining the source of wood supply in the future. Because human populations will expand substantially in the next century, considerable increases in the demands on native forests will occur. Currently, only a limited amount of tree plantations have been established specifically to provide firewood in developing countries. This may be the result, in part, of a world society that has evolved a general preference for pastures, but an aversion to tree plantations. In total, pastureland and tree plantations amount to 26 percent and 1 percent of the world's land base, respectively. Nevertheless, our actions today will determine whether children in the future collect firewood from natural stands or from tree plantations. It is estimated that a substantial afforestation program could increase the amount of tree plantations to equal 5 percent of the world's land base by the year 2050. Ten billion dollars (United States) is a rough estimate of the annual costs for such a program (assuming no overhead or administrative costs). Most of the wood needs could be met from tree plantations. However, if the world's society wants most of its wood in 2050 to come from natural stands (>80 percent), then tree plantations can be limited to just 1 percent of the land base.

Introduction

The world population for 1999 is about 6 billion. If the population continues to increase at a rate of 1.5 percent/yr, the population will double before the year 2100 (Figure 1). Some estimates are that the world population will be 11.2 billion by the year 2100 (World Resources Institute 1992). If the per capita consumption does not change, the world consumption of wood could be greater than 7 billion m^3/yr by the end of the twenty-first century. This wood will come from an increase in harvest of natural stands and/or from plantations. The amount of wood that is harvested from natural stands in the future will depend directly on the amount of plantations we establish today. Instead of planting more trees to provide wood for a population of 10 billion, many countries are reducing their rates of tree planting. For this

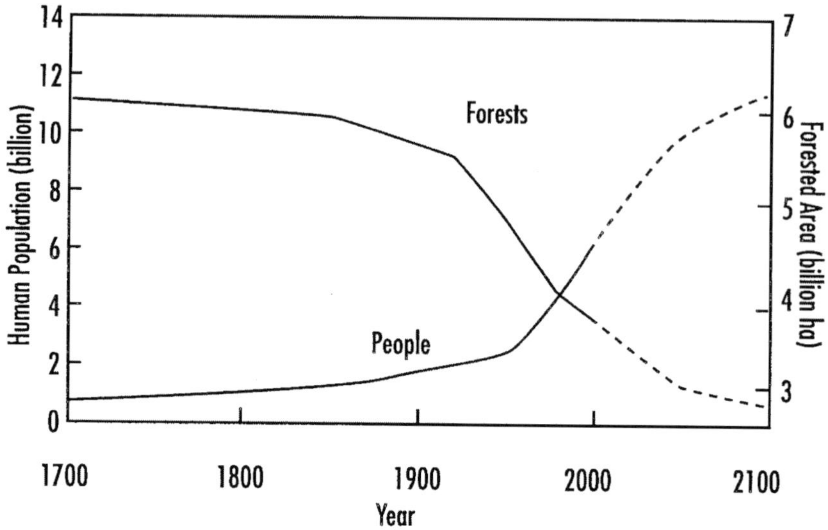

Figure 1. Projected decrease in the world's forests in comparison to projected increase in the world population (World Resources Institute 1992). Some estimate the world's forests for 1990 at about 3.4 billion ha (FAO 1995).

reason, the following hypothetical word problem is provided to illustrate how a society's views regarding land management can affect the wood supply for future generations.

Word problem

In the Hypothetical Ocean, there are three islands that are identical in size and natural resources. Each island is 12 ha in size and is covered with 10 ha of forest (the beach covers about 1 ha and there is 1 ha of high ground). Each island is inhabited by a different society. Each of these societies has the same ideas about population growth, food consumption, and wood consumption. However, each society has a different view about how to manage the land for future wood needs. Initially, at year zero, each society has ten people, and follows a strict ethic regarding population growth. Therefore, population growth is limited to only two persons per year per island.

Each society obtains all the food it needs from the ocean. However, these societies need wood for cooking and heating. Firewood is consumed at a rate of 1 m^3/person/yr. The 10 ha of forest produce about 24 m^3/yr, and this wood can be collected on a sustainable basis without reducing the growing stock. However, the trees are small and, therefore, the standing volume of the forest never exceeds 2.4 m^3/ha. To clarify, the 10 ha forest has 24 m^3 of wood on

January 1st and, during the year, 24 m^3 of firewood (e.g., dead branches) can be collected. On December 31st, the standing volume of the forest is still 24 m^3. If the firewood is not collected, the branches fall off the trees and either decay or are occasionally washed off the island. However, firewood can be collected and stored indefinitely (in a dry, high place) without wood decay. If harvest reduces the standing volume to less than 24 m^3, the annual firewood production will only equal the standing volume on January 1st (for ease of calculation).

Each society has the same major objectives: to keep cooking food and to keep at least 5 ha of forest as a preserve. Although firewood can be collected from the preserve, the objective is to not reduce the standing volume on these 5 ha.

The three societies differ in the way they decide to manage their islands. The society on Island A decides to keep all 10 ha of forest in a natural state; all the firewood is collected from the forest (24 m^3), and any unused firewood is stored for future needs. On January 1st (year 1), the society has 14 m^3 of firewood in storage.

On Island B the society decides to cut down 5 ha of forest and to put the wood in storage. On the cleared land, a 5 ha plantation is established with an exotic species (*Salix* spp.). The willow plantation is harvested each year and produces 10 m^3 of firewood/ha/yr. On January 1st (year 1), there is 64 m^3 of firewood in storage (50 m^3 from the plantation, 12 m^3 from cutting down 5 ha of woods, and 2 m^3 of extra firewood collected from the preserve). The society plans to continue managing the 5 ha willow plantation on a sustainable basis (adding kelp for fertilizer to maintain production at 10 m^3/ha/yr).

The society on Island C does the same as the society on Island B. On January 1st, (year 1) this society also has 64 m^3 of stored firewood. However, the society decides to invest time and effort into research. Although some mistakes are made, new ways are discovered to increase the volume production from the willow plantation. Amazingly, each year the researchers figure out a way to increase willow production by 0.2 m^3/ha/yr. For example, the first willow crop was 50 m^3, the second was 51 m^3, the third was 52 m^3, and so forth.

With the information given, how long does it take before each society depletes the storage of wood? Which society is most likely to protect the 5 ha preserve for the longest period of time?

Answer

Island A can only sustain a population of 24 people (i.e., 24 m^3/yr). By storing firewood collected from the natural forest, this society's supply of stored

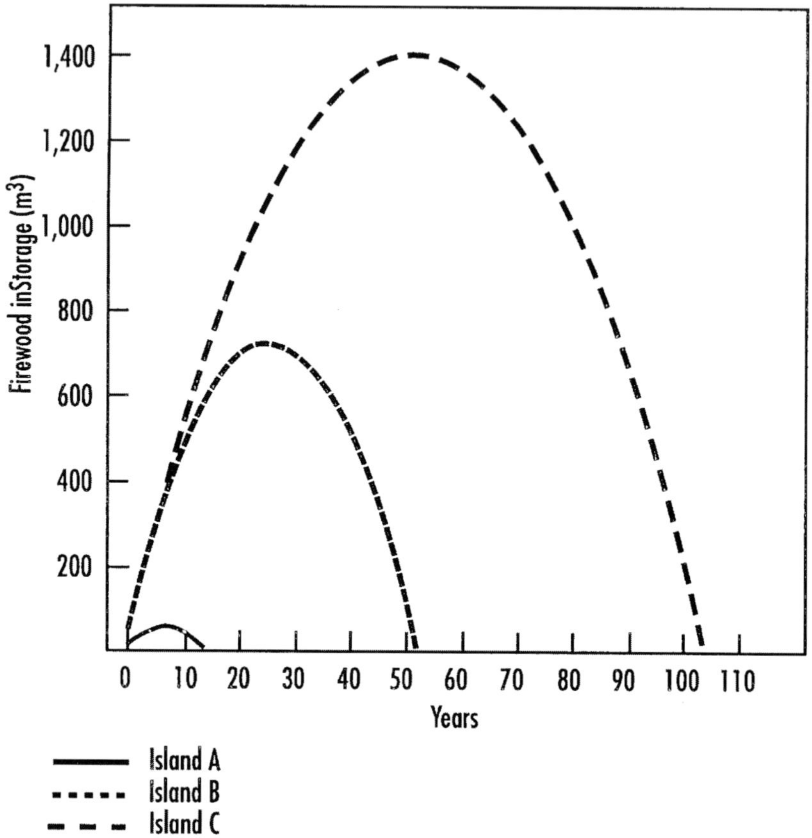

Figure 2. The amount of firewood in storage for hypothetical Islands A, B, and C.

wood runs out during year 14 (Figure 2). On January 1st (year 15), there are 40 people on the island and during the year they consume 24 m^3 of growth for the year. They end up cutting down 6.7 ha of the natural forest. In year 16, there are only 3.3 ha of forest remaining, which is not enough to supply ten people (much less 42). The low productivity of the nonpreserve land limits the ability to protect the preserve.

The society on Island B increases the production of firewood with a willow plantation. With both wood from the plantation and wood from the preserve, this society is able to produce 62 m^3/yr. Therefore, Island B can sustain a population of 62 people (two and a half times higher than Island A). When combined with storage, the society lasts for 51 years before the supply runs out. The preserve was protected for more than three decades longer than on Island A.

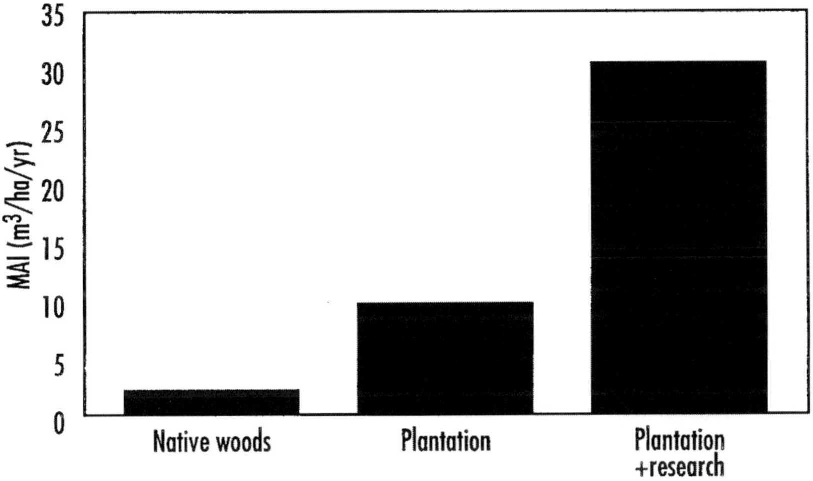

Figure 3. Mean annual increment (MAI) for the native woods on Island A, the exotic plantation on Island B (with no research), and the exotic plantation on Island C (after 100 years of research).

By investing time and effort in research, the society on Island C was able to find ways to increase volume production/hectare. As a result, the preserve was protected for more than a century. By conducting research aimed at improving yield, this society was able to have a 5 ha preserve and sustain at least 190 people (a yield of 190 m^3/yr could be obtained in year 116). However, at year 104 the population is 218 and the amount of wood in storage is down to 12 m^3. For the next year this society decided that some of the 5 ha "preserve" would need to be converted to a willow plantation in order to supply enough wood for everyone. The plantation was expanded by 2 ha to meet the needs of 220 people.

The ability to protect the preserve varied with the management regimes adopted by the society. Societies that relied on plantations to supply part of their wood needs were able to keep a preserve intact for much longer than the society that relied entirely on native woods. One could conclude that high productivity on nonpreserve land (Figure 3) can help to protect the preserve by meeting the day-to-day consumption needs of the population. I conclude that high productivity from intensively managed lands can help protect wilderness or nonutilized lands from exploitation. Likewise, agronomists have concluded that increasing productivity on the current agricultural lands can feed 10 billion people without increasing the amount of farmland (Waggoner 1994).

None of the societies can sustain a population growth of two persons/yr (Figure 4). Therefore, the real problem for all three societies is population

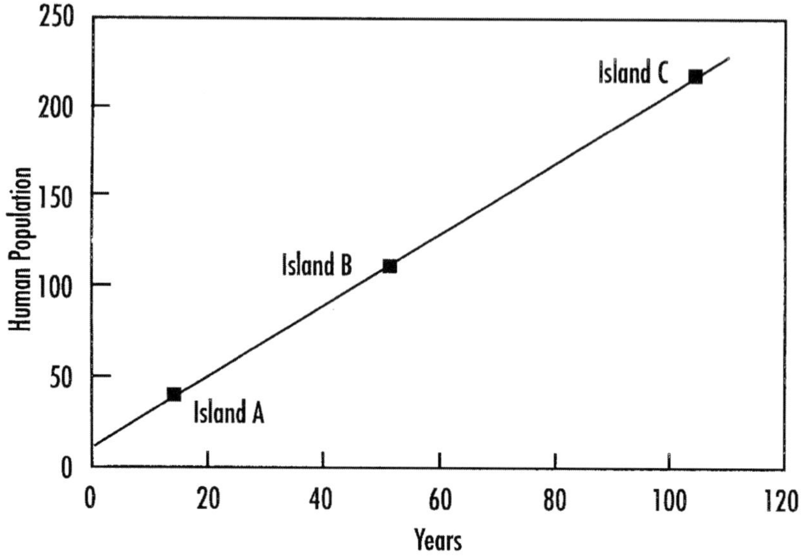

Figure 4. The population growth curve and the respective year when the society on each island cuts down part of the preserve.

growth. Adoption of new technology can delay the inevitable (Figure 2), but for areas with limited resources, societies must adopt appropriate ethics about population growth if they hope to be sustainable during the next several millennia. From this exercise, it also can be concluded that for any of the three islands to have a sustainable ecosystem and an unconstrained population growth either (1) per capita wood consumption must decline in direct proportion to the population growth (through recycling, product substitution, or reduced standard of living), or (2) technology must devise ways to continually improve yields.

In this example, even when population growth is linear, limited resources cannot sustain unlimited population growth. I realize this statement does not agree with others (Barnett and Morse 1963; Manthy 1977; Simon 1977, 1981, 1992). I imagine that Julian Simon would propose a fourth island where researchers were able to increase growth (each year) by 0.6 m^3/ha/yr. This rate of improvement would be more than enough to support the population growth and would eventually allow the natural preserve to expand in area. Simon (1981) has stated that "if the past is any guide, natural resources will progressively become less scarce, and less costly, and will constitute a smaller proportion of our expenses in future years. And population growth is likely to have a long-run *beneficial* impact on the natural-resource situation." Although I agree with much of Simon's philosophy, I do not accept a cause and effect relationship between population growth and the supply of firewood.

Just because wood supply increased from year 0 to year 50 as the population of Island C increased, we should not assume that wood supply will continue to increase for the next 50 years (Figure 2).

Real world examples

Although the islands in the word problem are fictitious, real islands and/or countries that illustrate each management type do exist. Island A is similar to Easter Island. Around 400 AD, people arrived on an island that was covered with trees and palms. A society emerged that is well known for statue building. However, as the population increased, the consumption of natural resources increased. It is estimated that forest clearing began around 700 AD, and that the last remnants of the forest were gone by about 1400 AD (Bahn and Flenley 1992). Initially, the society had wood to build large canoes, but by 1722 wood was so limited that the few remaining canoes were small (3 m long), frail, and leaky. By 1770, Gonzalez wrote, "Not a single tree is to be found capable of furnishing a plank so much as six inches in width." Although the society apparently did not plant trees for their children, they often promised to send their children a tree from the afterlife (Bahn and Flenley 1992). Bahn and Flenley believe that the social catastrophe that occurred in the 1700s was a direct result of overexploitation of the forest. I do not believe the population growth on Easter Island had a long-run "beneficial" impact on the natural resources of the island. Although the islanders may have invested research into how to move large statues, they likely did not invest time in determining how to propagate nonfood trees.

Although Alabama (in the southern United States) is not an island, it can be used as an example of a society which is similar to that on Island B. Prior to 1900 AD, this society relied heavily on wood as a fuel source, but today fossil fuels are used to cook food and heat homes. Large tree planting programs during the late 1950s and late 1980s converted old agricultural land to pine plantations. Today, approximately 10 percent of the land base is in plantations (1.4 million ha) and 56 percent is in native woodlands (McWilliams 1992). Currently, about one-third of the softwood harvested is from plantations. Although more plantations are expected to increase in the future (Figure 5), only a few forest scientists in Alabama are currently exploring ways to increase the wood yields (i.e., mean annual increment). With a population of over 4 million, Alabama has only about eight researchers actively looking for ways to improve plantation yields. Currently, production of pine plantations is about 10 m^3/ha/yr.

New Zealand can be used as a society representative of that on Island C. Currently, these islanders have established tree plantations (1.3 million

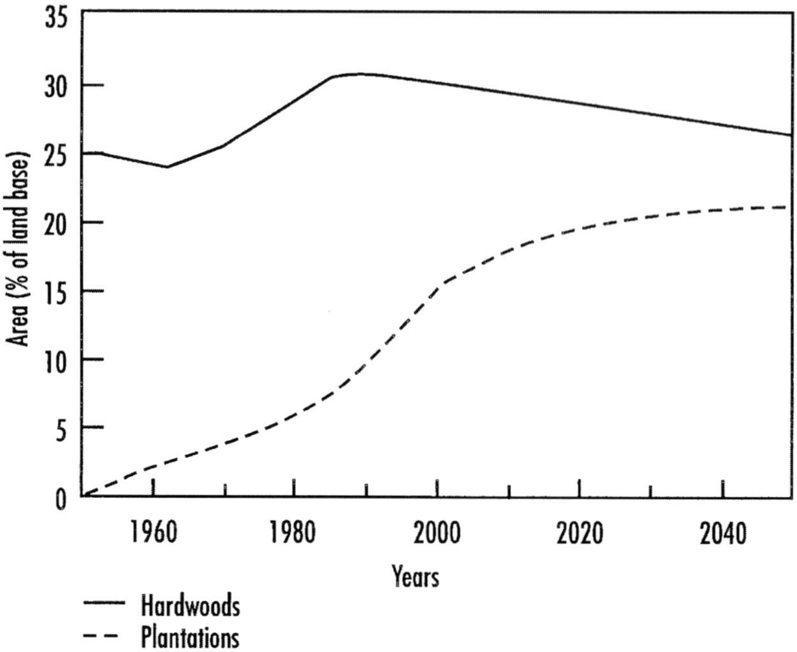

Figure 5. Projected increases in pine plantations in comparison with the amount of natural hardwood stands (bottomland plus upland) for the State of Alabama. (*Source*: USDA Forest Service 1988).

ha) on about 5 percent of the land base and have retained 23 percent of the land in native woodlands. Today, about 98 percent of the wood harvested is from plantations. This reliance on wood from plantations is mostly a result of large tree planting programs conducted during the 1930s and again from 1970 to the late 1980s. With a population of 3.8 million, New Zealand has approximately 30 scientists working to improve plantation growth. Currently, production of pine plantations is often about 24 m^3/ha/yr.

The purpose of the word problem and the "real world" examples is to illustrate the "yin-yang" aspects associated with tree monocultures and native woodlands. Some people believe that a dualistic philosophy is superior to the "all or none" philosophy touted by others. The important question, I think, for a population of 10 billion, is how much land do we want to devote to each of the major cover types (e.g., wilderness, native forests, tree farms, pastureland, farm crops, roads, and cities)? This seems more relevant than debating the pros and cons of either having all natural forests or all industrial plantations.

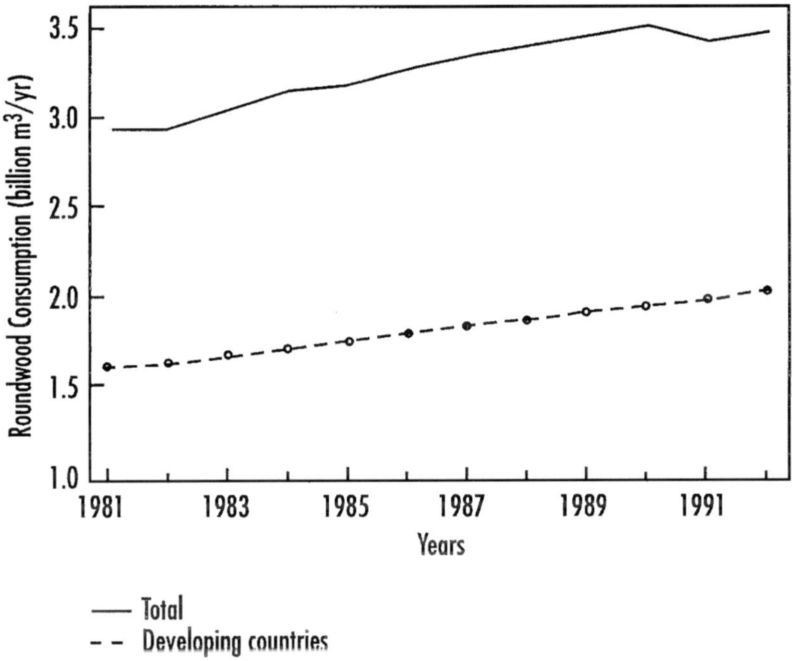

Figure 6. The global increase in roundwood consumption from 1981 to 1991 indicates a steady increase in wood consumption from developing countries (*Source*: FAO 1992).

The world as an island

Some ecologists believe the concept of carrying capacity is directly relevant to the realistic evaluation of the future size and impact of human populations (Pulliam and Haddad 1994). Perhaps now is the time for foresters to make an attempt to define the "carrying capacity" for wood production. As a society, foresters are concerned about population growth and its effects on natural resources (see Appendix). Although estimates of annual wood consumption are increasing, do we know how to provide a sustainable wood supply for a population level of 10 billion people? In just 11 years, the consumption of roundwood in developing countries has increased by about 25 percent (Figure 6). In just 15 years, the world's roundwood consumption will approach 5 billion m^3/yr (FAO 1995). If this rate of wood use per capita is maintained, and the population continues to increase as expected, then by the year 2050 the world might be consuming about 7 billion m^3/yr. This level of consumption is about 75 percent greater than that today (1995) and, assuming a low level of plantation establishment, this rate of harvest will likely exceed the maximum level of sustainability. Historically, deforestation occurs as rural populations increase (Clawson 1979; Inman 1993).

As a forester, I realize that the net annual increment (NAI) is not the best scientific estimate of how much wood our world can supply indefinitely. However, because I believe we are not planting enough trees for the future, I will use the ratio of NAI to annual harvest as a rough indicator of "sustainability." If the NAI for developed countries is 2.4 billion m^3/yr (FAO 1995) and we assume that the NAI for developing countries is 2 billion m^3/yr (not all forests are "production" forests), then a very rough estimate for the total NAI for 1990 is near 4.4 billion m^3. If the demand for wood in 2015 is 5 billion m^3 (FAO 1995), then deforestation will likely result in developing countries where the annual harvest will exceed the NAI, mostly as a result of a high demand for firewood. The demand in developing countries for fuelwood and charcoal alone is estimated to reach 2.4 billion m^3/yr by the year 2010. By the year 2050, the demand for firewood (perhaps not the supply) could reach 3.3 billion m^3/yr.

What type of world society (A, B, or C) will help to ensure that our children do not end up exploiting the world's forests? Assuming that we consume wood only from sustainable systems, where will the extra wood be produced? Do we want the extra 3 billion m^3 to be logged from natural stands, or do we want most of the additional wood to come from plantations? If most individuals are from a society similar to that on Island A, then the "politically correct" policy would be to harvest wood mainly from native woods. Some individuals from this society would not favor the expansion of plantations, and would likely protest the allocation of 5 percent of the land base to plantations.

However, individuals from societies where plantations comprise 5 percent or more of the land base (e.g., Alabama, Japan, New Zealand, and Sweden) would see no problem with using 5 percent of the land base for tree farms and fiber farms. Individuals from these societies would believe the "correct" policy would be to establish plantations to provide a renewable resource to sustain the society for a much longer time period.

Global planting rate

Estimates of the total amount of land in tree plantations can vary widely. However, a recent estimate is 128 million ha (Brooks 1993). Over the past 15 years, plantations have been expanding at a rate of 2.4 million ha/yr. If this rate continues for the next 60 years, then the world will have 272 million ha of plantations in 2050. This is equivalent to 2 percent of the world's land base. Assuming that 20 years is an average rotation length for plantations, a total planting rate (for 1990) would be approximately 9 million ha/yr and a final planting rate (2050) would be 16 million ha/yr. Estimates for annual planting rates vary from 4 to 14.5 million ha/yr (World Resources Institute 1986, 1992; Sedjo 1995).

If the world's society decides to set a goal of 630 million ha of plantations by the year 2050 (about 5 percent of the land base), the expansion rate for plantations will need to be about 9 million ha/yr (as opposed to a current expansion rate of 2.4 million ha/yr). The total annual planting rate will initially need to be approximately 15 million ha/yr. At year 2050, the annual planting rate might be closer to 33 million ha/yr. In contrast, if our society decides to limit plantations to just 1 percent of the world's land base, then (assuming an average rotation length of 20 years) the annual planting rate would drop to about 6.5 million ha/yr.

Currently, the United States has about 1.5 percent of the land base in plantations (Table 1) and the annual planting rate is about 1 million ha. If the average rotation length in the United States is 30 years, and the planting rate remains at 1 million ha/yr, then we should eventually reach a plantation acreage of 30 million ha (about 3.3 percent of the land base). In the southern United States, there may be 20 million ha of plantations by the year 2030 (U.S.D.A. Forest Service 1988). Because of the use of plantations, it is expected that roundwood consumption in the United States will increase by 40 percent even though the amount of exploitable timberland will decrease by 4 percent (Figure 7). If U.S. society wants to produce 960 million m^3/ha/yr from plantations by the year 2050, then about 96 million ha of plantations will be required. Plantation acreage will need to be expanded by 1.4 million ha/yr. Initially, this will require an increased planting rate of 1.8 million ha/yr, but will end up at a planting rate of about 3.2 million ha/yr. Because almost 1.4 million ha were planted in 1988 (Figure 8), planting 28 percent more should not be too difficult. If these goals are to be met, the United States would have about 11 percent of the land base in plantation by 2050. According to current planting rates, some predict that 20 percent of Alabama's land base will be in plantations by the year 2030 (Figure 21.5). If all additional U.S. plantations were to be established on pastureland, then the amount of pastureland in the United States in 2050 would decrease to about 16 percent of the land base.

Pastures and plantations

In the past, the world's society has preferred to establish pastures rather than tree plantations. Except for Japan and Sweden, the ratio of pastures to plantations is very high in most countries (Figure 9). In just 33 years (from 1955 to 1988), pastureland increased globally by about 1 billion ha (Figure 10), while tree plantations increased by about 0.1 billion ha (a ratio of 10 to 1). Pasturelands have been allowed to increase from 18 percent of the

Table 1. Population and plantation areas for 1990 (Brooks 1993), estimated populations for the year 2050, and proposed plantation areas for the year 2050.[a]

Country/region	Population		Plantation area: Million ha		Plantation area: % of land base		Expansion rate[b] (million ha/yr)
	1990	2050	1990	2050	1990	2050	
World	5,316	10,000	128	629	1	5	9.1
United States	249	400	13.4	96	1.5	11	1.5
Canada	27	50	1.5	12	0.2	1	0.2
Europe	385	930	33	84	6	15	0.9
Japan	124	150	10.2	14	27	36	0.1
Africa	648	2,100	2.1	105	0.07	4	1.9
China	1,140	1,900	30	95	3.2	10	1.7
Asia (other)	1,845	3,660	31.5	183	1.8	10	1.2
Latin America	448	810	6	40	0.3	2	0.6

Note: This assumes a per capita consumption rate of 2.4 m^3 of wood for Canada and United States; 0.9 m^3 for Europe and Japan; and 0.5 m^3 for Africa, Asia, and Latin America.

[a] Calculations are made with the assumption that 93 percent of the wood need is met from plantations.

[b] Plantation expansion rate does not include replanting of harvested plantations.

world's land base in 1955 to about 26 percent in 1990. In contrast, it seems unlikely that the world's society would allow tree plantations to increase from 1 percent of the world's land base in 1990 to 9 percent in just 33 years (from 1995 to 2028). The number of printed words (mostly in newspapers and magazines) objecting to establishing tree plantations seems much greater than the number objecting to the establishment of pastures. Because of both increasing population pressures and the preference for eating red meat, it is likely that more natural forests will be converted to pastures than will be converted to tree plantations during the next 55 years. This will occur even though overgrazing can cause soil degradation, and tree plantations can be used to restore degraded land. According to the World Resources Institute (1992):

> Overgrazing by livestock decreases vegetation, exposing the soil to water and wind erosion. In addition, livestock trample and thereby compact the soil, reducing its capacity to retain moisture. Overgrazing is the most pervasive cause of soil degradation, affecting 679 million ha (35 percent of all degraded land). In Africa and Australia, overgrazing causes 49 percent and 80 percent, respectively, of soil degradation, mainly in semiarid and arid regions.

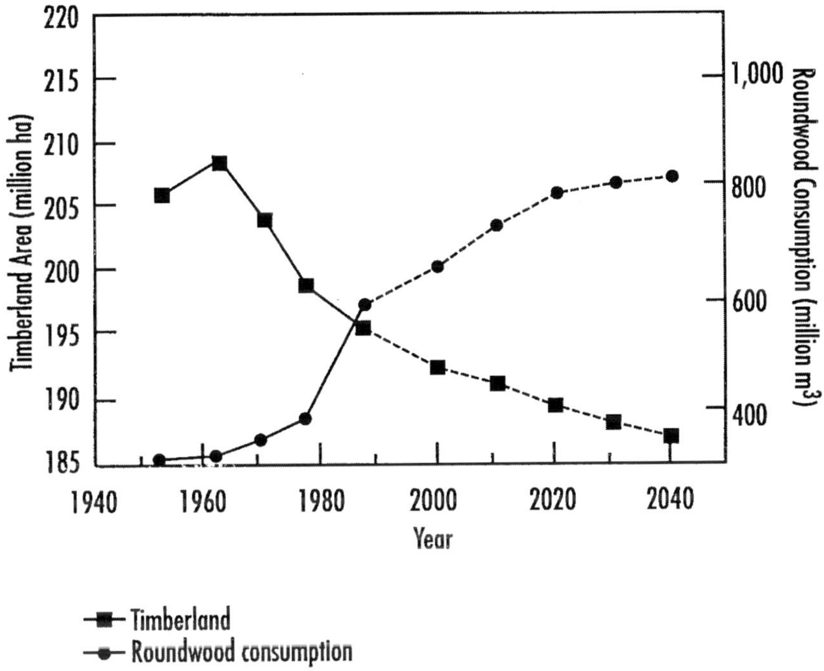

Figure 7. The expected decrease in timberland in the United States and the expected increase in roundwood consumption (*Source*: USDA Forest Service 1989).

Two management scenarios

Let's examine two possible management scenarios for the world. The "native stand scenario" would keep tree plantations at the current level of about 130 million ha (1 percent of the world's land base). This would allow pastures to increase to about 28 percent of the land base. Although most of the wood would be harvested from native stands, about 18 percent would be supplied by plantations (Table 2). Over half the wood consumption would be used for firewood.

In contrast, the "plantation scenario" would increase the amount of plantations to about 5 percent of the land base (Table 1), but would decrease the amount of pastureland to 22 percent of the land base. Most of the wood produced from this scenario would be harvested from plantations. Only about 7 percent of the wood would be supplied by managed native forests. Although about 130 million ha of afforestation would be for biomass energy (Sampson et al. 1993; Wright and Hughes 1993) gain, about 260 million ha would be for firewood production in developing countries.

All of the "new" plantations could be established on pastureland (i.e., afforestation). Removing 500 million ha of degraded pastureland would still

Table 2. Two possible wood production scenarios for the world in the year 2050.

Land use	MAI[c]	Native stand scenario[a]			Plantation scenario[b]		
	(m^3/ha/yr)	Forest area (billion ha)	World's land base (%)	Annual harvest (billion m^3)	Forest area (billion ha)	World's land base (%)	Annual harvest (billion m^3)
Protected forest	0	0.28	2	0	0.65	5	0
Unmanaged native stand	0	0	0	0	2.08	16	0
Managed native stand	2	3	23	6	0.25	2	0.5
Plantation	10	0.13	1	1	0.65	5	6.5
Total	–	3.4	26	7	3.63	28	7
Pasture	–	–	28	–	–	22	–

[a] The native stand scenario estimates that 86 percent of the wood is harvested from managed native stands.
[b] The plantation scenario estimates that 93 percent of the wood is harvested from plantations.
[c] MAI = mean annual increment.

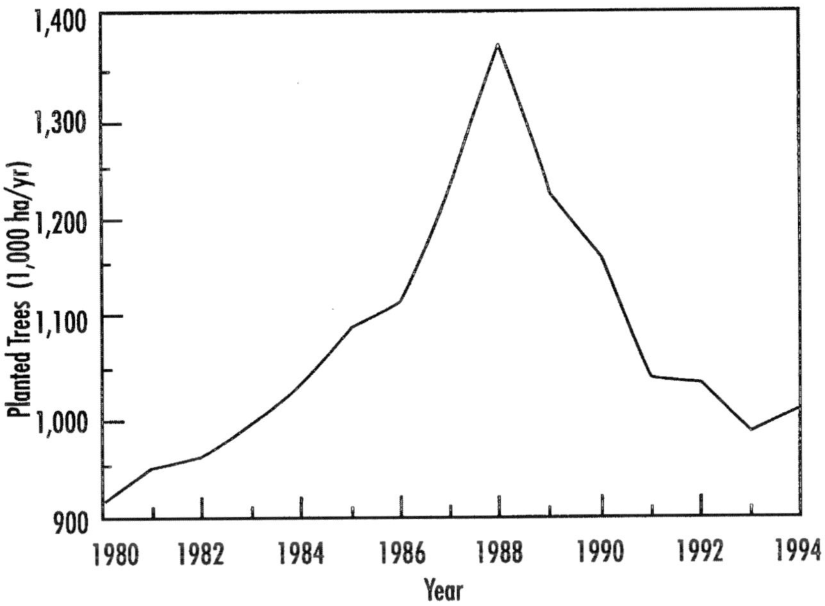

Figure 8. The trend in tree planting in the United States from 1980 to 1994 (*Source*: Moulton et al. 1995).

leave 22 percent of the world's land base in pasture (about the 1965 level). This might reduce the production of pasture fed meat by 15 percent or less. However, it is likely that some societies will not support the establishment of plantations on pastureland. Some will claim that the economics favor production of red meat instead of trees. Some will object to plantations if they displace the traditional sources of livelihood for local people (Barber et al. 1994). Some may argue that a monoculture of trees is more harmful to the environment than a monoculture of monocots.

If productivity of plantations is substantially higher than for native woodlands, then an increase in plantations should remove some of the pressure from natural stands. I have assumed that tree plantations can average about 10 m^3/ha/yr and natural woodlands will produce about 2 m^3/ha/yr (Table 2). In the southeastern United States, native hardwood stands produce about 2.5–3.0 m^3/ha/yr (U.S.D.A. Forest Service 1988). For the Commonwealth of Independent States, an average mean annual increment is estimated at 1.4 m^3/ha/yr (Richards 1987). Some tropical forests in the Asia-Pacific region may produce 1 m^3/ha/yr on a sustainable basis (Sedjo and Lyon 1990). Although higher values for individual regions can be obtained, 2 and 10 m^3/ha/yr are rough estimates of average production on a global basis. They, therefore, should not be viewed as accurate estimates.

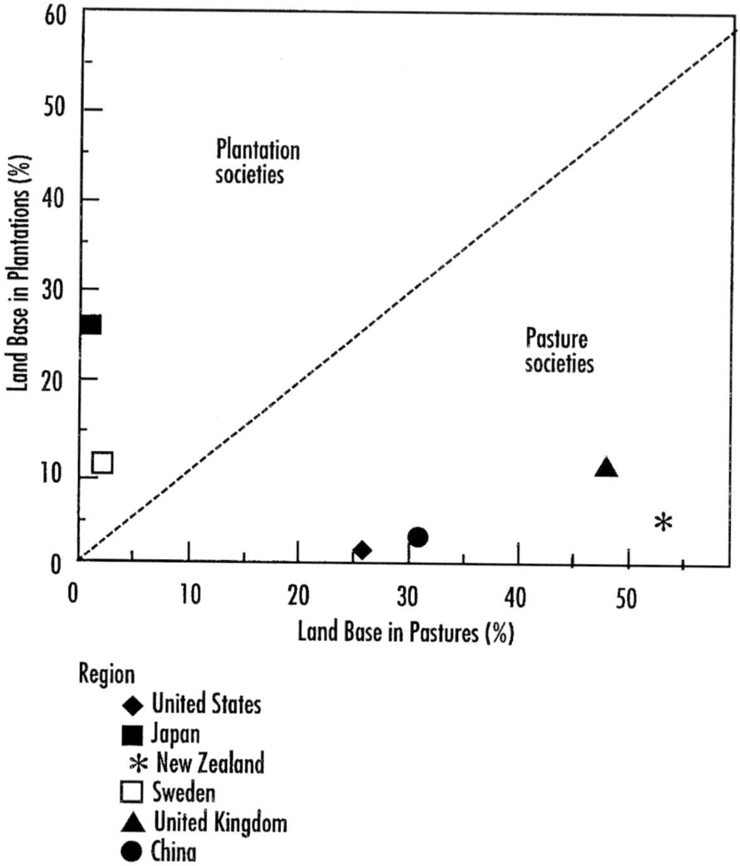

Figure 9. A comparison of land uses (tree plantations versus pastures) for six countries. Countries with more tree plantations than pastures include Japan and Sweden.

Global costs

Approximate costs for establishing plantations are provided in Table 3. The costs vary as a result of differences in labor costs, seedling costs, and site preparation costs. If a cost of $830/ha is used as a median value, then a planting program aimed at establishing 12 million ha/yr would cost 10 billion dollars. This is only half the amount that some have recommended for afforestation and soil conservation (World Bank 1992). This would be less than $2/person/yr. In comparison, a U.S. nuclear submarine can cost 2.5 billion, and the U.S. Senate is considering building 30 new nuclear-powered submarines over the next 20 years. Currently, the payment on interest for the U.S. debt amounts to about $775/person/yr. The annual cost of filling out tax forms for the United States is estimated at $20 billion annually. I therefore

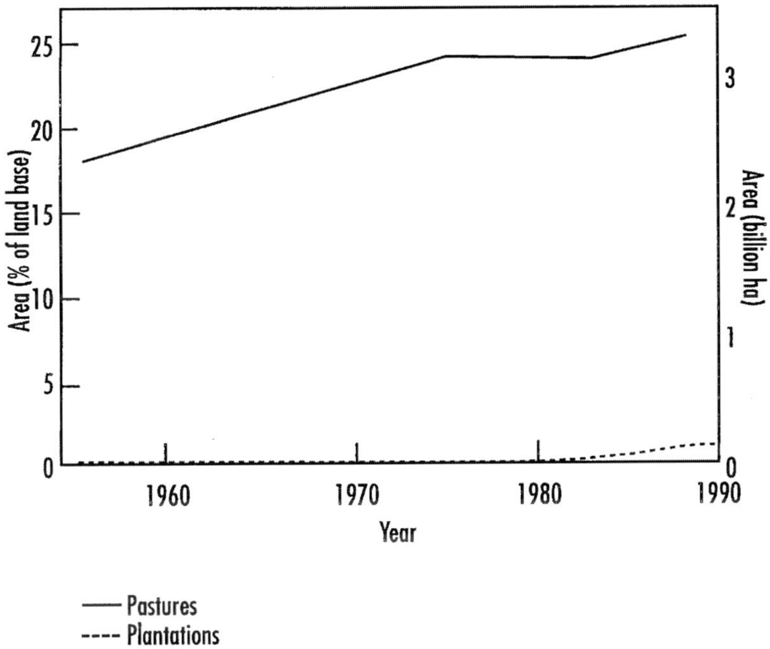

Figure 10. A comparison of the increase in tree plantations and pastures since 1955 (*Source*: World Resources Institute 1986, 1992).

conclude that global funds for planting 12 million ha/yr are not a limiting factor. The limiting factor is the priorities of our society. Some people believe we have a moral obligation to make sure that future generations have an equal or even greater supply of wood than we have claimed for ourselves (Schulz 1993). However, planting trees today to benefit our children and grandchildren may not be a high priority for many societies in the world.

Recommendations

Recommendations for beginning to deal with the situation include the following:

- Take steps to increase the planting of trees for fuel and fiber.
- Set a global planting goal to increase plantations to 630 million ha by the year 2050 (250 million ha for industrial roundwood; 250 million ha for firewood in developing countries; 130 million ha for biomass-fuelwood in developed countries).
- Encourage afforestation of pasturelands.
- End tax subsidies for parents having more than one child.
- Just do it.

Table 3. Approximate costs of establishing plantations and associated Mean annual increments, MAI (*Adapted from*: Winjum and Schroeder 1991).

Country/region	Afforestation	Cost/ha ($)	MAI (m^3/ha/yr)
Southern United States	Yes	270	11
Southern United States	No	370	9
South Africa	Yes	370	15
New Zealand	Yes	490	20
Brazil	Yes	600	23
India	Yes	800	30
Finland	No	900	3
Canada	No	1,000	6
United Kingdom	Yes	1,100	14
Tasmania	No	1,280	14
Congo	Yes	1,400	30
Western United States	No	1,730	14

Appendix: Management of Natural Resources in the Face of an Increasing Human Population

A Position Statement of the Society of American Foresters. Adopted by the Council of the Society of American Foresters on May 2, 1984, and renewed on November 10, 1987, November 13, 1990, December 7, 1994 and December 5, 1995. This policy expired December 17, 1997.

The relationship of human populations to forest land resources is a critical factor in achieving the full benefits of those resources. As human populations continue to increase substantially, increased demands on forest land resources will result.

The United States has the capacity to provide leadership in this global population challenge – as it has done in the conservation movement. Our legislative measures to ameliorate air and water pollution and toxic wastes, and to protect endangered species and wildlands have established a world standard. Yet these measures treat only the symptoms of uncontrolled population growth. This primary conservation issue has yet to be seriously addressed by the nation.

Professional foresters are concerned about the destruction and degradation of habitat for both humans and wildlife. Mounting population pressures not only lower the quality of life for humans but also contribute to the extinction of plant and wildlife species. The parallels of current population trends to wildlife management principles are obvious – in some places people are overrunning their own habitat and that of other life forms and making natural-resource management ineffectual. The best science and technology we can devise will not extricate us from the absolute limitations of the carrying capacity of our environment.

The relationship of human populations to forest land resources is a critical factor in optimizing forest benefits. If human populations expand substantially in the future, considerable increases in the demands on forest land resources will occur. Although we recognize that the much debated political aspects of population policy are peripheral to the expertise of professional forest land managers, we also recognize that the long-term effectiveness of forest management and conservation efforts depends on the resolution of this major domestic and global challenge. Therefore, the Society encourages efforts to place before the public scientific information on the dangers of unlimited population expansion and the land-management options that will have to be faced.

References

Bahn, P. and Flenley, J. 1992. Easter Island, Earth Island. Thames and Hudson, Inc., New York.

Barber, C.V., Johnson, N.C. and Hafild, E. 1994. Breaking the Logjam: Obstacles to Forest Policy Reform in Indonesia and the United States. World Resources Institute, Washington, DC.

Barnett, J. and Morse, C. 1963. Scarcity and Growth: The Economics of Natural Resource Availability. Johns Hopkins University Press for Resources for the Future, Inc., Baltimore.

Brooks, D.J. 1993. U.S. Forests in a Global Context. USDA Forest Service General Technical Report RM-228. USDA Forest Service, Washington, DC.

Clawson, M. 1979. Forests in the long sweep of American history. Sci. 204: 1168–1174.

FAO (Food and Agriculture Organization of the United Nations) 1992. Yearbook of Forest Products. Forestry Series 25. FAO, Rome.

FAO (Food and Agriculture Organization of the United Nations) 1995. Forest Resources Assessment 1990: Global Synthesis. FAO Forestry Paper 124. FAO, Rome.

Inman, K. 1993. Fueling expansion in the third world: Population, development, debt, and the global decline of forests. Soc. Nat. Resour. 6: 17–39.

Manthy, R.S. 1977. Scarcity, renewability, and forest policy. J. For. 75: 201–205.

McWilliams, W.H. 1992. Recent Trends in Afforestation and Reforestation of Nonindustrial Private Pine Forests in Alabama. Resource Bulletin SO-164. USDA Forest Service, New Orleans, LA.

Moulton, R.J., Lockhart, F. and Snellgrove, J.D. 1995. Tree Planting in the United States 1994. Cooperative Forestry, USDA Forest Service, Washington, DC.

Pulliam, H.R. and Haddad, N.M. 1994. Human population growth and the carrying capacity concept. Bull. Ecol. Soc. Am. 75(3): 141–157.

Richards, E.G. 1987. Forestry and the Forest Industries: Past and Future. Martinus Nijhoff, Dordrecht, The Netherlands.

Sampson, R.N., Wright, L.L., Winjum, J.K., Kinsman, J.D., Benneman, J., Kursten, E. and Scurlock, I.M.O. 1993. Biomass management and energy. Water, Air, Soil Pollut. 70: 139–159.

Schulz, H. 1993. The development of wood utilization in the 19th, 20th and 21st centuries. For. Chron. 69: 413–418.

Sedjo, R. 1995. The World's Forests: Conflicting Signals. Competitive Enterprise Institute, Washington, DC.

Sedjo, R. and Lyon, K.S. 1990. The Long-Term Adequacy of World Timber Supply. Resources for the Future, Washington, DC.

Simon, J.L. 1977. The economics of Population Growth. Princeton University Press, Princeton, NJ.

Simon, J.L. 1981. The Ultimate Resource. Princeton University Press, Princeton, NJ.

Simon, J.L. 1992. Population Matters: People, Resources, Environment, and Immigration. Transaction Publishers, New Brunswick, NJ.

USDA (United States Department of Agriculture) Forest Service 1988. The South's Fourth Forest: Alternatives for the Future. Research Report 24. USDA Forest Service, Washington, DC.

USDA (United States Department of Agriculture) Forest Service 1989. An Analysis of the Land Base Situation in the United States: 1989–2040. General Technical Report RM-181. USDA Forest Service, Washington, DC.

Waggoner, P.E. 1994. How Much Land Can 10 Billion People Spare for Nature? Task Force Report No. 121. Council for Agricultural Science and Technology, Ames, IA.

Winjum, J. and Schroeder, P. (Eds) 1991. International Workshop on Large-Scale Reforestation. EPA/600/9-91/014. Environmental Protection Agency Environmental Research Laboratory, Corvallis, OR.

World Bank 1992. Development and the Environment. Oxford University Press, Oxford.

World Resources Institute 1986. World Resources 1986. Oxford University Press, Oxford.

World Resources Institute 1992. World resources 1992–93. Oxford University Press, Oxford.

Wright, L.L. and Hughes, E.E. 1993. U.S. carbon offset potential using biomass energy systems. Water, Air, Soil Pollut. 70: 483–497.

New Forests **17:** 213–226, 1999.

Applications of biotechnology for forest regeneration

STEVEN C. GROSSNICKLE and BEN C.S. SUTTON
[1] *Forest Biotechnology Centre, BCRI, Vancouver, BC, Canada*

Key words: tissue culture, molecular genetics, microbial inoculations, ecophysiology, *Picea*, somatic embryo

Abstract. The Forest Biotechnology Centre is an interdisciplinary research group dedicated to the development and application of advanced technology for the enhancement of forest regeneration. The Centre carries out contracts on behalf of clients in forest-related industries and government agencies. In addition, there are a number of long-term, in-house projects aimed at the development of proprietary technologies in genetics and propagation, and seedling production and establishment. Technical capabilities include: tissue culture, molecular genetics, pathology and microbial inoculants, and ecophysiology. These techniques are also being used to improve nursery culture regimes, disease assessment, planting regimes, and new product development for a variety of conifer species. Additional programs relate population genetics to adaptive traits, and develop clonal testing within elite families from tree-breeding programs.

Introduction

Successful regeneration relies on the application of work from many forestry disciplines toward a common goal. At the center of any successful regeneration program is the production of high-quality seedlings that have good performance on reforestation sites. Performance on a reforestation site depends on seedling growth potential and the degree to which field site conditions allow this potential to be expressed. Seedling growth potential is influenced by the inherent genetic make-up of source material and the culture used during nursery development. If these attributes can be directed toward improving seedling growth on a reforestation site, then the potential productivity of planted forests will be increased. Disciplines that are oriented toward improving these facets of producing high-quality seedlings are the main focus of the Forest Biotechnology Centre (FBC).

The FBC is an interdisciplinary research group dedicated to the development and application of advanced technologies for improved regeneration practices within planted forests (Figure 1). The FBC carries out long-term, in-house projects and contracts on behalf of clients in forestry and related industries. Technical capabilities applied by the FBC to forest regeneration programs include:

- defining species variation through molecular genetic markers and ecophysiological parameters,
- creating new genotypes through gene transfer technologies,
- developing advanced propagation systems through somatic embryogenesis tissue culture technology,
- improving seedling quality through disease diagnosis, remediation, and incorporation of plant growth-promoting bacteria, and
- applying ecophysiological assessment techniques in support of seedling production, improved quality, and reforestation site performance.

The following sections describe the four major research groups within the FBC – tissue culture, molecular genetics, pathology and microbial inoculants, and ecophysiological assessment. These research groups focus on applying their respective technologies to forest regeneration programs.

Tissue culture

The main emphasis of the tissue culture program is to use somatic embryogenesis to develop propagation systems for conifers. These systems allow the mass propagation of elite families from tree-breeding programs, and, in addition, the selection of superior clones that can be stored and propagated in a sustained manner. Over the last several years, the FBC has made significant advances in the maturation, germination, and acclimatization phases of spruce somatic embryogenesis (Sutton et al. 1993). Progress in producing high-quality somatic embryos and resulting seedlings has enabled the development of an operational production system for spruce. Recently, the FBC has developed protocols to produce and deliver high-quality pine somatic embryos (Cyr 1998). Commercial production has been initiated by an associated company, Silvagen, Inc.

Molecular genetics

The molecular genetics program involves an integrated approach in creating the necessary molecular tools for various short- and longer-term applications to operational forestry. These have ranged from development and use of species-specific probes for analysis of natural hybrids to the development of highly polymorphic DNA markers (DNA fingerprints) for monitoring seed orchards (Sutton et al. 1991a, 1991b, 1994). It is now possible to index clonal material unambiguously, and to monitor pollen contributions, pollen contamination, and in-breeding. In addition, these same tools can be used

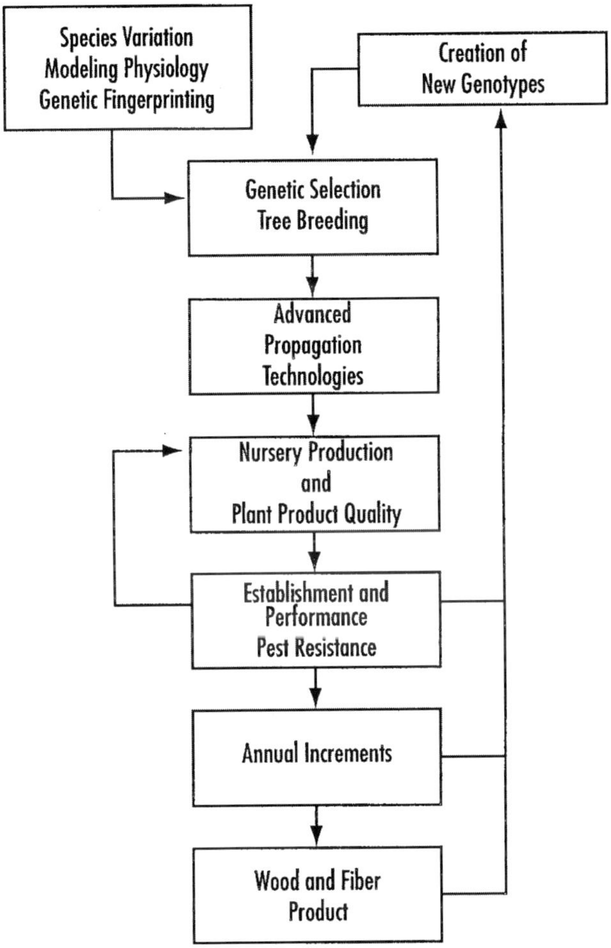

Figure 1. Technology factors influencing plantation forestry.

to characterize natural populations for planning tree-breeding units and seed orchards.

Longer-term work in gene expression and genetic transformation has also been carried out, thus allowing for considerable progress toward the goal of providing routine genetic engineering of improved clonal material for value-added traits. The FBC reported the first successful genetic engineering of spruce carrying spruce budworm resistance (Ellis et al. 1993).

Pathology and microbial inoculants

Fungal disease problems can cause major losses in forest nurseries and on reforestation sites. The FBC has a program to investigate causes of these problems, and finds means to limit their impact. The approaches taken include both biological and cultural approaches to enhance conifer seedling health and growth. For example, the influence of fungal inoculum sources and nursery cultural practices on root infection of conifer seedlings is evaluated. This has resulted in the development of improved seed treatments that considerably reduce the incidence of *Fusarium* contamination (Axelrood et al. 1995). Through a major in-house research initiative, the Centre has identified a number of microbial strains for biological disease control and growth enhancement of conifer seedlings. These are being developed further for commercial use in association with Agrium Inc.

Ecophysiological assessment

Ecophysiology research projects are carried out on behalf of a variety of clients in the forest and agrochemical sectors. Over the past few years the FBC has developed an assessment program for the laboratory, farm field sites, and reforestation sites that integrates a number of physiological and morphological measurements, including drought and frost tolerance, gas exchange capability, and growth capacity (Grossnickle et al. 1991a, 1991b, 1991c; Grossnickle and Arnott 1992; Major et al. 1994). These techniques have resulted in a stock quality assessment approach that is used to identify optimal nursery culture and planting regimes for a variety of species (Grossnickle and Folk 1993). In addition, these assessment methods are used to describe the effects of various precommercial products on the performance potential of various plant species (e.g., abscisic acid analogs, Grossnickle et al. 1996a).

A methodology is under development for the physiological selection of desirable genotypes from a variety of species. Selection methodology will be generated by collecting data over a wide range of environmental conditions, including light intensity, evaporative demand, and soil temperature and moisture. This assessment procedure is used to forecast genotype performance relative to specific site environmental conditions (e.g., Grossnickle and Fan 1998, Fan and Grossnickle 1998).

Specific examples of FBC programs

The following are two specific examples of how the FBC has integrated various scientific disciplines to produce information or propagation systems that are used to improve forest regeneration programs for planted forests. The examples include the relationship between nuclear DNA markers and physiological parameters for Sitka × interior spruce populations, and somatic embryogenesis in interior spruce.

Introduction to the relationship between nuclear DNA markers and physiological parameters for Sitka × interior spruce populations

Special problems exist for reforestation of introgression zones. One specific area is the Nass Skeena transition, where there is a substantial and increasing demand for successful reforestation. In the Nass Skeena transition, a large introgression zone occurs between Sitka spruce (*Picea sitchensis* [Bong.] Carr.) and interior spruce (*Picea glauca* [Moench] Voss × *Picea engelmannii* Parry ex. Engelm.) (Little 1953; Daubenmire 1968; Roche 1969). Sitka spruce occurs naturally in wet, maritime climates, whereas interior spruce occurs across continental areas which experience summer droughts and severe winters. As a result, there are risks associated with using seed from the Sitka × interior transition zone with unknown genetic characterization. Because the exact extent of this zone is not clear and information on the genotypes represented within it is insufficient, the use of seed orchard seed to reforest areas distant from original parent tree locations has the potential to place progeny off site. The result can be poor survival and/or growth of planted seedlings.

Approach

Research based on chloroplast DNA, to estimate species components of hybrid seedlots, indicates a good agreement between growth patterns and morphological development of nursery grown seedlings (Sutton et al. 1991a). In fact, DNA probes have been developed to analyze species contribution to the mitochondrial and nuclear genomes of hybrids (Sutton et al. 1991b, 1994). Nonetheless, no information has been collected on how DNA analysis relates to seasonal changes in phenology, morphology, and physiology of field-planted Sitka × interior genotypes. By combining ecophysiological and genetic characterization, a better species deployment strategy could be developed for the Sitka × interior spruce transition zone.

Sitka spruce and interior spruce were sampled from a zone of Sitka/interior spruce introgression in British Columbia, Canada. Restriction fragment length polymorphisms of the nuclear ribosomal RNA genes (rDNA) defined species-specific patterns for Sitka and interior spruce populations (Figure 2). Hybridization was estimated from an index based on the relative abundance of polymorphic rDNA bands for each population. The same Sitka and interior spruce populations were assessed for the physiological parameters of drought tolerance, freezing tolerance, and gas exchange (Grossnickle et al. 1996).

Results

Sitka × interior spruce seed sources had an interior spruce (Si) rDNA index ranging from 0.04 (Lower Nass) from the most westerly collected seed

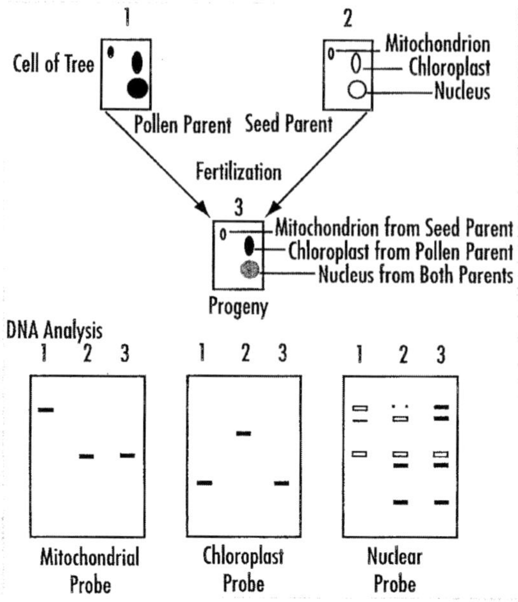

Figure 2. Spruce hybrid analysis using chloroplast, mitochondrial, and nuclear DNA probes. NOTE: Plant cells contain subcellular organelles (chloroplasts and mitochondria) that carry out metabolic functions and also contain genetic material (DNA). Research using probes for each of these organelles in spruce has shown that species-specific patterns can be found for both chloroplasts and mitochondria. By looking at the progeny of crosses between two species it was found that chloroplasts are inherited from the pollen parent and mitochondria are inherited from the seed parent (Sutton et al. 1991b). The nuclear DNA, consisting of two sets of chromosomes, is inherited equally from both parents (one set of chromosomes from each parent). Thus, each organelle represents a maternal or paternal lineage of the ancestors of a tree, whereas the nucleus can display the tree's overall genetic makeup. In the example above, the DNA patterns represent what is seen using the species-specific probes where tree 1 is Sitka spruce and tree 2 is interior spruce. The intermediate pattern displayed by the progeny with the nuclear probe can be quantified to generate and index of hybrid fraction range from 0 (Sitka spruce) to 1 (interior spruce). In this simple case, where the two species have been crossed, the index for the progeny (tree 3) is 0.5.

source to 0.97 from the Bulkely Valley low-elevation, easterly source. Spruce seedling populations exhibited similar Si rDNA index values when compared to mature trees previously surveyed from the same approximate locations (Sutton et al. 1994).

Sitka × interior hybrid seedlings had typical seasonal patterns of shoot water relation parameters (i.e., osmotic potential at saturation, and turgor loss point, relative water content at turgor loss point, and total turgor) (Grossnickle et al. 1996). During shoot elongation in the spring, drought tolerance decreased, whereas in the summer and throughout the fall, after budset, drought tolerance increased in a manner that is very typical of spruce species

(Grossnickle 1989). During all times of the year, there were linear relationships between the Si rDNA index and shoot water relations parameters; as Si rDNA index increased, drought tolerance, represented by osmotic potential at turgor loss point (Ψ_{tlp}) just after summer budset, increased in a predictable manner (Figure 3A).

Gas exchange patterns measured just after summer budset, under optimum conditions, indicated that Sitka, compared to interior spruce populations, has higher gas exchange capability, represented by net photosynthesis, P_n (Figure 3B; Fan et al. 1997). When seedlings were soil droughted (predawn shoot water potential = -1.5 MPa), P_n rates declined overall but interior spruce populations maintained higher P_n rates relative to coastal populations. Thus, under optimum conditions, populations with a greater Sitka spruce component had greater P_n capability. On the other hand, during drought conditions, populations with a greater interior spruce component had greater P_n capability, which was related to their greater drought tolerance (shown in Figure 3A).

During the fall, Sitka $\times$ interior hybrid seedlings had a seasonal increase in freezing tolerance that was dependent upon species hybridization (Grossnickle et al. 1996). Seedlings with a higher Si rDNA index had greater freezing tolerance, represented by the lower lethal temperatures that caused 50 percent needle electrolyte leakage (LT_{50}) throughout the fall (Figure 3C, measurements taken on November 10th). Populations with a greater interior spruce component had greater development of freezing tolerance during fall acclimation.

Conclusions

The following conclusions can be drawn from this program. First, DNA analysis is a useful tool for accurate identification of natural Sitka and interior spruce hybrids in the Nass Skeena Transition Zone. Second, seed source Si rDNA index was directly related to the degree of drought tolerance throughout most of the year, to freezing tolerance during fall acclimation, and to gas exchange patterns during the summer season. Third, incorporation of genetic marker procedures, when correlated with ecophysiological patterns, should improve the seed transfer guidelines of Sitka and interior spruce populations for forest regeneration programs.

Introduction to somatic embryogenesis in interior spruce

Somatic embryogenesis is a tissue culture method for asexual propagation. The term "somatic" refers to the fact that embryos develop asexually from vegetative (somatic) tissue rather than as a result of fertilization. This propagation method allows for the multiplication of superior families, as identified

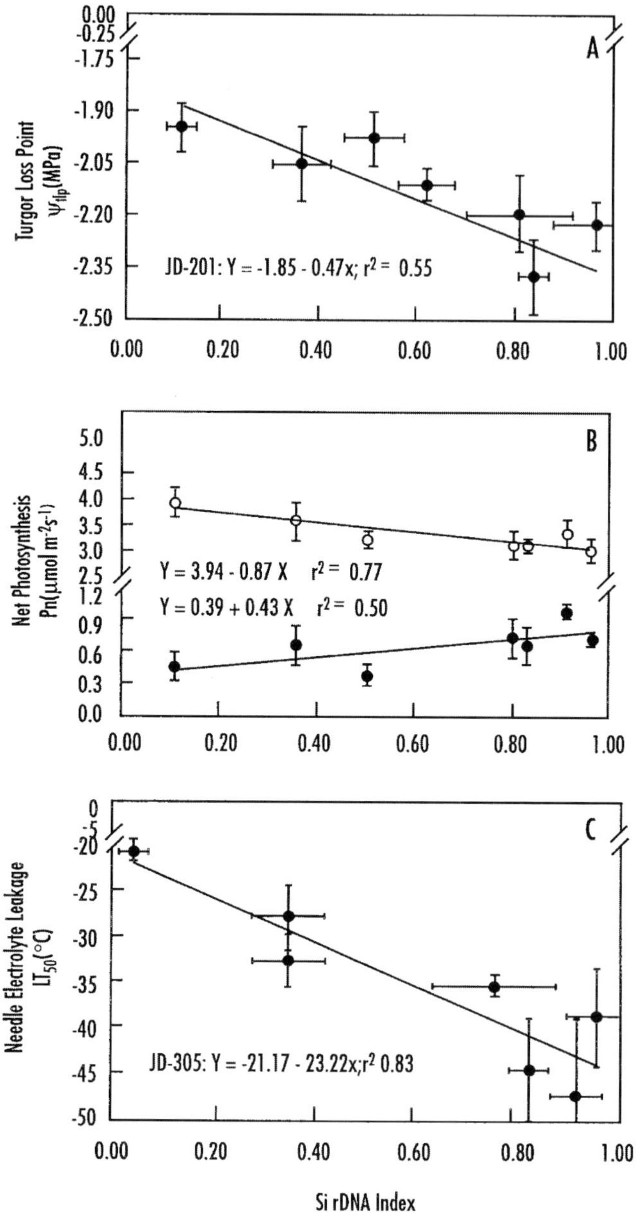

Figure 3. Relationship between Sitka × interior spruce seed sources in relation to their Si rDNA index (mean ± SE) and: (A) drought tolerance, osmotic potential at turgor loss point, Ψ_{tlp} (mean ± SE), of seedlings just after summer budset (Julian Day 201); (B) net photosynthesis, P_n, under both optimum (open symbol) and drought (filled symbol) soil moisture conditions of seedlings just after summer budset; and (C) freezing temperature resulting in 50 percent needle electrolyte leakage, LT_{50} (mean ± SE), measured on November 10th (Julian Day 305).

in tree improvement programs, and the selection of elite clones to capture a greater portion of the gain from additive, dominance, and epistatic variation (Libby and Rauter 1984; Mullin and Park 1992). In addition, somatic embryogenesis is a propagation technique that allows for cryopreservation of propagated clones, thus allowing for maintenance of culture lines while testing is ongoing for selection of elite clones to be used within a deployment strategy.

Value-added traits that could be captured and propagated through somatic embryogenesis include yield, wood quality, and pest and disease resistance. Particular opportunities also exist where seed supply or germination is inherently poor. The somatic embryogenesis propagation system is also ideal for genetic engineering of improved clonal material with value-added traits.

Propagation technique

Somatic embryogenesis is a successfully implemented tissue culture method for the asexual propagation of interior spruce (*Picea glauca* × *Picea engelmannii*). Somatic embryos are derived from excised seed embryos which are placed on the proper medium to produce a culture composed of many proembryos (i.e., early stage somatic embryos), similar in appearance to zygotic embryos soon after fertilization (Hakman and von Arnold 1985; Webb et al. 1989). Each culture can produce essentially an unlimited number of proembryos, each proembryo being a clone of the original explant. In order to produce plants, cultures are placed on a different medium where proembryos stop proliferating and proceed through more advanced stages of embryogenesis, this resulting in the formation of cotyledonary embryos similar to a mature seed (Roberts et al. 1990a; Flinn et al. 1991). Somatic embryos are germinated in enclosed containers to produce somatic seedlings which resemble young seedlings (Roberts et al. 1990b; Cyr et al. 1991). Somatic seedlings are transferred to styrofoam blocks, acclimatized to *ex vitro* conditions, and placed in the nursery (Webster et al. 1990).

Protocols for cryopreservation of interior spruce suspension cultures have been developed (Kartha et al. 1988), and interior spruce cultures from a range of genotypes are currently being preserved for long-term storage. In addition, applying somatic embryogenesis to clonal propagation requires genetic stability. An examination with isozyme patterns and culture morphology of the genetic stability of these embryos has revealed no evidence of somaclonal variation (Eastman et al. 1991).

Operational testing

Somatic seedlings have been tested during all phases of an operational forest regeneration program. Somatic seedlings from nine clones representing three families were germinated *in vitro* concomitantly with the germination of genetically related seed in the nursery. Somatic seedlings were grown alongside control seedlings in an operational nursery and overwintered in frozen storage. Somatic and regular seedlings were tested with a comprehensive stock quality assessment approach just prior to planting, and spring planted on a reforestation site in the interior of British Columbia (52°28′ N, 122°41′ W) with field performance monitored over a 2-year period. Results from the operational testing are briefly described below with details of the program reported elsewhere (Grossnickle and Major 1994a, 1994b; Grossnickle et al. 1994).

Somatic, compared to regular, seedlings have slower growth during the initial phase of growth in the nursery. Somatic seedlings initially require higher humidity and lower light levels during acclimation in the nursery. Thereafter, height and root growth are similar between somatic and regular seedlings throughout the growing season. Recent nursery performance of somatic seedlings have shown that a proper nursery cultural environment during the initial establishment stage results in normal morphological development of seedlings (Grossnickle et al. 1996b). During fall acclimatization, somatic and regular seedlings have similar dormancy, freezing tolerance, and root growth patterns, and meet all standards for the successful lifting and storage of interior spruce.

Just prior to planting, somatic and regular seedlings were assessed for stock quality. Somatic seedlings met all testing standards for interior spruce seedlings used in operational regeneration programs in British Columbia. In addition, both stock types were tested for physiological performance and morphological development under environmental conditions that simulated potential reforestation site conditions. Somatic and regular seedlings performed similarly in simulated low temperature and drought conditions. This indicated that somatic seedlings have good field performance potential.

On a reforestation site, somatic and regular seedlings had comparable patterns of summer seasonal water relations and gas exchange responses, and were comparable in response to damaging winter conditions. Somatic and regular seedlings had a similar rate of incremental height and diameter growth across two growing seasons. Regular seedlings had larger shoot systems and root development at the end of the first growing season, and this reflects their larger shoot and root systems at the time of planting (Figure 4). At the end of the second growing season, somatic and regular seedlings had comparable root development. Although regular seedlings had larger shoots than somatic seedlings at the end of the second growing season, they had a similar shoot-to-root balance. Similar morphological balance was reflected in their comparable gas exchange and water relations patterns. At the end of the second growing season, survival was 83 percent for somatic seedlings

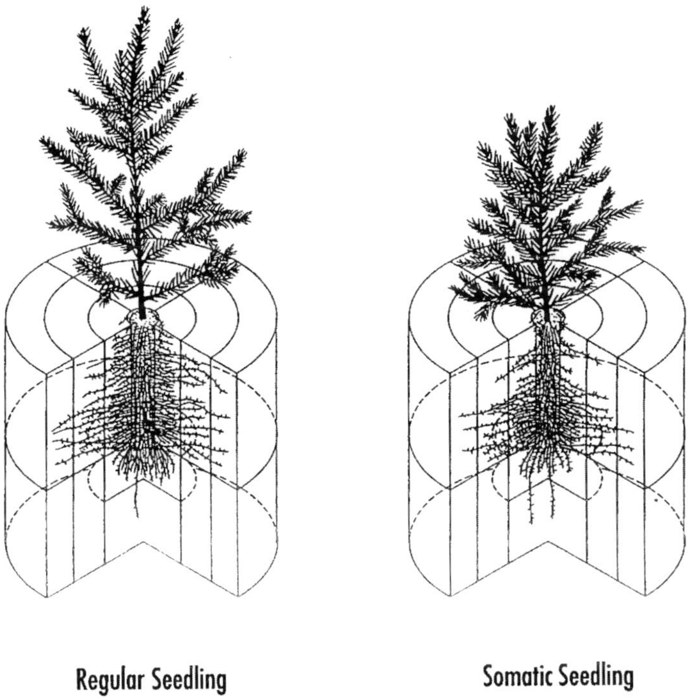

Figure 4. Diagrammatic representation of morphological development after one growing season on a reforestation site for regular and somatic interior spruce seedlings.

and 77 percent for regular seedlings. Long-term (i.e. seven years) field trials indicate that somatic seedlings are capable of sustaining good shoot development when grown under plantation conditions (Grossnickle 1998). Somatic seedlings had all of the traits desired of stock used in successful forest regeneration programs.

Program development

The FBC has successfully implemented a somatic embryogenesis tissue culture production system with the following elements:

- Establishment of a diverse array of embryogenic culture lines (1,400 to date) from superior seed families.
- Application of a reliable long-term capability to store cultures in liquid nitrogen.
- Rapid bulk up of embryogenic cultures followed by bioreactor-based bulk embryo production.
- Drying of embryos in a manner suitable for subsequent storage or germination.
- Capacity for mass germination.

- Semiautomatic planting of germinants for growing in the nursery.
- Development of nursery cultural protocols to produce quality somatic seedlings.
- Development of early clonal selection capability.

To date approximately 100,000 somatic seedlings have been produced for trials. Commercial production has begun through an associated company, Silvagen, Inc., which delivered over 300,000 interior spruce somatic seedlings for commercial nursery production in 1997 and 1998. The major emphasis of this commercial program is the delivery of genetic material carrying insect resistance and increased growth rate.

Initiatives are underway in the following areas to improve the somatic seedling program. First, scaling-up of production capability to 1,000,000 over the next 3 years. Second, further establishing a diverse array of embryogenic culture lines from superior seed families. These lines are now undergoing field performance trials that will select elite lines for deployment in reforestation programs. Third, improving the cultural protocols for nursery production of somatic seedlings from a wide array of clonal lines. Fourth, developing early selection capability to identify superior families and lines. This early selection capability will be used to develop profiles of lines that will be deployed in reforestation programs.

Conclusion

The FBC is an interdisciplinary research group that develops and applies advanced technologies to enhance forest regeneration programs. The FBC has developed major research groups in the areas of tissue culture, molecular genetics, pathology and microbial inoculants, and ecophysiological assessment. These research groups work together to apply various disciplines toward a common goal of developing high-quality seedlings. By producing high-quality seedlings, improved seedling growth on reforestation sites will increase the productivity of planted forests.

References

Axelrood, P.E., Neuman, M., Trotter, D., Radley, R., Shrimpton, G. and Dennis, J. 1995. Seedborne *Fusarium* on Douglas-fir: Pathogenicity and seed stratification method to decrease *Fusarium* contamination. New For. 9: 35–51.

Cyr, D. 1998. Cryopreservation of embryogenic cultures of conifers & its application to clonal forestry. In Somatic Embryogenesis in Woody Plants Volume 4. Eds. Jain, S.M., Gupta, P.L. and Newton, R.J. Kluwer Academic Publishers, Dordrecht, Boston & London (in press).

Cyr, D.R., Webster, F.W. and Roberts, D.R. 1991. Biochemical events during germination and early growth of somatic embryos and seed of interior spruce (*Picea glauca/engelmannii* complex). Seed Sci. Res. 1: 91–97.

Daubenmire, R. 1968. Taxonomic and ecological relationships between *Picea glauca* and *Picea sitchensis* and their ecological interpretation. Can. J. Bot. 46: 787–798.

Eastman, P.K.A., Webster, F.B., Pitel, J.A. and Roberts, D.A. 1991. Evaluation of somaclonal variation during somatic embryogenesis of interior spruce (*Picea glauca/engelmannii* complex) using culture morphology and isozyme analysis. Plant Cell Rep. 10: 425–430.

Ellis, D.D., McCabe, D.E., McInnis, S., Ramachandran, R., Russell, D.R., Wallace, K. M., Martinell, B.J., Roberts, D.R., Raffa, K.F. and McCown, B.H. 1993. Stable transformation of *Picea glauca* by particle acceleration. Biotechnol. 11: 84–89.

Fan, S. and Grossnickle, S.C. 1998. Comparison of gas exchange parameters and shoot water relations of interior spruce (*Picea glauca* (Moench) Voss × *P. engelmannii* Parry ex engelm.) clones under repeated soil drought. Can. J. For. Res. 28: 820–830.

Fan, S., Grossnickle, S.C. and Sutton, B.C.S. 1997. Relationships between gas exchange adaptation of Sitka × interior spruce genotypes and ribosomal DNA markers. Tree Physiol. 17: 115–123.

Flinn, B.S., Roberts, D.R. and Taylor, I.E.P. 1991. Evaluation of somatic embryos of interior spruce. Characterization and developmental regulation of storage proteins. Physiol. Plant. 82: 624–632.

Grossnickle, S.C. 1989. Seasonal shoot phenology and water relations of *Picea glauca*. Can. J. For. Res. 19: 1287–1290

Grossnickle, S.C. 1998. Performance of conifer stock produced through somatic embryogenesis. In Somatic Embryogenesis in Woody Plants. Volume 4. Eds. Jain, S.M., Gupta, P.K. and Newton, R.J. Kluwer Academic Publishers, Dordrecht, Boston & London (in press).

Grossnickle, S.C. and Arnott, J.T. 1992. Gas exchange response of western hemlock seedlings from various dormancy induction treatments to reforestation site environmental conditions. For. Ecol. Manage. 49: 177–193.

Grossnickle, S.C. and Fan, S. 1998. Genetic variation in summer gas exchange patterns of interior spruce (*Picea glauca* (Moench) Voss × *P engelmannii* Parry ex engelm.). Can. J. For. Res. 28: 831–840.

Grossnickle, S.C. and Folk, R.S. 1993. Stock quality assessment: Forecasting survival or performance on a reforestation site. Tree Planters' Notes 44: 113–121.

Grossnickle, S.C. and Major, J.E. 1994a. Interior spruce seedlings compared to emblings produced from somatic embryogenesis. II) Stock quality assessment prior to field planting. Can. J. For. Res. 24: 1385–1396.

Grossnickle, S.C. and Major, J.E. 1994b. Interior spruce seedlings compared to emblings produced from somatic embryogenesis. III) Physiological response and morphological development on a reforestation site. Can. J. For. Res. 24: 1397–1407.

Grossnickle, S.C., Cyr, D. and Polonenko, D.R. 1996b. Somatic embryogenesis tissue culture for the propagation of conifer seedlings: A technology comes of age. Tree Planters' Notes 47: 48–57.

Grossnickle, S.C., Major, J.E., Arnott, J.T. and LeMay, V.M. 1991a. Stock quality assessment through an integrated approach. New For. 5: 77–91.

Grossnickle, S.C., Arnott, J.T., Major, J.E. and Tschaplinski, T.J. 1991b. Influence of dormancy induction treatment on western hemlock seedlings. 1) Seedling development and stock quality assessment. Can. J. For. Res. 21: 164–174.

Grossnickle, S.C., Arnott, J.T. and Major, J.E. 1991c. Influence of dormancy induction treatments on western hemlock seedlings. 2) Physiological and morphological response during the first growing season on a reforestation site. Can. J. For. Res. 21: 175–185.

Grossnickle, S.C., Major, J.E. and Folk, R.S. 1994. Interior spruce seedlings compared to emblings produced from somatic embryogenesis. I) Nursery development, fall acclimation and over-winter storage. Can. J. For. Res. 24: 1376–1384.

Grossnickle, S.C., Sutton, B.C.S., Folk, R.S. and Gawley, B.J. 1996. Relationship between nuclear DNA markers and physiological parameters for Sitka × interior spruce populations. Tree Physiol. 16: 547–555.

Grossnickle, S.C., Folk, R.S., Abrams, S.R., Dunstan, D.I. and Rose, P.A. 1996a. Performance of interior spruce seedlings treated with abscisic acid analogs. Can. J. For. Res. 26: 2061–2070.

Hakman, I. and von Arnold, S. 1985. Plantlet regeneration through somatic embryogenesis in *Picea abies* (Norway spruce). J. Plant Physiol. 121: 149–158.

Kartha, K.K., Fowke, L.C., Leung, N.L., Caswell, K.L. and Hakman, I. 1988. Induction of somatic embryos and plantlets from cryopreservation cell cultures of white spruce (*Picea glauca*). J. Plant Physiol. 132: 529–539.

Libby, W.J. and Rauter, R.M. 1984. Advantages of clonal forestry. For. Chron. 60: 145–149.

Little, E.L. 1953. A natural hybrid spruce in Alaska. J. For. 51: 745–747.

Major, J.E., Grossnickle, S.C. and Arnott, J.T. 1994. Influence of dormancy induction treatments on the photosynthetic response of field planted western hemlock seedlings. For. Ecol. Manage. 63: 235–246.

Mullin, T.J. and Park, Y.S. 1992. Estimating genetic gains from alternative breeding strategies for clonal forestry. Can. J. For. Res. 22: 14–23.

Roberts, D.R., Flinn, B.S., Webb, D.T., Webster, F.B. and Sutton, B.C.S. 1990a. Abscisic acid and indole-3-butyric acid regulation of maturation and accumulation of storage proteins in somatic embryos of interior spruce. Physiol. Plant. 78: 355–360.

Roberts, D.R., Sutton, B.C.S. and Flinn, B.S. 1990b. Synchronous and high frequency germination of interior spruce somatic embryos following partial drying at high relative humidity. Can. J. Bot. 68: 1086–1090.

Roche, L. 1969. A genecological study of the genus *Picea* in British Columbia. New Phytol. 68: 505–554.

Sutton, B.C.S., Flanagan, D.J. and El-Kassaby, Y.A. 1991a. A simple and rapid method for species determination of spruce seedlots using restriction fragment length polymorphism. Silva Gen. 40: 119–123.

Sutton, B.C.S., Flanagan, D.J., Gawley, R., Newton, C.H., Lester, D. and El-Kassaby, Y.A. 1991b. Inheritance of chloroplast and mitochondrial DNA in *Picea* and composition of hybrids from introgression zones. Theor. Appl. Genet. 82: 242–248.

Sutton, B.C.S., Grossnickle, S.C., Roberts, D.R., Russell, J.H. and Kiss, G.K. 1993. Somatic embryogenesis and tree improvement in interior spruce. For. 91: 34–38.

Sutton, B.C.S., Pritchard, S.C., Gawley, J.R., Newton, C.H. and Kiss, G. 1994. Analysis of Sitka × interior spruce introgression in British Columbia using cytoplasmic and nuclear DNA probes. Can. J. For. Res. 24: 278–285.

Webb, D.T., Webster, F., Flinn, B.S., Roberts, D.R. and Ellis, D.D. 1989. Factors influencing the induction of embryogenic and caulogenic callus from embryos of*Picea glauca* and *P. engelmannii.* Can. J. For. Res. 19: 1303–1308.

Webster, F.B., Roberts, D.R., McInnis, S.M. and Sutton, B.C.S. 1990. Propagation of interior spruce by somatic embryogenesis. Can. J. For. Res. 20: 1759–1765.

New Forests **17:** 227–238, 1999.

Vulnerability of planted forests to biotic and abiotic disturbances

PETER D. GADGIL[1] and JOHN BAIN[2]
[1]*Forest Health, Forest Research Institute, Rotorua, New Zealand;* [2]*Forest Research Institute, Rotorua, New Zealand*

Key words: disease, forest health, insects, sustainability

Abstract. Losses through reduced growth or mortality attributable to pests, diseases, and abiotic disturbances are generally less in intensively managed, planted forests than in natural forests. Among plantation forests, the most healthy forests are those of exotic species growing a long way from their native range (e.g., *Pinus radiata* D. Don in New Zealand) and the least healthy ones are those of indigenous species (e.g., *Eucalyptus nitens* (Deane & Maiden) Maiden in Tasmania). The better health of plantations is a reflection of the generally good growing conditions, the proper matching of site to species that is possible in artificial culture and in the case of exotic species, the freedom from the many pests and diseases present in the native habitat of exotics.

Introduction

For the purposes of this paper, the term "planted forests" is taken to mean man-made forests, either planted or regenerated, primarily of one or very few species, grown in blocks of varying sizes of a single age class with production of wood as one of the main purposes. Most of the planted production forests fit this definition and are often referred to as "monocultures." The term monoculture is rather misleading because planted forests usually contain many other plant species, ranging from grasses to a heavy understory of woody shrubs, in addition to the main crop species.

Monocultures have generally had a bad press, beginning with Hartig (1894) and including Boyce (1954) and Gibson and Jones (1977). However, alternative views have also been expressed (Bain 1981, Chou 1981, Whitehead 1981). The objections to plantation forestry were perhaps best summed up by Ciesla and Donaubauer (1994) who said, "single species plantations, especially of exotics, which may represent a narrow genetic base, are generally considered to be at high risk of damage by pests and diseases." In this paper we consider how far these fears are justified.

Table 1. Silvicultural practices in plantation forestry.

Species	Country	Initial stocking (stems/ha)	Thinnings	Final stocking (stems/ha)	Author
		Exotic species			
Picea sitchensis (Bong.) Carr.	United Kingdom	2,500	2[a]	Variable	Kerr 1990
Pseudotsuga menziesii (Mirb.) Franco	France	1,000	2–3	250	Bailley 1990
Pinus radiata D. Don	New Zealand	800	2–3	200–250	James 1990
Pinus radiata D. Don	Australia	1,000–1,500	2–3	200	Shepherd 1990
		Indigenous species			
Picea abies (L.) Karst.	Germany	2,500	2–3	Variable	Kramer 1990
Pinus sylvestris L., *Picea abies* (L.) Karst.	Sweden	1,500–2,500	1–3	Variable	Knuttell 1990
Pinus taeda L., *Pinus elliottii* Engelmann	United States	1,700	1	Variable	Greene 1990

[a] Production.

Characteristics of intensively managed plantations, whether trees are of indigenous or exotic origin, are relatively low initial stocking (800–2,500 stems/ha); early thinning to improve wind stability and resistance to snow damage, and to remove malformed, dead, or diseased trees, followed by a variable number of later, usually production thinnings; pruning of lower branches in one or more lifts; and a low final stocking (250–400 stems/ha). Fertilizer is often applied, particularly at planting to aid establishment and sometimes later in the rotation to cure mineral nutrient deficiencies. The harvest pattern varies from the felling of small coupes of a few hectares to clearcutting entire compartments up to 250–300 ha. Silvicultural practices in a number of countries are summarized in Table 1.

Plantation forests are also likely to be composed of selected stock bred for special characters such as growth rate, stem straightness, branch diameter and angle, wood density, and resistance to pests and diseases. Such breeding programs are bound to reject a substantial number of parents with silviculturally undesirable properties, and plantations containing the progeny from a breeding program will have a narrower genetic base than the wild populations of the same species. Even when natural regeneration is used to propagate an indigenous species, the use of a limited number of seed trees results in less diversity in the regenerated stand compared to the original stand.

Vulnerability of plantations as affected by management

Low initial stocking, a succession of thinnings, and the pruning of lower branches lead to an open, well-ventilated stand with few, if any, dead or dying trees. There is little competition for moisture and nutrients. In less intensively managed plantations, such as those grown on pulpwood regimes, these conditions are, of course, different. Clearcutting creates a large number of stumps, and may result in the accumulation of woody debris. These conditions influence the incidence and severity of pests and diseases.

Pests

It is generally accepted that fast-growing, vigorous trees are less susceptible to insect damage than suppressed or less vigorous ones. This means that fewer insect problems should be encountered in intensively managed stands. In general this is so, although there are exceptions. The most notable exceptions are insects, such as *Pissodes strobi* in North America and *Hypsipylla* spp. throughout much of the tropics, that damage tips and buds. These insects favor open-grown trees, and common strategies to minimize their damage include planting at closer spacings and ensuring that the stand reaches canopy closure as soon as possible. The slash produced by thinning in intensively managed regimes provides ideal breeding material for many insects, particularly bark beetles. It is often contended that insects, such as various species of *Ips*, emerge in large numbers from this material and attack and kill neighboring trees (e.g., Berryman 1986). This is not the case with *Ips grandicollis* in Australia, except when stands are under severe drought stress, or in the southern United States with several species of *Ips* (Mason 1969). The beetles prefer to infest the slash and, of course, in intensively managed stands there is an ongoing supply of logging slash.

Sirex noctilio is a good example of an insect that predominates in unthrifty or suppressed trees. In intensively managed *Pinus radiata* D. Don forests in New Zealand it causes no problems (Nuttall 1989); in fact, it is very rare. In some stands grown on pulpwood regimes (of which there are very few), that is, with relatively high stocking rates and little or no thinning, *Sirex* can kill quite a high proportion of suppressed or even codominant trees. Periodically in New Zealand there are outbreaks of *Pseudocoremia sauvis* on *Pinus radiata* and *Pseudotsuga menziesii* (Mirb.) Franco. These are invariably related to drought stress, and defoliation is always greater in overstocked stands (M.K. Kay, personal communication).

Diseases

Propagules of leaf pathogens generally require the presence of free water on leaf surfaces for germination and infection. Because of the open nature of plantations, foliage dries quickly and provides less favorable conditions for infection, repeated thinnings allow individual trees particularly susceptible to infection by the local pathogens to be removed, and the pruning of lower branches lessens the amount of infected foliage and hence the inoculum available for reinfection. Leaf pathogens, therefore, tend to be less severe in managed stands. For example, pruning at the right time can delay the necessity for chemical control of *Dothistroma* needle-blight for several years (Kershaw et al. 1988). Alternatively, management practices have no influence on foliage pathogens which require only a short leaf wetness period for infection. Hood and Sandberg (1979) found no difference in needle retention or in *Phaeocryptopus gaeumannii* infection levels in stands with 3,000, 740, or 220 stems/ha, and Manly (unpublished 1985) reports that there were no differences in the growth loss attributable to *P. gaeumannii* infection between thinned and unthinned stands. Stem pathogens, such as *Diplodia pinea* which can only attack trees suffering from moisture stress, are less of a problem in managed plantations. Obligate parasites, such as stem rusts, can be managed effectively in plantations by the removal of diseased trees as thinnings. Root diseases, however, present a different problem, and these may be exacerbated through the practices of thinning and clearcutting. Species of *Armillaria* are common root pathogens throughout the temperate regions (Hood et al. 1991). Their common mode of local spread is through rhizomorphs; long-distance spread is by means of airborne basidiospores. In New Zealand, *Armillaria* spp. are among the normal components of the indigenous fungal flora of native forests. When such forests are converted to exotic plantations, *Armillaria* spp., which have colonized the many stumps created by felling, parasitize the newly planted stock and can cause considerable damage (van der Pas 1981). There is no mortality attributable to *Armillaria* spp. in first-rotation plantations established on previously unforested land. However, stumps created by the harvesting of the first crop and those created by thinning become infected by *Armillaria* spp. to various degrees, and, although very little mortality is seen in the second rotation crop because the trees are larger and more resistant to infection by the time the colonized stumps produce rhizomorphs, chronic, nonlethal infection can persist throughout the rotation, and cause a loss of 6–13 percent of potential volume for a 28-year sawlog regime (MacKenzie 1987).

Abiotic factors

Wind

In planted forests the opportunity exists to minimize the effects of wind damage by having control over the quality of planting stock, planting pattern (Turner 1989), stocking density (Olsen 1989), the timing of thinning (Carter 1989; Turner 1989), and the choice of tree species (Carter 1989). Plantations, therefore, are less susceptible to wind damage than are natural forests.

Fire

All forests are vulnerable to fire, but the risk of suffering major damage from fire is less in managed plantations than it is in natural forests. Plantation management plans usually make provisions for the measurement of fire risk, for early detection of fires, and for ready availability of fire-fighting equipment and fire crews. The network of roads in a plantation provides good access to site of the fire, and, coupled with the strategic location of water points (tanks or ponds), makes effective fire fighting possible. Restrictions on public access and on the lighting of fires during periods of high fire risk are easier to enforce in plantations. Reduction of fuel by controlled burning is also a common management practice in some countries, such as Australia, where long, hot, and dry summers create conditions of high fire hazard.

Frost, hail, snow, and other natural phenomena

Damage from weather-related causes to plantations can be avoided, or at least minimized, by careful site selection. Many of the failures of exotic plantation species can be attributed to a poor understanding of the site requirements of the species. The importance of site selection is discussed in a separate section.

Vulnerability as affected by genetic selection

Use of tree breeds developed specifically for resistance to certain pests and diseases is possible in plantations. Examples where such breeds have been successfully employed are the breeding of *Pinus taeda* L. and *P. elliottii* Engelmann for resistance to fusiform rust (*Cronartium quercuum* f. sp. *fusiforme*) in the southeastern United States (Powers and Kraus 1983), and the breeding of *Eucalyptus* spp. for resistance against eucalypt canker (*Cryphonectria cubensis*) in Brazil (Ferreira 1989). However, tree breeding invariably leads to a narrowing of the genetic base, and the vulnerability of the products of breeding programs remains a much debated, but as yet unresolved, question. The lower vulnerability of a genetically diverse population to diseases and pests is generally accepted (Wolfe 1985). In agriculture, one means of reducing disease is the use of multilines or cultivar mixtures, thus maintaining a genetically heterogeneous host population. Stem rust in wheat has been managed successfully in Australia for many years in this way (Watson 1970), and there are many other examples (Simmonds 1962). McCracken and Dawson (1994) found that when *Salix* spp. clones were grown in intimate mixtures of six different clones, there were significant increases in yield compared to that from equivalent monoclonal stands. At least part of the increased yield was the result of restriction of disease. Theoretically, the greatest disease restriction would be achieved by interspersing as many different mixture components as possible. However, practical plantation management considerations, associated with silviculture and harvesting, require that the number of components be kept to the minimum consistent with reasonable restriction of disease progress. Unfortunately, there are few experimental data to determine what this minimum number should be. It should be pointed out that disease restriction by the use of a genetically heterogeneous population is most effective against specialist pests or pathogens (e.g., rusts), and confers no protection against organisms with a wide host range (e.g., most of the root pathogens, such as *Armillaria* spp.).

Vulnerability of plantations of exotic and indigenous species

The vulnerability of a plantation species to natural agencies depends to some extent on whether the species used is exotic or indigenous to the country where it is planted.

Pests and pathogens

An exotic species is normally free from its indigenous pests and pathogens when it is first introduced. It is vulnerable to such pests and pathogens, which may be able to adapt to it in its new country, and to others, which may be introduced later.

An indigenous species will have its (usually) large complement of pests and pathogens, and, equally with the exotic species, is vulnerable to introduced organisms. Generally, plantations of indigenous species suffer from more pests and pathogens than those of exotic species (Table 2).

It can be argued that native pests and diseases will eventually "catch up" with their hosts in exotic locations, and, although this possibility cannot be denied, it has not happened to any great extent particularly in those countries that have stringent quarantine regulations that are vigorously implemented. New Zealand and, to a lesser extent, Australia are good examples of this.

It is just as likely that new host associations that result from pests and diseases being introduced into a new environment will occur in natural forests and it is very likely that predicting the outcome of these occurrences will be far more difficult than predicting the outcome of a host being reunited with a pest or disease from its native range. It is easier to prepare contingency plans and devise strategies to minimize the effect of these associations than it is for unknown or less predictable associations.

Site selection

In establishing plantations, choosing the right species for a site is important if the species vulnerability to natural agencies is to be minimized. Mismatching of site and species happens more often with exotic than with indigenous species, generally because much more is known about the site requirements of native flora.

Ciesla and Donaubauer (1994), in a global overview of decline and dieback of forests, list and discuss six examples of decline or dieback of exotic plantations. Five of these are thought to be caused by the mismatching of tree species and site: *Casuarina equisetifolia* in Africa, *Eucalyptus globulus* in Columbia and Peru, *Eucalyptus* spp. in Brazil, and *Pinus patula* in Tanzania.

Species of *Eucalyptus* have been widely planted as exotics in many parts of the world (Hillis and Brown 1984), often with spectacular success, but with some notable failures. As a group, eucalypts are well adapted to a wide range of climatic and soil conditions, but what is often not generally appreciated is that many species have very specific site requirements and much care is needed in matching species and sites. A good example is the decline of *Eucalyptus regnans* and *E. delegatensis* in the central North Island of New Zealand. This region has a similar climate and annual precipitation to those parts of Australia where these two species naturally occur, but the distribution of rainfall is markedly different. The Australian locations have a distinct winter rainfall pattern, whereas in the central North Island there is a uniform pattern with occasional summer peaks. Is has been postulated that this difference in rainfall pattern is the main reason for dieback of these species in the central North Island (Rawcliffe, unpublished data).

Table 2. Major pests and pathogens of some plantation species.

Species	As an Indigenous Species			As an Exotic Species		
	Geographic area	Indigenous pests and pathogens	Exotic pests and pathogens	Geographic area	Indigenous pests and pathogens	Exotic pests and pathogens
Pinus radiata	California	*Coleosporium madiae*	*Dothistroma pini*	New Zealand	*Armillaria* spp.	*Dothistroma pini*
		Cronartium comptoniae	*Fusarium subglutinans*		*Pseudocoremia sauvis*	*Cyclaneusma minus*
		Cyclaneusma minus	*Rhyacionia frustrana*			*Diplodia pinea*
		Endocronartium harknessii				
		Arceuthobium campylopodum		Chile	None	*Dothistroma pini*
		Armillaria spp.				*Diplodia pinea*
		Heterobasidion annosum				*Rhyacionia buoliana*
		Toumeyella pinicola				
		Ips paraconfusus				
Pseudotsuga menziesii	Northwest North America	*Phaeocryptopus gaeumannii*	None	New Zealand	*Pseudocoremia sauvis*	*Phaeocryptopus gaeumannii*
		Rhabdocline pseudotsugae				*Phacidiopycnis pseudotsugae*
		Arceuthobium douglasii				
		Phacidiopycnis pseudotsugae		Germany	*Heterobasidion annosum*	*Phaeocryptopus gaeumannii*
		Armillaria spp.				*Rhabdocline pseudotsugae*
		Heterobasidion annosum				*Phacidiopycnis pseudotsugae*
		Phellinus weirii				*Adelges cooleyi*
		Leptographium wageneri				
		Dioryctria abietivorella				
		Choristoneura occidentalis				
		Orygia pseudotsugata				
		Buprestis aurulenta				
		Dendroctonus pseudotsugae				

Table 2. Continued

Species	As an Indigenous Species			As an Exotic Species		
	Geographic area	Indigenous pests and pathogens	Exotic pests and pathogens	Geographic area	Indigenous pests and pathogens	Exotic pests and pathogens
Pinus taeda	Southeast United States	*Coleosporium vernonae*	None	Australia	*Armillaria* spp.	None
		Cronartium comandrae				
		Cronartium comptoniae				
		Scirrhia acicola				
		Cronartium quercuum f. sp. *fusiforme*				
		Cronartium strobilinum				
		Fusarium subglutinans				
		Armillaria spp.				
		Heterobasidion annosum				
		Dioryctria zimmermani				
		Rhyacionia frustrana				
		Neodiprion excitans				
		Neodiprion taedae linearis				
		Dendroctonus frontalis				
		Dendroctonus terebrans				
		Hylobius pales				

Concluding statement

It is economics that largely drives plantation forestry. Most plantation forests are established and managed for profit. Even if forest managers did not subscribe to the views expressed in this paper, we would predict that they would still manage their production forests on a monocultural basis, provided they thought the risks were acceptable and could be managed. There is a risk with plantation forestry (as there is with any business venture), but there are no reasons to conclude that it is any greater than risks associated with managing natural forests. Indeed a lot of the evidence indicates the contrary.

References

Bailley, A. 1990. Evolution and current status of wide spacings and heavy thinnings in France, pp. 43–46. In: James, R.N. and Tarlton, G.L. (Eds) New Approaches to Spacing and Thinning in Plantation Forestry. Forest Research Institute Bulletin 151. IUFRO and Ministry of Forestry, Rotorua, New Zealand.

Bain, J. 1981. Forest monocultures – how safe are they? An entomologist's view. N. Z. J. For. 26(1): 37–42.

Berryman, A.A. 1986. Forest Insects: Principles and Practice of Population Management. Plenum Press, New York.

Boyce, J.S. 1954. Forest Plantation Protection Against Diseases and Pests. Food and Agriculture Organization of the United Nations (FAO) Forest Development Paper No. 3. FAO, Rome.

Carter, P.C.S. 1989. Wind damage in Kinleith Forest – effects of cyclone Bola, pp. 17–19. In: Somerville, A., Wakelin, S. and Whitehouse, L. (Eds) Workshop on Wind Damage in New Zealand Exotic Forests. Forest Research Institute Bulletin 146. Forest Research Institute, Rotorua, New Zealand.

Chou, C.K.S. 1981. Monocultures, species diversification, and disease hazards in forestry. N.Z. J. For. 26(1): 20–36.

Ciesla, W.M. and Donaubauer, E. 1994. Decline and Dieback of Trees and Forests. A Global Overview. Food and Agriculture Organization of the United Nations (FAO) Forestry Paper 120. FAO, Rome.

Ferreira, F.A. 1989. Patologia Florestal – Principas Doenças Florestais no Brasil. Sociedade de Investigações Florestais, Viçosa, Brazil.

Gibson, I.A.S. and Jones, T. 1977. Monoculture as the origin of major forest pests and diseases, pp. 139–161. In: Cherret, J.M. and Sagar, G.R. (Eds) Origins of Pest, Parasite, Disease and Weed Problems. Blackwell Scientific Publications, Oxford.

Greene, W.D. 1990. Use of wide spacings and heavy thinnings in the southeastern United States of America, pp. 54–57. In: James, R.N. and Tarlton, G.L. (Eds) New Approaches to Spacing and Thinning in Plantation Forestry. Forest Research Institute Bulletin 151. IUFRO and Ministry of Forestry, Rotorua, New Zealand.

Hartig, R. 1894. Textbook on Diseases of Trees. Macmillan and Co, London.

Hillis, W.E. and Brown, A.G. (Eds). 1984. Eucalypts for Wood Production. Academic Press, New York.

Hood, I.A. and Sandberg, C.J. 1979. Changes within tree crowns following thinning of young Douglas-fir infected by *Phaeocryptopus gaeumannii*. N. Z. J. For. Sci. 9: 177–184.

Hood, I.A., Redfern, D.B. and Kile, G.A. 1991. *Armillaria* in planted hosts, pp. 122–149. In: Shaw, C.G. and Kile, G.A. (Eds) Armillaria Root Disease. Agriculture Handbook No. 691. USDA Forest Service, Washington DC.

James, R.N. 1990. Evolution of silvicultural practice towards wide spacing and heavy thinnings in New Zealand, pp. 13–20. In: James, R.N. and Tarlton, G.L. (Eds) New Approaches to Spacing and Thinning in Plantation Forestry. Forest Research Institute Bulletin 151. IUFRO and Ministry of Forestry, Rotorua, New Zealand.

Kay, M.K. 1995. Forest Research Institute, Rotorua, New Zealand. Personal communication.

Kerr, G. 1990. Evolution and current status of wide spacing and heavy thinning in Great Britain, pp. 21–27. In: James, R.N. and Tarlton, G.L. (Eds) New Approaches to Spacing and Thinning in Plantation Forestry. Forest Research Institute Bulletin 151. IUFRO and Ministry of Forestry, Rotorua, New Zealand.

Kershaw, D.J., Gadgil, P.D., Ray, J.W., van der Pas, J.B. and Blair, R.G. 1988. Assessment and Control of *Dothistroma* needle-blight. Forest Research Institute Bulletin No. 18. Forest Research Institute, Rotorua, New Zealand.

Kramer, H. 1990. Evolution and current status of spacing and thinning in Germany, pp. 28–35. In: James, R.N. and Tarlton, G.L. (Eds) New Approaches to Spacing and Thinning in Plantation Forestry. Forest Research Institute Bulletin 151. IUFRO and Ministry of Forestry, Rotorua, New Zealand.

Knuttell, H. 1990. Evolution and current status of spacing and thinning in Sweden, pp. 47–53. In: James, R.N. and Tarlton, G.L. (Eds) New approaches to spacing and thinning in plantation forestry. Forest Research Institute Bulletin 151. IUFRO and Ministry of Forestry, Rotorua, New Zealand.

MacKenzie, M. 1987. Infection changes and volume loss in a 19-year-old *Pinus radiata* stand affected by Armillaria root-rot. N. Z. J. For. Sci. 17: 100–108.

Manley, B.R. 1985. Forest Research Institute, Rotorua, New Zealand. Personal communication.

McCracken, A.R. and Dawson, W.M. 1994. Experience in the use of mixed-clonal stands of *Salix* as a method of reducing the impact of rust diseases. Norw. J. Agri. Sci., Supplement 18: 101–109.

Mason, R.R. 1969. Behaviour of Ips populations after summer thinning in a loblolly pine plantation. For. Sci. 15: 390–398.

Nuttall, M.J. 1989. *Sirex noctilio* F., sirex wood wasp (Hymenoptera: Siricidae), pp. 299–306. In: Cameron, P.J., Hill, R.L., Bain, J. and Thomas, W.P. (Eds) A Review of Biological Control of Invertebrate Pests and Weeds in New Zealand, 1874 to 1987. CAB International Institute of Biological Control, London.

Olsen, P.F. 1989. Wind risk – a consultant's perspective, p. 26. In: Somerville, A., Wakelin, S. and Whitehouse, L. (Eds) Workshop on Wind Damage in New Zealand Exotic Forests. Forest Research Institute Bulletin 146. Forest Research Institute, Rotorua, New Zealand.

Powers, H.R. and Kraus, J.F. 1983. Developing fusiform rust-resistant loblolly and slash pines. Plant Dis. 67: 187–189.

Rawcliffe, D.J. 1985. Forest Research Institute, Rotorua, New Zealand. Personal communication.

Shepherd, K.R. 1990. Trends in plantation silviculture in Australia – the immediate past and future, pp. 58–65. In: James, R.N. and Tarlton, G.L. (Eds) New Approaches to Spacing and Thinning in Plantation Forestry. Forest Research Institute Bulletin 151. IUFRO and Ministry of Forestry, Rotorua, New Zealand.

Simmonds, N.W. 1962. Variability in crop plants: its use and conservation. Biol. Rev. 37: 422–465.

Turner, R.M. 1989. Canterbury Plains and Lake Taupo forests, pp. 10–12. In: Somerville, A., Wakelin, S. and Whitehouse, L. (Eds) Workshop on Wind Damage in New Zealand Exotic Forests. Forest Research Institute Bulletin 146. Forest Research Institute, Rotorua, New Zealand.

van der Pas, J.B. 1981. A statistical appraisal of Armillaria root rot in New Zealand plantations of *Pinus radiata.* N. Z. J. For. Sci. 11: 23–36.

Watson, I.A. 1970. Changes in virulence and population shifts in plant pathogens. Annu. Rev. Phytopathol. 8: 209–230.

Whitehead, D. 1981. An ecological overview of plantation forestry. N. Z. J. For. 26(1): 14–19.

Wolfe, M.S. 1985. The current status and prospects of multiline cultivars and variety mixtures for disease resistance. Annu. Rev. Phytopathol. 23: 251–273.

New Forests **17:** 239–262, 1999.

Environmental impacts of forest monocultures: water use, acidification, wildlife conservation, and carbon storage

MELVIN G. R. CANNELL
Institute of Terrestrial Ecology, Edinburgh Research Station, Bush Estate, Penicuik, Midlothian, Scotland

Key words: planted forests, sustainability

Abstract. A broad assessment is given of the contentions that plantation forests are high consumers of water, increase acidification, sustain a low diversity of wildlife, and store more carbon than do unmanaged forests. The following conclusions are drawn: (1) Evapotranspiration from planted forest monocultures is greater than from short vegetation, as a result of greater interception loss. Water loss from conifer forests is usually greater than from deciduous hardwoods, but evapotranspiration from *Eucalyptus* in the dry tropics is often no greater than from native hardwoods. (2) Compared to short vegetation, forests can significantly increase the transfer of acidifying pollutants from the air to the soil and surface waters, and conifers are more likely to enhance acidification than are hardwoods. (3) There are normally sufficient plantation management options available to make most plantation landscapes the homes of a rich diversity of flora and fauna. (4) An area covered with a plantation managed for maximum volume yield will normally contain substantially less carbon than the same area of unmanaged forest.

Introduction

The purpose of this paper is to examine four claims about the effects of plantation forests on the environment. Three of the claims are negative: that they are high consumers of water, increase acidification, and sustain a low diversity of wildlife compared to unmanaged forests and short vegetation. The fourth is positive: that more carbon can be stored in harvested plantations than in unmanaged forests. For each claim, I shall draw a conclusion that seems to be generally true, based on the evidence cited.

Water use

Claim: Planted forest monocultures are high consumers of water. Total water use is the sum of transpiration (mainly through stomata) and evaporation of intercepted water (from leaf and woody surfaces). In the following, I shall consider these two components separately and then together, comparing conifer plantations both with unmanaged forests of conifers or hardwoods and with areas of short vegetation.

Transpiration from dry canopies

Transpiration losses from any vegetation are determined (1) by climatic demand – radiation, saturation deficit, wind speed, turbulence, and temperature, and (2) by what may be regarded as "supply" factors, which include canopy conductance, leaf area index, canopy structure, and the availability of soil water in the rooting zone. This discussion refers to vegetation in temperate regions, where water use is determined largely by climatic demand, moderated by supply factors. A brief discussion is given at the end of the section on conditions in the dry tropics, where water use is determined largely by supply factors.

The Penman-Monteith equation (Monteith 1981) expresses the dependence of the transpiration rate, E, of vegetation on H, the net amount of heat available in the system (mainly from radiation); D, the saturation vapor pressure deficit; g_c, the surface conductance (determined by the stomata); and g_a, the conductance of the aerodynamic boundary layer between the canopy surface and the air above. The equation is as follows:

$$E = \frac{\Delta H + \rho c_p g_a D}{L[\Delta + \gamma(1 + g_a/g_c)]}$$

where (in consistent units) L is latent heat of evaporation; ρc_p is the volumetric specific heat of air; γ is the psychrometric constant; and Δ is the change in vapor pressure per unit change in temperature.

The canopies of all tall forests are well coupled to the atmosphere above. The aerodynamic conductance, g_a (to the transport of momentum, force, energy, and mass), between the canopy and the air above is relatively large because the canopy surface is rough, thus generating vigorous mixing in the relatively high winds experienced at tree top height. Because the leaves and air readily exchange energy and water vapor, the saturation deficit at the canopy surface is similar to that of the ambient air above. Consequently, for forests growing in temperate regions (where the climatic demand is not large relative to soil water supply), the term $\rho c_p g_a D$ in the Penman-Monteith equation is very large compared to ΔH, such that the latter can be ignored. Furthermore, the term g_a occurs in both the numerator and the denominator, so the equation can be simplified to the following:

$$E = \frac{\rho c_p}{L_\gamma} g_c D \quad \text{or} \quad E \propto g_c D$$

That is, the transpiration rates of such well-coupled canopies are strongly controlled by canopy conductance, g_c (i.e., stomatal behavior), and the saturation deficit, D (Whitehead and Jarvis 1981; Jarvis and McNaughton 1985).

Now, it is well established that stomatal closure occurs in response to increasing saturation deficit in most species, including trees, so that canopy conductance falls as the saturation deficit increases. When the stomatal sensitivity to saturation deficit is high, as in *Picea sitchensis* (Bong.) Carr. (Sitka spruce), the transpiration rate may be low relative to other vegetation, because the canopy conductance is low (Jarvis and Stewart 1979). In other words, such conifer plantations have low transpiration rates because their stomata close in response to increasing saturation deficits, which are readily imposed at the canopy surface, such that canopy conductances and saturation deficits are never both large at the same time. The transpiration rate of *P. sitchensis* in the United Kingdom is commonly one-half or one-third that of short vegetation, and transpiration may use only a fraction of the energy available (i.e., the Bowen Ratio is greater than 1, often 2–4).

Most types of short vegetation have relatively smooth canopies and small aerodynamic conductances. They are consequently described as being poorly coupled to the air above; they have transpiration rates that are dictated more by the amount of net radiation received than by their canopy conductances. When well supplied with water, short vegetation may use most or all of the available energy in transpiration (i.e., the Bowen Ratio is 0 or less than 1), and will consequently lose more water by transpiration than do conifer forests such as *P. sitchensis*.

The transpiration rates of different forest types in temperate regions differ, depending upon their stomatal sensitivity to saturation deficit, species differences in maximum canopy conductance, and differences in leaf area index, regardless of whether they are managed plantations or natural forests. The maximum canopy conductances of stands of conifer species are often quite low, and, although their leaf area indices can be large, their maximum transpiration rates are often constrained within 2–4 mm/day by the closure of stomata in response to increasing saturation deficits (Whitehead and Jarvis 1981).

The transpiration rates of broadleaved tree species (hardwoods) per unit leaf area are typically two-three times greater than those of conifers during the growing season, because hardwoods have either more stomata per unit leaf area, or larger or more exposed stomatal pores and a better vascular supply (Korner et al. 1979). However, total transpiration/ha may be the same or less than that of conifer stands, because hardwoods have a smaller leaf area (Ladefoged 1963; Swank and Milner 1968). Thus, we can conclude that planted forests in temperate regions will often lose less water by transpiration than will short vegetation, and that planted forests may transpire as much, more, or less water than unmanaged forests on the same site, depending on the tree species.

Evaporation of intercepted water

Annual evaporative losses of intercepted water (interception loss) are determined by (1) the rate at which that water evaporates, and (2) the amount of water that can be stored on the leaf and branch surfaces during and after each period of precipitation (the storage or canopy capacity). The distinguishing feature of tall forests is that the average evaporation rate from their wet surfaces is typically two-five times greater than that from wet short vegetation such as grass. Evaporation rates from wet *P. sitchensis* canopies in the United Kingdom are commonly 0.1–0.3 mm/h – equaling or exceeding typical transpiration rates, and using all or most of the available energy (Jarvis and Stewart 1979). The reason has already been explained – i.e., the large surface roughness and high wind speeds, which result in effective turbulent mixing. The conductance to water vapor exchange can be so high that water evaporates readily into moist air during, as well as after, periods of rainfall (Calder 1990). In fact, with prolonged gentle rain, most interception loss occurs while it is raining, and the storage capacity is less important than the aerodynamic resistance. By contrast, in areas with short, heavy rainstorms, most interception loss occurs after it has stopped raining, and the amount lost is then governed principally by the canopy storage capacity.

Figure 1 shows the relationship between the rate of evaporation of intercepted water and the boundary layer conductance (g_a) at a range of saturation deficits (D) with a net radiation of 100 W/m^2. Clearly, as the saturation deficit increases, the evaporation rate from forest canopies (with large boundary layer conductances) becomes progressively greater than the evaporation rate from short vegetation (with small boundary layer conductances).

The storage capacities of forests and other vegetation are greatest when their leaf and branch indices are large. They also depend on species-specific characters such as leaf wettability and leaf orientation, as well as the intensity of rainfall and shaking in the wind. Rain forests can hold up to 8 mm of water (Herwitz 1985). The leaf area indices and seasonal durations of leaf retention (i.e., the leaf area durations) of conifer forests are both about twice those of

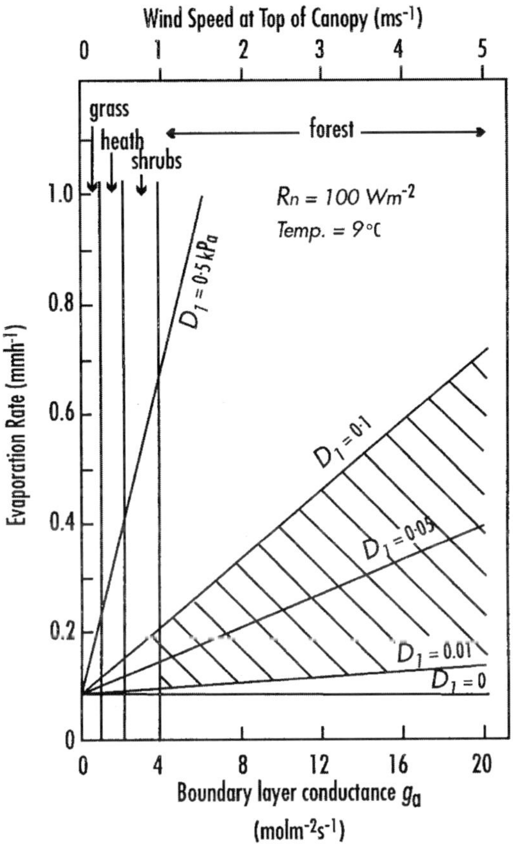

Figure 1. Ranges of likely boundary layer conductances for grass, shrub, and forest canopies, indicating the relationship between rate of evaporation of intercepted water and boundary layer conductance at a range of saturation deficits (D) (*Source*: Jarvis 1985, reproduced with permission).
Note: Shaded areas mark the range of values for wet forest canopies.

deciduous broadleaved woodlands on the same sites (Gholz et al. 1976). As a result, the storage capacities of conifer forests are much larger than those of broadleaved woodlands; in the summer, typical values for conifers and broadleaves are 1.0–2.5 mm and 0.5–1.0 mm, respectively. This difference in storage capacity can, under some circumstances, lead to larger interception losses from conifer plantations than from deciduous woodlands which they may replace (Swank and Douglass 1974). The range of values recorded for the percentage of total rainfall lost by interception is 20–85 percent for conifer stands and 10–50 percent for foliated hardwoods. The wide variation in values is the result of species characteristics and the intensity and duration of rainfall events (Benecke 1976). We can conclude that interception losses are almost

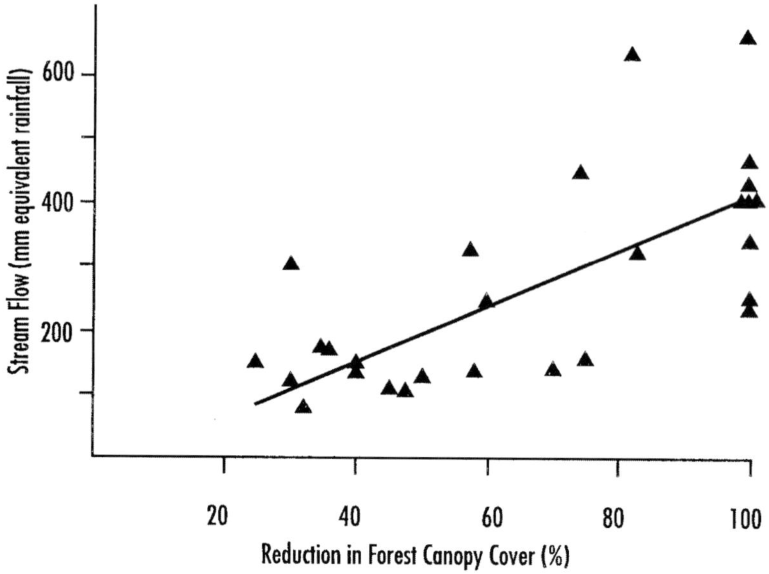

Figure 2. Relationship between the percentage reduction in canopy cover as a result of clearcutting within catchments in the United Kingdom and the change in stream flow (*Source:* Bosch and Hewlett 1982, with permission ©Elsevier Science).

certain to be increased when conifer plantations replace short vegetation, and are likely to be increased when conifer plantations replace broadleaved woodlands.

Total water use in temperate regions

Clearly, if plantations replace short vegetation, it can be expected that total water use (transpiration + interception loss) will be increased, owing to increased interception losses, especially if the canopies are wet for a large proportion of the year. There is an extensive literature showing that afforestation of nonforest land generally decreases the flow of water from catchments, i.e., water yield (Hibbert 1967; Nutter 1979). In the United Kingdom, the rule of thumb for upland coniferous forests, based on observations made on paired catchments, is that water yields are decreased by 1.5–2.0 percent for every 10.0 percent of a catchment that is afforested. Conversely, clearcutting increases water yield from catchments (Figure 2), although there is probably a threshold level of forest cover, or forest removal, before changes in water yield can be detected. The most serious effects of these decreases in water yield are on low flows during droughts and on the generating output from hydroelectric schemes (e.g., Loch Doon and Loch Tarn in Scotland; Maitland et al. 1990).

If conifer plantations replace hardwood forests, the probability is that total water use will be increased. Penman (1967), in a theoretical study, estimates that conifers in the United Kingdom may use 13 percent more water than do hardwoods in summer and 100 percent more during the winter. Also, Swank and Douglass (1974) report decreases in stream flow of 230 mm and 200 mm as a result of interception by conifers and hardwoods, respectively. By contrast, Ladefoged (1963) reports little difference in the water loss from conifer and hardwood stands in Sweden, perhaps related to the cool moist climate. But Hibbert (1967) concludes that plantation management practices that increase forest cover, leaf area index, and growth rate of the forest will generally increase water loss, particularly if there is a change from hardwood to conifer species. We should also note that, when natural hardwoods (or other vegetation) are replaced by conifer plantations, not only may the total stream flow be decreased, but also the peak flows may be increased as a result of plowing, draining, and building roads and trails.

Total water use in the dry tropics

In the dry zones of the tropics, transpiration can be primarily determined by the water supply rather than by the climatic demand. Restrictions imposed by soil water availability can be so severe that they place an upper limit on the amount of transpiration that can take place. If the climatic demand is far in excess of the water supply, then neither the saturation deficit nor the net radiation receipt can be said to control the annual transpiration loss (although they will, of course, control the rate of loss at any time). The controlling variables are the amount of soil water available, the extent to which tree roots access the soil water, stomatal conductance, and leaf area (Calder 1992).

If natural forests are replaced by plantations of trees with similar rooting depths, stomatal behavior, and leaf area indices, then the transpiration losses will remain unchanged. In the dry zone of India (ca 800 mm/yr) on soils 3 m deep, the water use of young *Eucalyptus* plantations is similar to that of the indigenous dry deciduous forest; the annual water use of both forests is about equal to the annual rainfall (Calder 1992).

However, plantations of *Eucalyptus* and other species could use more water than indigenous forests (1) if their roots penetrate deeper and access water at greater depth than those of indigenous species – which seems intuitively unlikely, or (2) if their stomata do not close so readily in response to increasing saturation deficits or soil drying. Colquhoun et al. (1984) found that some *Eucalyptus* species (e.g., *E. marginata* Smith) show little stomatal control, whereas others (e.g., *E. wandoo* Blakely) have high stomatal control – but there are no field measurements to show that such differences have resulted in high water use by *Eucalyptus* plantations.

Of course, the situation is quite different when comparisons are made between *Eucalyptus* plantations and agricultural land. Cropped fields will usually use less water than forest plantations because (1) they do not have a complete canopy cover throughout the year, (2) more water may be lost to groundwater below the crop rooting depth, and (3) their interception loss is less. Afforestation of nonforested land in the tropics, as in temperate regions, will increase water losses, regardless of the tree species. However, in the dry tropics, because water loss is so dependent on water "supply", rather than on atmospheric "demand", there is the additional possibility that trees planted on former agricultural land will access water at depth. Calder (1992) found that young *Eucalyptus* planted on 8 m deep soils seemed to be "mining" water as the roots penetrated greater depths, so that annual evapotranspiration exceeded the rainfall (3,400 mm and 2,100 mm, respectively). Greenwood et al. (1985) found that annual evapotranspiration from *Eucalyptus* at a site in Australia exceeded the annual rainfall by a factor of four, owing to groundwater extraction. Similarly high water use probably occurs where *Eucalyptus* has been planted alongside watercourses; indeed *Eucalyptus* has been used to "drain" marshland and to deliberately lower the water table where saline water has risen to the surface (Calder 1992).

It may be added that the relatively high water use by forests compared to agricultural crops transfers water to the atmosphere with consequent effects on the local climate, which may be important in the tropics. Harding (1992) shows that, in Karnataka, India, a forest which is over 50 km in extent should have important, positive effects of increasing humidity and decreasing temperature. There has been similar work showing that extensive deforestation in Amazonia may have the reverse, negative effect of decreasing humidity and increasing temperature (e.g., Gash and Shuttleworth 1991).

Water use efficiency

It is beyond the scope of this paper to review the factors affecting the ratio of water lost to carbon fixed by forests and other vegetation (Stanhill 1986). It is important to state that, wherever interception losses are large, the water use efficiency of the stand will inevitably be low. However, differences in transpiration among forest types, including *Eucalyptus*, are generally positively related to volume production – that is, given comparable climatic and nutritional conditions, the more water transpired, the greater the growth rate of the trees. The water use efficiencies of trees, like those of other plants, decrease with increase in saturation deficit and when trees are subject to nutrient or severe water stress. Large plantation forests, which increase the humidity of the air above them and are also kept well supplied with nutrients, may be expected to have water use efficiencies at least as great as those of unmanaged forests.

Conclusion

Total water loss from planted-forest monocultures will normally be greater than from short vegetation, owing to greater interception loss. Also, total water loss from conifer forests will usually be greater than from deciduous hardwood forests, because conifer canopies usually have a greater water storage capacity. However, total water loss from *Eucalyptus* in the dry tropics is often no greater than from native hardwoods, although it is greater than from agricultural crops, especially if water is accessed at depth by the *Eucalyptus* roots.

Acidification

Claim: Planted forest monocultures increase soil and water acidification. There is good evidence that afforestation of previously nonforested land can increase the concentrations of acidity and aluminum in surface waters, sometimes to such an extent that salmonid fish and invertebrate populations are adversely affected (Stoner and Gee 1985; Ormerod et al. 1989; Kreiser et al. 1990). Such effects occur where the region has appreciable concentrations of pollutants in the air and base-poor soils overlying rocks with few weathered minerals. In Wales, it has been proposed that conifer planting be restricted in areas where the drainage water already has less than 25 mg/l of calcium carbonate, and that no planting be done in areas with less than 12 mg/l of calcium carbonate (Hornung et al. 1990).

Direct effects of forests on acidification

Forests can increase the acidity of soils and surface waters "directly" (1) by increasing base cation uptake from the soil and the production in the soil of organic acids from decomposing needles (when conifers replace broadleaves or other vegetation), and (2) by altering hydrology, as a result of soil drying and site disturbance, which leads to increased solute concentrations and the mineralization of organic sulfur and nitrogen with the production of mobile anions. However, these processes are generally either slow or of only local importance. The dominant mechanism producing increased soil and water acidity is normally the "indirect" effect that the forest canopy has on increasing the transfer of pollutants from the air to the ground, i.e., the effect of the forest on acid deposition.

In polluted regions with moderate or high rainfall, about half of the total inputs of acidic substances are likely to be received in rain, mostly as sulfate (SO_4^{-2}), nitrate (NO^{-3}), and the hydrogen ion (H^+). The presence of forests has little effect on this form of acid input. Nor do forests have much impact on the deposition of those gases that are not absorbed on the external surfaces of plants, i.e., which are absorbed by dissolution or reaction within the substomatal cavities. These gases include nitrogen dioxide (NO_2) and, to a lesser extent, sulfur dioxide (SO_2), whose deposition is consequently determined by the bulk stomatal conductance of the canopy. As mentioned, conifer forest canopy conductances are normally smaller than or similar to those of other vegetation, so the rates of NO_2 deposition are likely to be less or similar (Fowler et al. 1989; Duyzer and Fowler 1994). A complication exists in the case of SO_2, because it is deposited on leaf surfaces more readily when they are wet and in the presence of ammonia.

By contrast, forests greatly increase the input of pollutants that are deposited and absorbed on leaf and branch surfaces, because they are transferred to those surfaces more rapidly in the turbulent air within aerodynamically rough canopies. These pollutants include nitric acid vapor (HNO_3), hydrochloric acid vapor (HCl), and ammonia (NH_3, which is also acidifying following transformation to $R.NH_2$ or NO_3 in the soil). These are the so-called "reactive" gases, which, for a forest 10 m tall with a wind speed of 5 m/s above the canopy, have "deposition velocities" approximately equal to the aerodynamic conductance of about 10 cm/s, compared to about 2–5 cm/s for short vegetation (Fowler et al. 1989).

Forests are also efficient at collecting cloud droplets, in the 5–40 μm diameter size range, by processes of impaction and sedimentation, particularly in wind speeds of 5–10 m/s. For cloud water, integrated across the droplet size spectrum, the rate of deposition (like that of the reactive gases) is similar to that of momentum (Fowler et al. 1991). Consequently, sulfate and nitrate aerosols that nucleate and accumulate in cloud droplets are deposited in large quantities onto some high-elevation forests. In the United Kingdom, the concentration of ions in cloud water can be two-ten times greater than those in rain, and some forests can be in cloud for 2,000 h/yr, so that cloud water deposition can increase precipitation by 12 percent and pollutant deposition by 40 percent (Crossley et al. 1992).

The extent to which planting forests on nonforest land will increase the input of the reactive gases and sulfate and nitrate ions will obviously depend on the pollution climate of the region and local meteorology. Calculations of the effect of converting a moorland to coniferous forest (15 m tall) in northern Britain (300 m above sea level, 1,500 mm annual rainfall), with typical rural air concentrations of pollutants, show that there could be a 4–5-fold increase

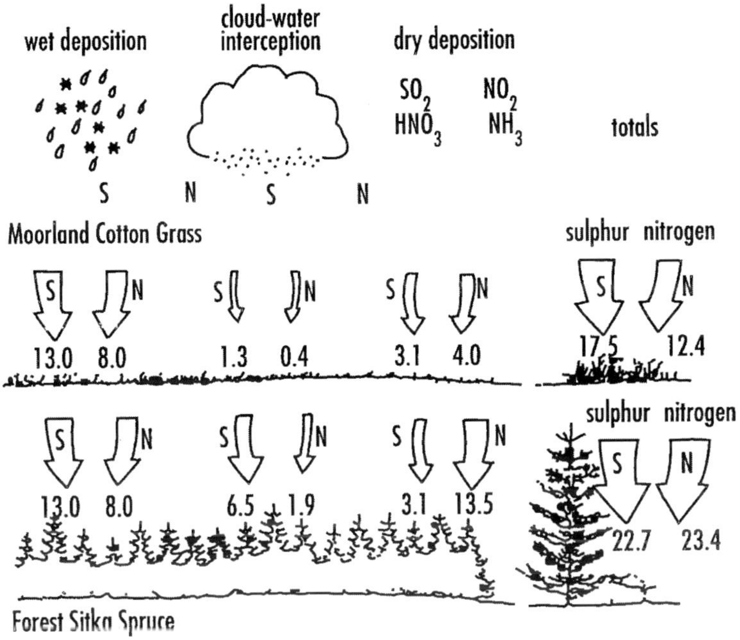

Figure 3. Estimated inputs (kg/ha/yr) of pollutant nitrogen and sulfur from the atmosphere to a moorland and forest in the United Kingdom[a] (*Source:* Fowler et al. 1989, reproduced with permission from Phil. Trans., Roy. Soc. London, 1989; 324B: 247–265.).
[a]The site is 60,000 ha in area, located 300 m above sea level, and receives 1,500 mm mean annual rainfall.

in sulfate and nitrate deposition in cloud water and a 3-fold increase in dry deposition of nitrogen (Figure 3; Fowler et al. 1989).

Effects of tree species on acid deposition

In general, acid deposition is enhanced more by conifer forests than by deciduous broadleaved woodlands, because conifers expose greater leaf areas for longer periods/yr. In the United Kingdom, excess deposition of sulfate-S (i.e., precipitation input minus throughfall plus stemflow) was estimated to be about 3, 5, 12, and 30 kg/ha/yr for oak, alder, pine, and spruce forests, respectively (Cape and Brown 1986). In Germany, inputs of sulfur to beech forests at Solling were 45–51 kg/ha/yr compared to 80–86 kg/ha/yr to an adjacent spruce forest, this attributable to greater deposition of both reactive gases and cloud water (Mayer and Ulrich 1978; Nys et al. 1990). Thus, other things being equal, the conversion of broadleaved woodlands to conifer plantations is likely to increase acid deposition.

Conclusion

Forests can significantly increase the transfer of acidifying pollutants from the air to terrestrial surfaces (and thence to groundwater) compared to short vegetation. This is a large effect where forests are exposed to high concentrations of "reactive" gases (HNO_3, NH_3, and HCl) and to polluted cloud water. Conifers are more likely to enhance acidification than are hardwoods.

Wildlife conservation

Claim: Planted forest monocultures sustain a low diversity of wildlife. People may object to planted forests because they are unlike the "natural" vegetation that would otherwise exist at the site (Margules and Usher 1981). The basis for this objection is clear when, say, natural scrub forests are replaced with *Eucalyptus* plantations, or when forests are planted on blanket bogs. The basis is less clear when, say, mixed-species, old-growth forest is replaced with plantations of one of those species. The arguments become finely balanced when forests are planted on former agricultural land or on deforested moorland. Whatever the basis, if naturalness is important, it may be impossible for plantations to be highly rated. In these circumstances, the best that can be done is to undertake some form of cost-benefit analysis of the land-use change, provided that foresters and conservationists can agree on a common currency.

The commonest objection to plantations, on conservation grounds, is that they contain a low diversity of wildlife. When considering ecosystems, diversity refers principally to species, and the conservation value is considered to be low when the total number and relative abundance of species is below that which would otherwise occur in natural, unmanaged ecosystems at the same site. Again, the arguments become complex when plantations replace nonforest vegetation. For the purposes of this discussion, let us concentrate on the situation in which plantations replace old-growth forests.

In order to evaluate the effects of afforestation on species diversity we need to identify the differences in forest attributes which have important effects on the numbers and types of plants and animals that can exist within the forests, and then to determine the ways in which plantations can be managed to mimic the essential attributes of unmanaged forests. Four important forest attributes are light transmission to the ground, duration of successional stages, change in the species composition, and diversity in forest structure.

Light transmission to the ground

The degree of shade (combined with the amount of smothering litter and competition for water and nutrients), rather than the tree species per se, is of overriding importance in determining the type and diversity of plant species that can exist in the understory. In temperate regions, a ground flora of vascular plants is eliminated once forests intercept 80–90 percent of the incoming radiation (Hill 1979), but, if light interception is kept to 80 percent or less, a similar ground flora can develop in forests with different overstory tree species, given similar soils and climate. Thus, in the United Kingdom, many of the same plant species occur in the ground vegetation of broadleaved plantations as do in that of native broadleaved woodlands (given similar soils and climate), and, surprisingly, conifer plantations can have ground floras similar to those of broadleaved plantations, provided the conifers are thinned or consist of species that cast light shade (Hill 1987; Hill and Wallace 1989).

However, if plantations are to be managed to maximize volume production, they must achieve maximum light interception during each rotation – or maximum water use (Cannell 1989). Apart from rides, roadsides, and streamsides, the forest floor will receive over 20 percent of incoming radiation only during the periods of recovery after each clearcutting. The total amount of light reaching the ground over a whole region over periods of several hundred years may be larger in plantations than in unmanaged forests, but the question is whether or not the species that can recolonize after each clearcutting are the same as those that persist with less frequent disturbance in unmanaged forests. If this is not the case, then attention must be given to rides, roadsides, streamsides, and unplanted areas as refuges for the less-mobile, light-demanding ground flora.

Duration of successional stages

Many natural forests experience disturbance by fire, wind, pests, or pathogens every few hundred years, after which there is a prolonged period of recovery, a prolonged period of maturation, and a period of natural decline (Figure 4). Plantation management truncates these successional (seral) stages: the period of recovery after felling is made as short as possible, growth is as fast as possible, and felling occurs well before natural decline. This difference between natural and plantation forests can have positive or negative impacts on wildlife diversity (Rochelle and Bunnell 1979).

The recovery period after disturbance or clearcutting is often one during which there is luxuriant growth of vegetation, supporting a rich or abundant wildlife. In the United Kingdom, the period between conifer planting and

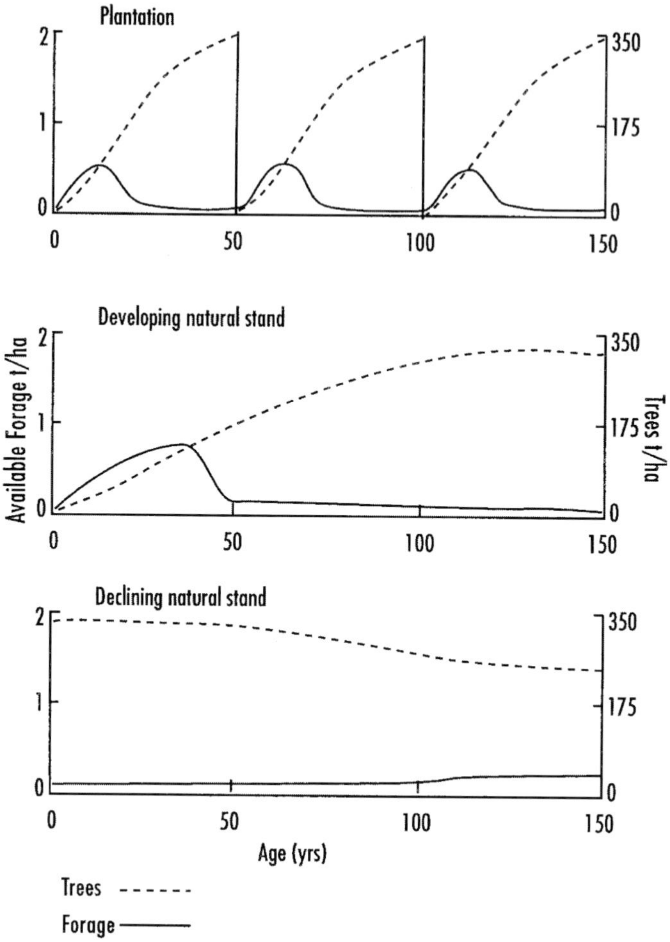

Figure 4. Idealized changes in the biomass of forest trees and forage biomass during plantation rotations compared to a natural stand (*Source:* Rochelle and Bunnell 1979, reproduced with permission).

canopy closure is often characterized by large populations of voles, which support large populations of owls, kestrels, and other raptors (Newton and Moss 1981). In the United States, populations of mountain beavers can expand to occupy recent clear-cut areas, and deer populations are typically greatest at this time. Of course, rodents, mountain beavers, and deer can all damage tree seedlings, and this is the stage at which wildlife abundance conflicts with the need for rapid forest regeneration. Populations may need to be controlled directly, or by forest management, such as altering the ratio of open to forest-covered land in time and space (e.g., Rocky Mountain elk; Thomas et al. 1976).

The late recovery period, when there is a mixture of saplings and brush/shrub vegetation, may be missing altogether in managed plantations, because of dense tree planting and vegetation (weed) control. Consequently, the wildlife associated with that stage may be depleted. Meslow and Wight (1975) report that, whereas 38 species of birds in western Oregon preferred the grass/forb stage of succession, 72 species preferred or were restricted to the brush/sapling stage. There seems little alternative but to provide habitat for such species in areas along forest edges, roads, streams, and in areas not subject to management for maximum volume production.

In the mature stage of natural forest succession, a large number of microhabitats develop and persist for a long time. This may never occur in plantations, which may therefore lack the associated flora and fauna. In the United Kingdom, many lichens, invertebrates, and epiphytes are found only in ancient woodlands, where their habitats have remained undisturbed for hundreds of years (Rose 1974). The existence of dead wood is particularly important for some lower plants, fungi, and invertebrates. Dead wood in the crowns of trees can have a flora and fauna different from similar wood lying on the ground (Elton 1966). It is also well known that some species of birds, such as the pileated woodpecker in Oregon and the red-cockaded woodpecker in the southeast United States, require large dead or dying trees (Bull and Meslow 1977). Such specialists could be accommodated in plantations by leaving a few old trees.

Finally, it may be noted that old-growth forests are needed to provide habitat for deer in winter in northern regions; the tall trees intercept snow, yet let through enough light to support understory vegetation. By contrast, plantation areas are either deep in snow or have little understory vegetation (Rochelle and Bunnell 1979).

Change in tree species composition

The conversion of a mixed species forest to an even-aged monoculture has obvious effects on the diversity of habitats, but how important is the restriction in tree species per se to the existence of fauna and other flora? This question has been examined in the United Kingdom, especially in relation to the introduction of non-native tree species. The main finding is that invertebrates, in particular, prefer certain host tree species. The non-native conifers, *Pinus contorta* Dougl. ex Loud. and *Picea sitchensis*, have fewer phytophagous insects living on them than does the native *Pinus sylvestris* L.; broadleaved tree species have many more invertebrate species than do conifers, but each tree species has a particular invertebrate fauna, especially those that eat woody tissues and associated saprophytic fungi (Kennedy and Southwood 1984; Welch 1986; Evans 1987). In general, other wildlife, such as mammals and birds, are more dependent on the presence of particular forest structures and ranges of food sources than on particular tree species (Haartman 1971).

Diversity in forest structure

As well as having a different successional structure (through time), plantation forests differ from old-growth forests in their physical structure (horizontally and vertically). However, the variety of physical structures that can be produced in a landscape by plantation management – by altering felling unit sizes, edge lengths, and aspects, and by creating a multiaged patchwork of stands – may be as great as or greater than that which would exist naturally. The only physical niches that may be lacking are those within multilayered high canopies. Otherwise, it may be possible to manage plantations to enrich the diversity of species in a small area. For instance, a range of felling unit sizes sustains the widest range of bird species in U.K. plantation forests (Bibby et al. 1986).

Conclusion

Although (1) planted forest monocultures may not be as highly valued for conservation as are unmanaged forests, (2) there are clear trade-offs between maximizing volume production and protecting wildlife diversity, and (3) there are some habitats that can only be provided by old-growth forests, there are normally sufficient plantation management options available to make most plantation landscapes the homes of a rich diversity of flora and fauna.

Carbon storage

Claim: More carbon can be stored in harvested plantations than in unmanaged forests.

Carbon storage in trees and litter

At first glance, it may be supposed that more carbon may be removed from the atmosphere by replacing old, slowly growing forests with young, fast-growing plantations, and then harvesting and storing the wood. But such reasoning ignores the fact that vastly more carbon is stored in old undisturbed forests than in plantations. Harmon et al. (1990) point out that a 450-year-old Douglas-fir/western hemlock forest in the Pacific Northwest contains at least twice as much carbon as a 60-year-old Douglas-fir plantation, and that this is

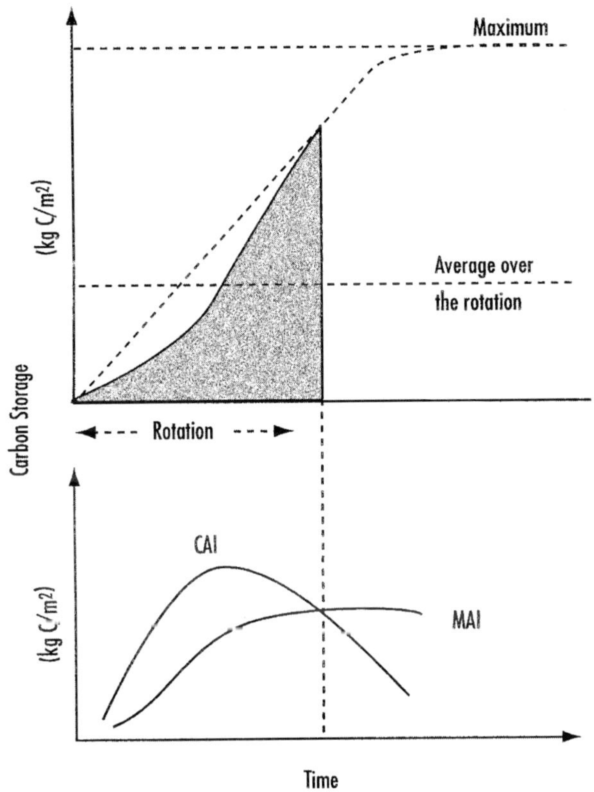

Figure 5. Relationship between the point of maximum mean annual increment (MAI) and the amount of carbon stored in a forest[a] (*Source:* Cannell 1995, reproduced with permission of EFI).

[a]Maximum MAI occurs when MAI equals the current annual increment (CAI), and is reached while the forest is still accumulating carbon, well before it reaches its maximum carbon content. The amount of carbon stored in a plantation (shaded), when averaged from planting to the point of maximum MAI, is only one-third the maximum carbon content that the forest would achieve if it were not harvested.

also true for a landscape of old-growth forests in different stages of recovery from natural disturbance compared to a landscape of plantations of different ages.

Plantations have a small average biomass, partly because they have a smaller detrital biomass on the forest floor than do mature forests. But, more importantly, plantations have a small time-averaged biomass because they are felled near the time of maximum mean annual volume increment (MAI), which we shall assume here is equivalent to MAI in carbon (Cooper 1983; Dewar 1991). Foresters identify the time of maximum MAI as the point at which the current annual increment (in any year) is equal to the MAI (averaged since planting). Figure 5 shows that this point occurs well before the forest reaches its maximum volume or carbon mass. If the cumulative amount of volume or carbon in a forest follows a typical sigmoidal pattern, as described by the logistic growth curve at the top of Figure 5, then the point of maximum MAI corresponds to the point at which a straight line from the origin is tangent to the upper limb of the growth curve. It can be shown that, for a variety of logistic growth curves approximated by most forest plantations, the time-averaged carbon storage from planting to harvest at maximum MAI (the average over the rotation) is only about one-third of the maximum amount of carbon stored in the forest at full maturity (Figure 5; Cooper 1983).

In other words, up to two-thirds of the carbon stored in the trees in a mature, old-growth forest is eventually "lost" (i.e., transferred to the atmosphere) when it is replaced by harvested plantations with the same growth characteristics. In reality, this theoretical "loss" is somewhat less than two-thirds, because old-growth forests are subject to periodic disturbances – so that their time-averaged carbon content is less than their maximum carbon content. Nevertheless, the loss of carbon is substantial. The main opportunities to compensate for the decrease in carbon in the trees when old-growth forests are replaced by plantations are (1) to transfer the wood harvested from the plantations into products that last much longer than the wood that decays on the forest floor in an old-growth forest, and (2) to plant tree species and tend the plantations so that they grow much faster than the old-growth forests that they replace. However, it is difficult to make up the loss of carbon, as the following example illustrates.

Figure 6 shows the simulated example of a mature, undisturbed, broadleaved forest in Europe (e.g., oak and beech) containing 38 kg C/m^2 (28, 2, and 7 kg C/m^2 in trees, litter, and soil, respectively), which is cut down and replaced by hardwood plantations of the same species with 80-year rotations (Cannell et al. 1992). These plantations were assumed to yield products with lifetimes of either 72 or 420 years, which are represented by the lower and upper lines for carbon in wood products in Figure 6. When the product lifetime is 72 years, the total equilibrium carbon storage reaches 23 kg C/m^2 after one or two rotations, and this represents a loss of 39 percent of the carbon from the original forest (which contained 38 kg C/m^2). The product lifetime of 420 years is that required to bring the total equilibrium carbon storage to 38 kg C/m^2. Clearly, it is unlikely that all of the wood, even from a hardwood forest, could be preserved for so long. Similar calculations were done to explore the option of replacing the mature woodland with fast-growing tree plantations. It was found that, in order to achieve an equilibrium carbon storage of 38 kg C/m^2, the rotation period would have to be decreased to 30

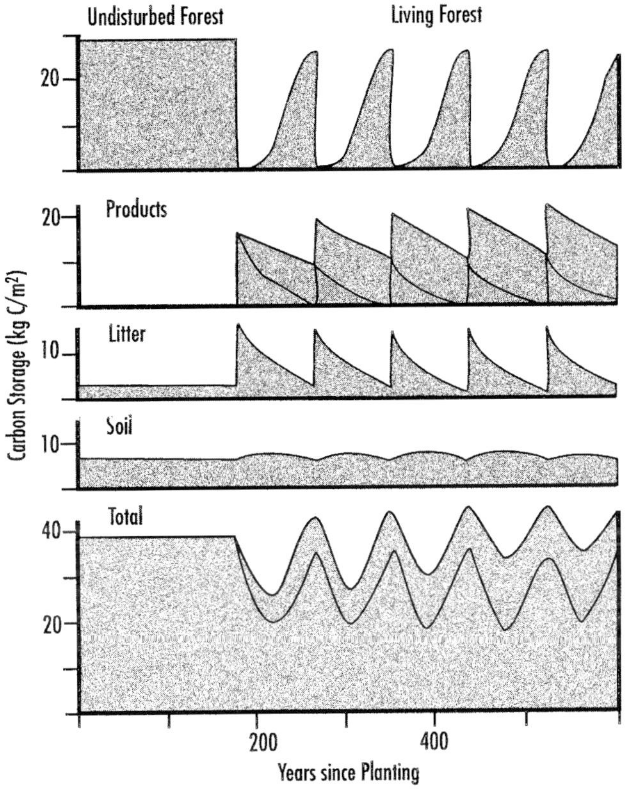

Figure 6. Simulated carbon storage in the trees, wood products, litter, and soil, when an undisturbed European broadleaved woodland (containing 38 kg C/m^2) is replaced by a plantation with similar growth characteristics over 80-year rotations (*Source:* Cannell et al. 1992, reproduced with permission).
NOTE: The upper and lower lines in the products and total compartments refer to wood product lifetimes of 420 and 72 years, respectively.

years, whereas the product lifetime remained at 72 years. It is conceivable that European broadleaved woodlands could be replaced by poplars (*Populus* spp.) with a rotation period of about 30 years, but it is unlikely, at present, that the average lifetime of the products from fast-growing species such as poplar could be extended to 72 years.

Carbon storage in soil organic matter

Concern has also been expressed that plantation forests, as well as storing less carbon in biomass than old-growth forests, may also store less carbon in the soil. In particular, it is feared that soil disturbance during site preparation and harvesting may accelerate the decomposition of soil organic matter. However,

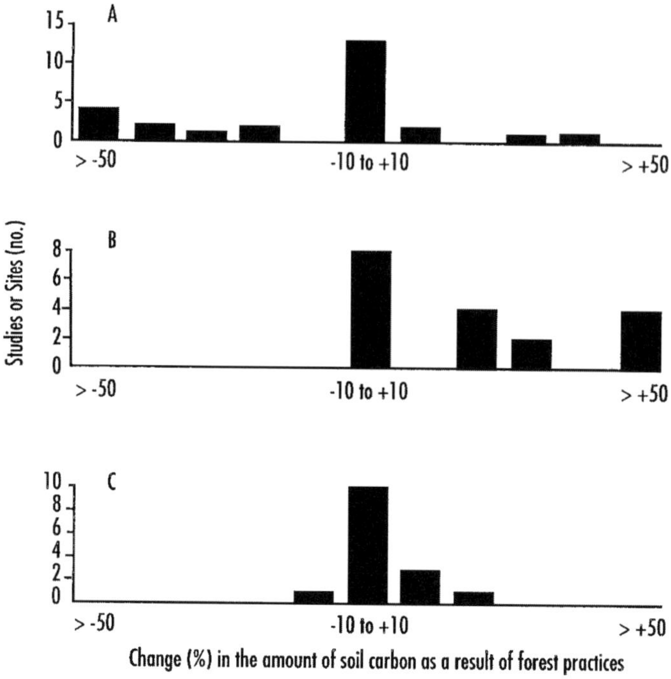

Figure 7. Numbers of studies showing gains (+) and losses (–) in soil carbon as a result of (A) harvesting, site preparation, and burning, (B) fertilization and presence of nitrogen-fixers, and (C) regular prescribed burning (*Source:* Johnson 1992, reproduced with permission from Water, Air, Soil Pollut. Kluwer Academic Publishers, 1992; 64: 83–120).

in a review of thirteen studies, Johnson (1992) found no general trend toward lower soil carbon with forest harvesting and site preparation; most studies showed no significant change in soil carbon, and some showed an increase (Figure 7). It would appear that any accelerated decomposition of soil organic matter is balanced by increased input of carbon from litter, perhaps because of increased net primary productivity and litter production by forests that are managed and harvested. One management activity that increases net primary productivity is the addition of fertilizers, and Johnson (1992) found clear evidence that soil carbon levels increased when stands were fertilized or had nitrogen-fixing tree species added (Figure 7).

Conclusion

Averaged over time (exceeding a rotation time), an area covered with plantations that are managed for maximum wood volume yield will normally contain substantially less carbon in the trees than the same area of unmanaged forest of species with similar growth characteristics. In most instances, the difference in carbon storage in the trees is so large that it is unlikely that it can be compensated by (1) the carbon stored in harvested wood products, unless they are very long-lasting, or (2) intensive management of the plantations. However, there is little evidence for a loss of soil carbon as a result of plantation management, and nitrogen fertilization generally increases soil carbon.

Conclusion

A broad assessment is given of the contentions that plantation forests are high "consumers" of water, increase soil and water acidification, sustain a low diversity of wildlife, but store more carbon than do unmanaged forests or areas of short vegetation. The following conclusions are drawn:

- Total water loss from planted forest monocultures will normally be greater than from short vegetation, because of greater interception loss. Also, total water loss from conifer forests will usually be greater than from deciduous hardwood forests, because conifer canopies usually have a greater water storage capacity. However, total water loss from *Eucalyptus* in the dry tropics is often no greater than from native hardwoods, although it is greater than from agricultural crops, especially if water is accessed at depth by the *Eucalyptus* roots.
- Forests can significantly increase the transfer of acidifying pollutants from the air to terrestrial surfaces (and thence to groundwater) compared to short vegetation. This is a large effect where forests are exposed to high concentrations of "reactive" gases (HNO_3, NH_3, and HCl) and to polluted cloud water. Conifers are more likely to enhance acidification than are hardwoods.
- Although (1) planted forest monocultures may not be as highly valued for conservation as unmanaged forests, (2) there are clear trade-offs between maximizing volume production and protecting wildlife diversity, and (3) there are some habitats that can only be provided by old-growth forests, there are normally sufficient plantation management options available to make most plantation landscapes the homes of a rich diversity of flora and fauna.
- Averaged over time (exceeding a rotation time), an area covered with-plantations that are managed for maximum wood volume yield will normally contain substantially less carbon in the trees than the same area of unmanaged forest of species with similar growth characteristics. In most instances, the difference in carbon storage in the trees is so large that it is unlikely that it can be compensated (1) by the carbon stored in harvested wood products, unless they are very long-lasting, or (2) by

intensive management of the plantations. However, there is little evidence of a loss of soil carbon as a result of plantation management, and nitrogen fertilization generally increases soil carbon.

References

Benecke, P. 1976. Soil water relations and water exchange of forest ecosystems. Ecol. Stud. 19: 101–131.

Bibby, C.J., Phillips, B.N. and Sneddon, A.J.E. 1986. Birds of restocked conifer plantations in Wales. J. Appl. Ecol. 22: 619–633.

Bosch, J.M. and Hewlett, J.D. 1982. A review of catchment experiments to determine the effect of vegetation changes on water yield and evapotranspiration. J. Hydrol. 55: 3–23.

Bull, E.L. and Meslow, E.C. 1977. Habitat requirement of the pileated woodpecker in northeastern Oregon. J. For. 75: 335–340.

Calder, I.R. 1990. Evaporation in the Uplands. John Wiley and Sons, Chichester, England.

Calder, I.R. 1992. Water use of eucalypt – a review, pp. 167–179. In: Calder, I.R., Hall, R.L. and Adlard, P.G. (Eds) Growth and Water Use of Forest Plantations. John Wiley and Sons, Chichester, England.

Cannell, M.G.R. 1989. Physiological basis of wood production: A review. Scand. J. For. Res. 4: 459–490.

Cannell, M.G.R. 1995. Forests and the global carbon cycle in the past, present and future. Research Report No 3. European Forest Institute, Joensuu, Finland.

Cannell, M.G.R., Dewar, R.C. and Thornley, J.H.M. 1992. Carbon flow and storage in European forests, pp. 256–271. In: Teller, A. and Jeffers, J.N.R. (Eds) Responses of Forest Ecosystems to Environmental Changes. Elsevier Applied Science, London and New York.

Cape, J.N. and Brown, A.H.F. 1986. The effects of different tree species on the amount and chemical composition of precipitation passing through the canopy. Report to the U.K. Department of Environment. Institute of Terrestrial Ecology, Edinburgh, United Kingdom.

Colquhoun, I.J., Ridge, R.W., Bell, D.T., Loneragan, W.A. and Kuo, J. 1984. Comparative studies in selected species of Eucalyptus used in rehabilitation of the northern jarrah forest, western Australia. I. Patterns of xylem pressure potential and diffusive resistance of leaves. Aust. J. Bot. 32: 367–373.

Cooper, C.F. 1983. Carbon storage in managed forests. Can. J. For. Res. 13: 155–166.

Crossley, A., Wilson, D.B. and Milne, R. 1992. Pollution in the upland environment. Envirn. Pollut. 75: 81–88.

Dewar, R.C. 1991. Analytical model of carbon storage in the trees, soils and wood products of managed forests. Tree Physiol. 8: 239–258.

Duyzer, J. and Fowler, D. 1994. Modeling land atmosphere exchange of gaseous oxides of nitrogen in Europe. Tellus 46B: 353–372.

Elton, C.S. 1966. The Patterns of Animal Communities. Methuen, London.

Evans, H.G. 1987. Sitka spruce insects: Past, present and future. In: Henderson, D.M. and Faulkner, R. (Eds) Sitka spruce. Proc. R. S. Edinburgh 93B: 157–168.

Fowler, D., Cape, J.N. and Unsworth, M.H. 1989. Deposition of atmospheric pollutants on forests. Philos. Trans. R. S. London 324B: 247–265.

Fowler, D., Duyzer, J.H. and Baldocchi, D.D. 1991. Inputs of trace gases, particles and cloud droplets to terrestrial surfaces. Proc. R. S. Edinburgh 97B: 35–59.

Gash, J.H.C. and Shuttleworth, W.J. 1991. Tropical deforestation: Albedo and the surface energy balance. Clim. Chang. 19: 123–133.

Gholz, H.L., Fitz, F.K. and Waring, R.H. 1976. Leaf area differences associated with old-growth forest communities in the western Oregon Cascades. Can. J. For. Res. 6: 49–57.

Greenwood, E.A.N., Klein, L., Beresford, J.D. and Watson, G.D. 1985. Differences in annual evaporation between grazed pasture and Eucalyptus species in plantations on a saline farm catchment. J. Hydrol. 78: 261–278.

Haartman, von, L. 1971. Population dynamics, pp. 391–459. In: Farner, D.J. and King, J.R. (Eds) Avian Biology (volume 1). Academic Press, London.

Harding, R.J. 1992. The modification of climate by forests, pp. 332–346. In: Calder, I.R., Hall, R.L. and Adlard, P.G. (Eds) Growth and Water Use of Forest Plantations. John Wiley and Sons, Chichester, England.

Harmon, M.E., Ferrell, W.K. and Franklin, J.F. 1990. Effects on carbon storage of conversion of old-growth forests to young forests. Sci. 247: 699–702.

Herwitz, S.R. 1985. Interception storage capacities of tropical rainforest canopy trees. J. Hydrol. 77: 237–252.

Hibbert, A.R. 1967. Forest treatment effects on water yield, pp. 527–543. In: Sopper, W.E. and Lull, H.W. (Eds) Forest Hydrology. Proceedings of an International Symposium on Forest Hydrology. Pergamon Press, New York.

Hill, M.O. 1979. The development of a flora in even-aged plantations, pp. 175–192. In: Ford, E.D., Malcolm, D.C. and Atterson, J. (Eds) The Ecology of Even-aged Forest Plantations. Institute of Terrestrial Ecology Edinburgh, United Kingdom.

Hill, M.O. 1987. Opportunities for vegetation management in plantation forests, pp. 64–69. In: Good, J.E.G. (Ed) Environmental Aspects of Plantation Forestry in Wales. Institute of Terrestrial Ecology, Edinburgh, United Kingdom.

Hill, M.O. and Wallace, H.L. 1989. Vegetation and environment in afforested sand dunes at Newborough, Anglesey. For. 62: 249–267.

Hornung, M., Le-Grice, S., Brown, N. and Norris, D. 1990. The role of geology and soils in controlling surface water acidity in Wales, pp. 55–66. In: Edwards, R.W., Gee, A.S. and Stoner, J.H. (Eds) Acid Waters in Wales. Kluwer, Dordrecht.

Jarvis, P.G. 1985. Transpiration and assimilation of tree and agricultural crops: The "omega factor", pp. 460–480. In: Cannell, M.G.R. and Jackson, J.E. (Eds) Trees as Crop Plants. Institute of Terrestrial Ecology, Edinburgh, United Kingdom.

Jarvis, P.G. and McNaughton, K.G. 1985. Stomatal control of transpiration. Adv. Ecol. Res. 15: 1–49.

Jarvis, P.G. and Stewart, J. 1979. Evaporation of water from plantation forest, pp. 327–349. In: Ford, E.D., Malcolm, D.C. and Atterson, J. (Eds) The Ecology of Even-aged Plantations. Institute of Terrestrial Ecology, Edinburgh, United Kingdom.

Johnson, D.W. 1992. Effects of forest management on soil carbon storage. Water, Air, Soil Pollut. 64: 83–120.

Kennedy, C.E.J. and Southwood, T.R.E. 1984. The number of species of insects associated with British trees: A re-analysis. Anim. Ecol. 53: 455–478.

Korner, C., Scheel, J.A. and Bauer, H. 1979. Maximum leaf diffusive conductance in vascular plants. Photosynthetica 13: 45–82.

Kreiser, A.M., Appleby, P.G., Natkanski, J., Rippey, B. and Batterbee, R.W. 1990. Afforestation and lake acidification: A comparison of four sites in Scotland. Philos. Trans. R. S. London 327B: 377–383.

Ladefoged, K. 1963. Transpiration of forest trees in closed stands. Physiol. Plant. 10: 378–414.

Maitland, P.S., Newson, M.D. and Best, G.A. 1990. The Impact of Afforestation and Forestry Practice on Freshwater Habitats. Nature Conservancy Council, Peterborough, United Kingdom.

Margules, C. and Usher, M.B. 1981. Criteria used in assessing wildlife conservation potential: A review. Biol. Conserv. 21: 79–109.

Mayer, R. and Ulrich, B. 1978. Input of atmospheric sulphur by dry and wet deposition to two Central European forest ecosystems. Atmos. Envirn. 12: 375–377.

Meslow, E.C. and Wight, H.M. 1975. Avifauna and succession in Douglas-fir forests in the Pacific Northwest, pp. 266–271. In: Management of Forest and Range Habitats for Non-game Birds. USDA Forest Service, USDA Forest Service, General Technical Report WO-1. Washington, DC.

Monteith, J.L. 1981. Evaporation and surface temperature. Q. J. R. Meteorol. Soc. 107: 1–27.

Newton, I. and Moss, D. 1981. Factors affecting the breeding of sparrowhawks and the occurrence of their song-bird prey in woodlands, pp. 125–131. In: Last, F.T. and Gardiner, A.S. (Eds) Forest and Woodland Ecology. Institute of Terrestrial Ecology, Edinburgh, United Kingdom.

Nutter, W.L. 1979. Effects of plantations on the quantity, quality and timing of water supplies, pp. 351–367. In: Ford, E.D., Malcolm, D.C. and Atterson, J. (Eds) The Ecology of Even Aged Plantations. Institute of Terrestrial Ecology, Edinburgh, United Kingdom.

Nys, C., Stevens, P.A. and Ranger, J. 1990. Sulphur nutrition of forests examined using a sulphur budget approach, pp. 356–372. In: Harrison, A. F., Ineson, P. and Heal, O.W. (Eds) Field Methods, Applications and Interpretation. Elsevier Applied Science, New York.

Ormerod, S.J., Donald, A.P. and Brown, S.J. 1989. The influence of plantation forestry on the pH and aluminium concentration of upland Welsh streams: A re-examination. Envirn. Pollut. 62: 47–62.

Penman, H.L. 1967. Evaporation from forests: a comparison of theory and observation, pp. 373–380. In: Sopper, W.E. and Lull, H.W. (Eds) Forest Hydrology. Proceedings of an International Symposium on Forest Hydrology. Pergamon Press, New York.

Rochelle, J.A. and Bunnell, F.L. 1979. Plantation management and vertebrate wildlife, pp. 389–411. In: Malcolm, D.C. and Atterson, J. (Eds) The Ecology of Even-aged Forest Plantations. Institute of Terrestrial Ecology, Edinburgh, United Kingdom.

Rose, F. 1974. The epiphytes of oak, pp. 270–273. In: Morris, M.G. and Perring, F.H. (Eds) The British Oak. Classey, Faringdon, United Kingdom.

Stanhill, G. 1986. Water use efficiency. Adv. Agron. 39: 53–85.

Stoner, J. and Gee, A.S. 1985. Effects of forestry on water quality and fish in Welsh rivers and lakes. J. Inst. Water Eng. Manag. 39: 27–45.

Swank, W.T. and Douglass, J.E. 1974. Streamflow greatly reduced by converting deciduous hardwood stands to pine. Sci. 185: 857–859.

Swank, W.T. and Milner, N.H. 1968. Conversion of hardwood-covered watersheds to white pine reduces water yield. Water Resour. Res. 4: 947–954.

Thomas, J.W., Miller, R.J., Black, H., Rodiek, J.E. and Maser, C. 1976. Guidelines for maintaining and enhancing wildlife habitat in forest management in the Blue Mountains of Washington and Oregon. Trans. North Am. Wildl. Nat. Resour. Conf. 41: 452–476.

Welch, R.C. 1986. What do we know about insects in Scottish woods, pp. 95–100. In: Jenkins, D. (Ed) Trees and Wildlife in the Scottish Uplands. Institute of Terrestrial Ecology, Edinburgh, United Kingdom.

Whitehead, D. and Jarvis, P.G. 1981. Coniferous forests and plantations, pp. 49–152. In: Kozlowski, T.T. (Ed) Water Deficits and Plant Growth (volume VI). Academic Press, London.

New Forests **17:** 263–306, 1999.

On the sustainable productivity of planted forests

ROBERT F. POWERS
Pacific Southwest Research Station, USDA Forest Service, Redding, CA, USA

Key words: history of plantations, site potential, nutrition, soil properties, plantation yield

Abstract. Planted forests have more than a millennium of history and represent the world's best hope for meeting global wood requirements in the twenty-first century. Advances in genetic improvement, nursery practices, stand establishment, and tending, harvesting, and manufacturing have boosted plantation yields to a higher level than at any point in history. Despite this, forest managers face a mounting challenge to demonstrate that plantation productivity is sustainable. Tackling this challenge requires a sound understanding of the principles of forest productivity, how they apply to a developing plantation, and how they are influenced by management. In this paper criticisms of plantation forestry are discussed from the basis of world experience, and examples of productivity decline are described. Obvious declines are rare, and can be attributed to poor soil management. However, ambiguities exist and controversy will continue until sustainable productivity can be demonstrated conclusively. Proposed programs aim to provide the technical base needed for sound soil management and sustainable plantation productivity.

A history of the planted forest

> Planting cuttings of the fir along the roads.
> Enjoying the cool air in the moonlight of the future.
> (Zhu Xi, Song Dynasty, A.D. 1130–1200)

Forests exploited

Artificial forest regeneration has been practiced for more than a millennium. The need for planting is rooted in the clearing of forests – a vital step in the advancement of social order that began 10,000 years ago. As early as 3000 B.C., forests of Cyprus were felled to provide wood for the smelting of copper and silver (Hermann 1976). Solomon, nearly 3,000 years ago, supplied 80,000 fallers and another 70,000 skidders to move cypress and cedar timbers from the forests of Lebanon for the construction of the temple in Jerusalem (Winters 1974). In China, extensive clearing commenced about 2700 B.C., and continued unabated for 1,500 years until the Chou Dynasty (1127–255 B.C.) created in essence the world's first "Forest Service." With the demise of the Dynasty, however, deforestation resumed (Hermann 1976).

Europe remained forested through Roman occupation, but the first wave of clearing began with the break-up of the Empire in the fourth century and continued to the reign of Charlemagne about 800 A.D. Much of England remained forested beyond the Norman Conquest in 1066 A.D. Vast areas were set aside by Anglo-Saxon kings as royal game reserves to be guarded zealously against poachers and trespassers. The royal forest reached its zenith – about a third of the country – during the reign of Henry II in 1154–1189 A.D. (Hoskins 1955). Despite severe fines for doing so, peasants continued to clear the best lands for fields and homesteads.

The second great wave of European forest clearing surged in the eleventh and twelfth centuries, as wilderness was converted to arable land and as Germanic populations spread eastward. The single greatest cause of clearing was the demand for charcoal by the iron industry – an assortment of nomadic producers that advanced progressively into the forest (Nef 1952). However, the Salzburg forest ordinance of 1237 prohibited clearing forests for salt mine timbers so that forests would have time to repair themselves (Fernow 1911). By the fifteenth century, forest exploitation nearly brought some regions to economic collapse. Fuelwood shortages caused a decline in European iron production by the sixteenth century (Cramer 1984), thus triggering a timber famine in England in the 1600s. By the eighteenth century, much of Europe faced a wood crisis. France, where forests once flourished on four-fifths of the land, was only 14 percent forested by 1789 (Postel and Heise 1988). Of the 6.2 billion ha of forest and woodland believed to exist on earth at the start of the Holocene epoch, a third has been lost to crop production, fuelwood gathering, livestock grazing, and desertification (Postel and Heise 1988).

Forests restored

Western literature traces forest renewal efforts back, at least two millennia, to the time of the Caesars when Mediterranean gardens were planted to commemorate temples to the gods and to provide respite from the midday sun in the open Roman landscape (Sereni 1974). Later, temples were sacked, and sacred and functional plantings were degraded by the barbaric tribes that followed the collapse of the Roman Empire. But the Middle Ages brought some social stability, and with it an awareness that forests were not infinite. Deliberate forest renewal was practiced in Central Europe about seven centuries ago when feudal lords and communal forest owners sought by edict to reverse the loss of wood supply and game habitat (Barrett 1949). The first records of artificial reforestation in Europe date to 1368, when the City of Nuremberg seeded several hundred hectares of burned lands to pine, spruce, and fir (Toumey and Korstian 1942). John Evelyn, a Surrey landowner in England, published a plea for reforestation in his 1664 *Sylva* (Hoskins 1955). However, progress was slow because forest rules were based largely on observations and folklore of huntsmen. But the wood shortage crisis and the advent of systematic study in the eighteenth century caused dramatic change in the practice of forest renewal. Sound, reliable forest planting methods developed quickly. Germany, Austria, France, and Switzerland became leaders in reforestation. By 1789, shelterbelt plantings were introduced in the Russian steppes to control wind erosion and protect farmlands.

Planting has a shorter history in North America (Toumey and Korstian 1942; U.S.D.A. 1949). The first successful effort was an experiment in Massachusetts in the 1740s to grow oak for ship timbers. In 1819, *Pinus rigida* Mill. was transplanted from natural forests in Massachusetts to sandy soils too poor for agriculture. For the most part, tree planting (mostly oaks) continued sporadically into the mid-nineteenth century. The Homestead Act of 1862 brought a wave of settlers to the prairie states, and with them came strong appreciation of trees as the substance for shelter and fuel on an otherwise bleak landscape. By 1869, Kansas, Nebraska, and the Dakota Territory had tax exemption laws encouraging tree planting to boost the agricultural economy of the region. The modern sense of forest care and renewal as a general goal traces to Michigan and the formation of the first state forestry commission in 1867. Other state commissions soon followed. By 1872, Arbor Day had been established as a national encouragement to plant trees.

Throughout history, tree planting progress has been slow without governmental help (Stoeckeler and Williams 1949). The Timber Culture Act of 1873 (later repealed) offered individuals title to 65 ha of public land provided that at least one-quarter of it were planted with trees. By 1881 Congress had established a Division of Forestry (later, the U.S.D.A. Forest Service), but little attention was paid to seeding or planting until the national forests were transferred from the Department of Interior to the Department of Agriculture in 1905. In the West – and partly in response to timber losses to wildfire – reforestation began in earnest through the establishment of federal tree nurseries. Planting was expanded to all ownerships through the Clarke-McNary Law of 1924, which provided for cooperation between the federal government and states in the production and distribution of tree nursery stock. By 1920, no more than 200 ha of successful plantations existed in the southern United States, but organized reforestation was accelerating by 1926. Today, the forest bases of the United States and Europe are fairly stable.

World attention has been drawn to the harvesting of the natural forests of the tropics, but less attention is paid to the fact that plantation forestry is on the rise globally. Even in the United States, where even-age silviculture is being de-emphasized on public land (only about 15 percent of all plantings in recent years), the area of all ownerships in new plantings has averaged 1.1 million ha annually for the last decade, and has ranged from a low of 979,000 ha in 1993 to a peak of 1.37 million ha in 1988 (Moulton et al. 1996). One-quarter of this is in the West. Domestic demand for wood products will continue to rise in the twenty-first century. Therefore, timber management deemphasis on public forests and the reclassification of much natural forest to protected status places an unprecedented burden (and opportunity) on planted forests – particularly industrial plantations of the South and Pacific Northwest – to meet the needs of a wood-demanding public.

Modern challenges

A hallmark of a maturing industrial nation is an awareness that natural resources are finite and have values beyond that of economic exploitation. Attaining a comfortable standard of living marks a transition to a growing sense of stewardship and responsibility to future generations. This comes, not from a blinding flash of mass cognition, but through the work of individuals who, by their persuasive power, form groups committed to a central ideal. Assuming that arguments are well intended and reasonable, these groups may generate political force that culminates in laws meant to sustain or to restore the condition of the natural environment for the benefit of the nation. At worst, it leads to obfuscation, turmoil, tension, and heated confrontation.

The path to creating and maintaining productive plantations is not as direct as it might seem. In the United States there is a core of opposition to the prospect of plantation forestry that is rooted partly in personal belief and partly in science. The former concerns a sense of desecration when forests of nature are harvested, groomed, and regenerated artificially. The scarcity of natural forests in much of the developed world imparts in them an element of mystery and spirituality. In the minds of many, converting natural forests to plantations is more than merely a loss of "wilderness." It is a loss of innocence, too.

Personal beliefs and values are not in the purview of science and technology. However, plantation forestry also has been criticized on technical grounds. Natural forests are seen by some as centers of biotic diversity that, through scientific discovery, may promote the well-being of future generations – a diversity that is narrowed by conversion to simplified plantations. Others view natural forests almost as an organism comprised of interconnected and interdependent parts that promote long-term stability and high productivity. Proponents reason that simplifying or severing these connections threatens forest health and future productivity. Thus, managers are faced with technical challenges concerning planted forests. Do plantations differ from natural forests in susceptibility to natural disturbances? Do practices that increase short-term productivity sacrifice long-term values? Can productivity be sustained indefinitely? Objective answers will affect public attitudes. And public attitudes will influence policies on how forests will be managed in the future, irrespective of ownership. The technical question of sustainable plantation productivity with principles of forest growth, experience with managed forests, and findings of recent research is considered in this paper. The thrust is on planted forests of the Pacific Northwest, where controversy seems to be centered. However, the principles presented have broad application.

The conceptual basis for sustainable productivity in plantations

Discussing the question of sustainable plantation productivity demands both a clear definition of terms and a basic understanding of how plantations develop and how biomass is partitioned over time. This conceptual foundation gives us sound footing in evaluating sustainable productivity from the perspective of world experience.

Defining productivity

In its most fundamental sense, productivity is the primary synthesis of organic matter from carbon dioxide, water, and nutrients through the absorption of light energy. Generally, this "gross primary production" (GPP) is expressed as units of carbon fixed/unit area/unit time. Largely, the amount depends on the amount of leaf area (and chlorophyll) available to capture sunlight. Plants vary greatly in their photosynthetic capacity to assimilate carbon dioxide into organic compounds. Corn and tropical grasses are the most efficient, and are capable of increasing their photosynthetic rates to very high levels of light intensity (Devlin and Barker 1971). Forest trees are less efficient, and reach a photosynthetic saturation at relatively low light intensities.

Not all of the carbon fixed in GPP appears as the protoplasm, cellulose, and lignin we recognize as plant biomass. Instead, roughly 50–70 percent of fixed carbon is lost to respiration of foliage, woody tissues, and roots, and to allocation to such symbionts as mycorrhizal fungi (Ryan 1991). That which remains accumulates as biomass or is consumed by animals or disease. This is called "net primary productivity" (NPP), a rate expressed as biomass production/unit area/unit time. For land plants, production is affected strongly by the availability of water – not so much because of the role of water in the synthesis process itself, but more to maintain plant turgor by replacing water lost through transpiration when stomata are open for carbon dioxide uptake. Productivity also is limited by temperature controls of biological processes and by the availability of soil nutrients needed as electron carriers, in enzymatic reactions, and in the diverse compounds of plant cells.

Therefore, a good working definition of productivity with very broad application is the net production of biomass/unit area/unit time. Forests produce the greatest rates of NPP of any vegetation unit. Closed canopy forests of the tropics average between 6 and 35 mg/ha/yr, forests of temperate regions average 2–25 mg/ha/yr, and grasslands, 2–20 mg/ha/yr (Lieth 1975). Tropical forests have shown the highest rates of NPP, and suggest a ceiling of about 40 mg/ha/yr when climate is nearly optimal (Lieth 1975; Waring and Schlesinger 1985; Lugo et al. 1988). However, recent measurements of periodic increment in young, mixed plantings of *Paraserianthes* (*Albizia*) and *Eucalyptus* in Hawaii (Binkley et al. 1992) suggest that the limit may exceed 50 mg/ha/yr. In their survey of many tropical plantations, Lugo et al. (1988) found that mean annual increments of NPP averaged between 12 and 25 mg/ha/yr for the first 10–20 years of stand development, depending on soil water availability (Lugo et al. 1988). Although plantation data are scarce in the Pacific Northwest, rates for young-growth natural forests often fall between 10 and 38 mg/ha/yr (Grier et al. 1989), which suggests that the potential may even be higher under intensive management.

Traditionally, forest site productivity has been synonymous with stand growth rates. In fact, most of our historical knowledge of forest production rates is based upon simple measurements of tree boles – the part of greatest commercial value. Such cultural treatments as spacing and weeding can enhance stand productivity by allocating fixed site resources set by climate, soil, and relief to a target vegetation – trees, and to a target component – tree boles. However, they have nothing to do with site potential itself. A more useful view is to consider the production of each component of the entire tree – or, even more fundamentally, all of the vegetation on a site, regardless of physiognomy. This has the advantage of providing a more complete view of total carbon capture and how it might be managed to produce products useful to society.

Forest development and the partitioning of biomass

The uninterrupted trend for biomass production in a plantation or any even-aged stand follows a general pattern of increase from stand establishment to maximal production near crown closure when leaf area peaks and trees are

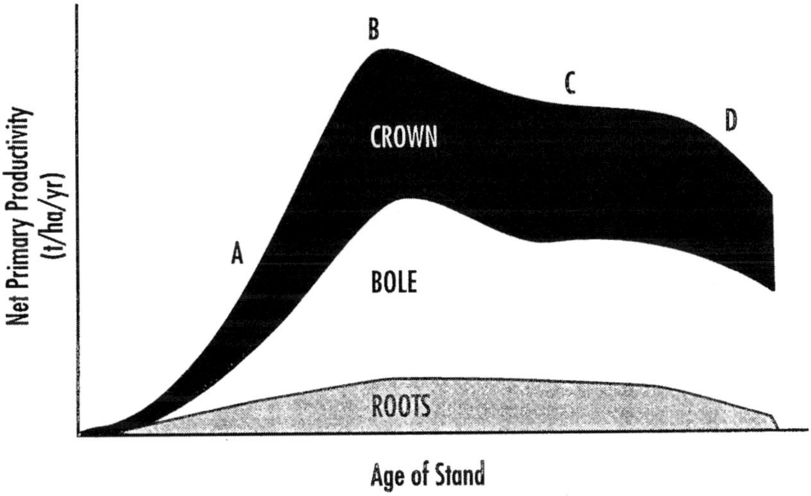

Figure 1. Typical pattern of even-aged stand development showing annual partitioning of productivity into roots, bole, and crown. Major phases are (A) rapid increases in productivity and nutrient demand as trees occupy site resources; (B) peak productivity and nutrient uptake at crown closure; (C) relatively stable productivity to maturity with increasing maintenance respiration (crown mass is fixed and much of the stand's nutrient demand is met through internal recycling); (D) rapid decline as stand senesces from natural causes (Waring and Schlesinger 1985).

fully exploiting the site (Figure 1). Throughout all phases of stand growth there is a close linkage between the mass of the crown and total production. Gross wood production is a linear physiological function of canopy light interception (Cannell 1989), as measured by the leaf area of a tree or stand. In general, the more leaf area, the more wood growth. Although it is modified by respiration (Gholz et al. 1990), the relationship between wood production and leaf area for a given species seems to be unaffected by water or nutrient stress. Although the relationship may be unaffected, however, a site's carrying capacity for leaf area or mass depends upon climate, soil moisture, and nutrient availability (Nambiar and Sands 1993). This carrying capacity is a fundamental property distinguishing one site from another. It is a precise measure of what foresters long have recognized as "site quality." Depending on the extent of limiting factors, it can be increased superficially by weed control or more profoundly by fertilization or drainage (Della-Tea and Jokela 1991), but not by thinning. Thus, climate, soil water supply, and nutrient availability determine site quality.

Crown closure also corresponds to the peak rate of nutrient uptake. Leaf area carrying capacity is reached, growth rates are high, and stands are fully taxing the site's ability to supply water and nutrients. After crown closure, water demand remains high because leaf area and the transpiring surface remains essentially constant. Despite a continuing high demand, the forest relies less on the soil for its nutrient supply. Once crown mass is fixed, between one-half and two-thirds of a plantation's annual needs for many nutrients, including nitrogen (N) and phosphorus (P), but not calcium (Ca), is met through internal recycling from older foliage to newer before leaf abscission (Miller 1984). Research in ponderosa pine (*Pinus ponderosa* Dougl. ex Laws.) plantations on a variety of sites in California (Figure 2) shows that roughly 60 percent of foliar N apparently has been retranslocated as amides and amino acids from needle fascicles when crown cover has reached 90 percent or beyond, thus leaving a residue at leaf fall of microbially resistant protein-polyphenol complexes. Compared to other nutrients in foliage, P is especially mobile, with about 80 percent apparently retranslocated as ionic phosphate and various organic compounds at crown closures of 50 percent and greater. That remaining at abscission consists mainly of phospholipids and insoluble calcium phosphate. In contrast with N and P, Ca accumulates in fascicles as insoluble pectates and oxalates even at very high crown closures. But, even for Ca, the proportion retained in senescent needles declines with increasing tree crown cover.

Internal recycling of nutrients is one of the main means by which forest trees cope with low availability of soil nutrients, and is a major nutritional distinction between perennial forest vegetation and annual agricultural crops. The fact that litterfall has higher nutritional quality at lower crown cover (Figure 2) suggests a more favorable dietary substrate for soil fauna during early stages of plantation development or in the period following a thinning. Total production of leaf litter and nutrient return would increase through crown closure, but the nutritional quality would lessen.

Typically, production rates are low when trees are young and crown leaf area is low. Much of the carbon assimilated annually is directed to production of leaves and the twig and branch system supporting them. As crown mass increases per unit area, production rates rise rapidly into an exponential phase that becomes sigmoid as the stand approaches the site's leaf area carrying capacity (Switzer and Nelson 1972; Waring and Schlesinger 1985). At crown closure, leaf area stabilizes and production rate peaks. Beyond crown closure, net production may decline slightly through maturity as an increasing proportion of photosynthate is used to maintain the respiring living matter accumulating in branches, bole wood, roots, and mycorrhizal networks. The pattern varies only by alterations in stocking from thinning or natural mortality, vigor reductions from fire, wind, insects, or disease, or from climatic vagaries. After maturity, productivity declines. This decline generally is ascribed to a lessened ability of a fixed crown mass to meet

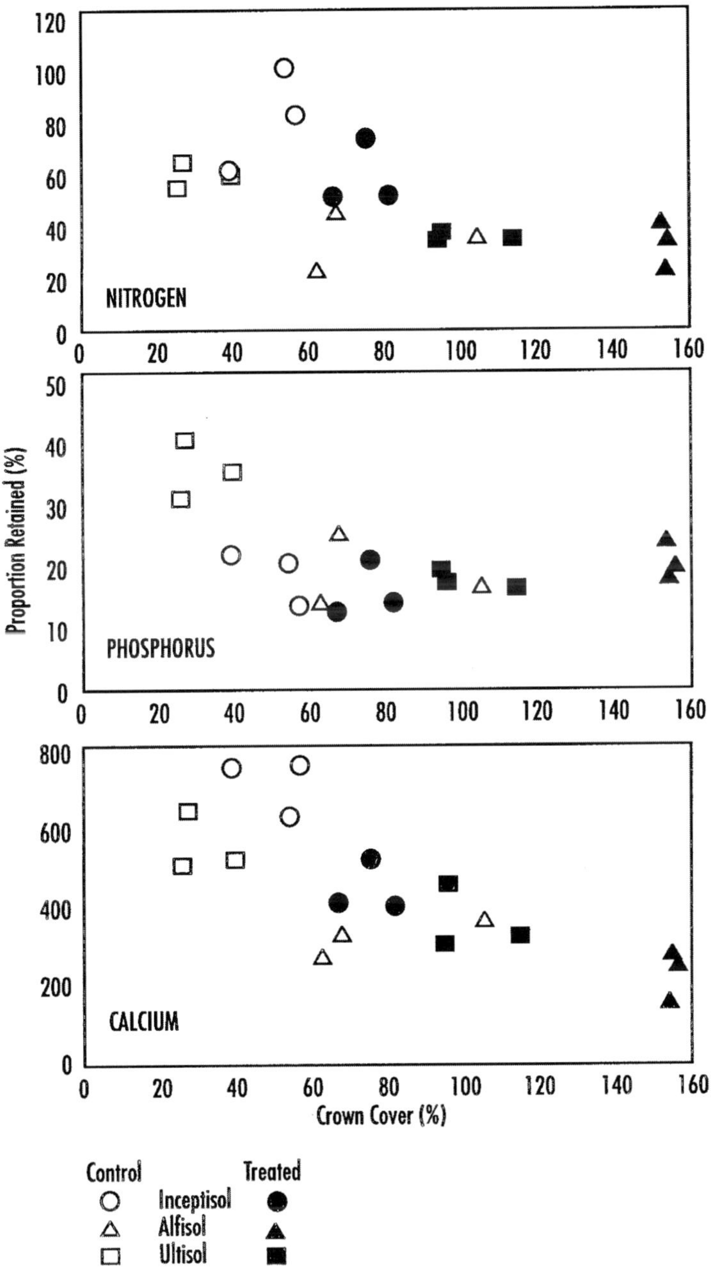

Figure 2. Percentage of initial foliar nutrient content remaining at leaf fall relative to percent crown cover of planted ponderosa pine. Results are from six treatment plots in each of three plantations on volcanic Inceptisols, Alfisols, and Ultisols in California. Open symbols are for control treatments. Filled symbols indicate repeated applications of herbicides and fertilizers.

the carbohydrate needs of respiring tissues in accumulating biomass (Waring and Schlesinger 1985). Thus, growth declines when maintenance respiration approaches assimilation. Because mycorrhizal fungi require steady supplies of carbohydrates from the host plant, mycorrhizal roots may decline because less carbohydrate is available for their maintenance. Reduced root surface leads to reduced water and nutrient uptake and to increasing stress, thereby reducing leaf area of individual trees. Ultimately, stand vigor declines, and insect and disease attacks become more severe. Canopy gaps then appear, leaf area decreases further, and production rate declines for the entire stand.

Ryan et al. (1997) discount the premise that increasing respiration is the principal cause of natural growth decline. Other explanations include decreasing nutrient availability, reduced assimilation because of increasing hydraulic resistance in tall trees, loss of leaf mass from wind abrasion in crowns, physiological aging of tissues, and increased reproductive output. Of these, increasing hydraulic resistance seems to be the most significant single factor (Ryan et al., 1997). All of these mechanisms may be important, and some can be influenced silviculturally. But, regardless of the actual causes, stand productivity clearly is dynamic (Figure 1). An important fact is that measurements made at different times during stand development may yield decidedly different values. Furthermore, measurements taken at a given time in stands of dissimilar stocking also will yield differing values – neither of which may equal the site's natural potential.

"Sustainable productivity," then, is the ability of managers to maintain NPP without a decline in rate. But the dynamic nature of plantation development demands a clear point of reference somewhere in the region B-C in Figure 1. A further need is to summarize NPP in all major forest components – not merely in tree boles. In principle, treatments which enhance short-term NPP in tree boles also may degrade the long-term productive potential of the site. This is important because a site's potential for NPP (its productive capacity) also reflects its potential for myriad uses and values. This productive capacity is a product of climate, soil, and topographic features acting upon vegetation that gives a site a unique potential. And as we shall see, it can be influenced by management.

Actual and potential productivity

What determines a site's natural potential? And, is it always achieved? Conceptually, productivity has two major elements. One is "current productivity," or the actual biomass produced by a forest over a recent period. Assuming that climate, soil, and genetic potential are not limiting, current productivity depends on stage of stand development (mature stands produce more than very young stands, Figure 1) and degree of stocking (fully stocked stands produce more than lightly stocked stands). Basically, this reflects differences in leaf area (Cannell 1989). Because dry matter production depends on photosynthesis, current productivity depends largely on leaf area of the vegetation. Current productivity can be measured at any time, but assessments at different stages of stand growth will lead to different conclusions.

The second important conceptual element is "potential productivity." Similar to current productivity, potential productivity relates to leaf area. However, it represents the site's potential for biomass production when the site is at full carrying capacity for leaf area. This occurs between crown closure and stand maturity (Figure 1, sectors B and C). A site's potential productivity is independent of stocking. It represents what could be produced if growth were constrained only by the factors of climate, soil, and genetic potential. Under the best of conditions, this may approach 50 mg/ha/yr in young, planted forests (Binkley et al. 1992).

Depending on management objectives and natural disturbances, a site's productive potential may or may not be achieved. Equally important is the fact that potential productivity is not immutable. Although it is a natural ceiling set by existing site resources, it can be raised or lowered through substantive changes in soil, climate, or genetics. Of these, climate and soil are the major factors. And, of these two, soil is most readily affected by management. Genetics has a lesser role in that some genotypes are adapted better to given site conditions than others. However, one should not assume that faster-growing genotypes increase the site potential. Instead, they merely may reach the site ceiling sooner than others (meaning that some genotypes are more adept at reflecting true site potential). Further, there can be genotypic variation in the way that photosynthate is partitioned into crown, bole, or roots (meaning that site potential is reflected differently among genotypes). But, popular impressions to the contrary, genetic improvement is not a panacea. It cannot compensate in any substantive way for poor climate or soil. What, then, are the consequences of management on plantation productivity?

Altering productivity

Both actual and potential productivity can be modified. Climate, soil, and genetic potential determine natural limits on site productivity, whereas stocking determines the degree to which this limit is achieved. For instance, the site potential for an understocked plantation is set by soil physical, chemical, and biotic components. However, low tree stocking or a high weed component keeps leaf area low and prevents the plantation from achieving its potential (Figure 3A). Improved stocking captures the site's leaf area carrying capacity so that current and potential productivity coincide (Figure 3B). Although

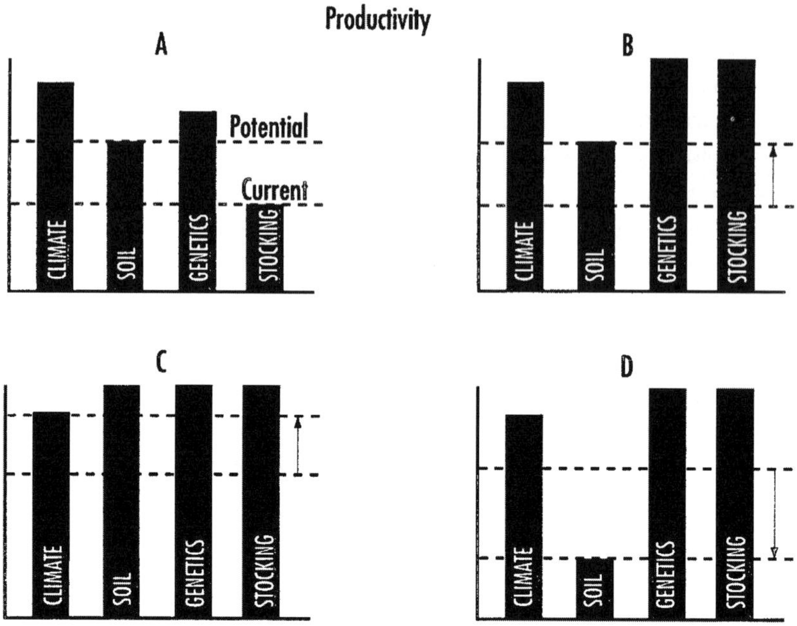

Figure 3. Relationship between current and potential productivity of a plantation as constrained by climate, soil, genetics, and stocking. (A) An understocked stand is performing at less than potential as limited by the natural properties of the soil. (B) Improvements in genetics and stocking increase productivity to the level constrained by the soil. (C) Soil amelioration (fertilization, drainage) raises productivity to a new potential set by local climate. (D) Both current and potential productivity are reduced through soil degradation.

genetics may have been improved as well, the limit remains set by the soil. Unless it can substantially change leaf area at full stocking or can improve photosynthetic efficiency in some fundamental manner, genetic improvement merely gets the plantation to the site limit faster. Conversely, diseases prevent a plantation from expressing its genetic potential. Improving soil properties, genetics, and stocking (Figure 3C) boosts potential site productivity to a higher plane constrained by climate. Alternatively, soil erosion, compaction, or nutrient drain may alter the site's potential so that productivity is degraded (Figure 3D). Superior conditions of climate, genetics, or stocking will not compensate for this, and operations are analogous to mining a nonrenewable resource.

Management often works within the fixed limits of natural potential productivity (Figure 4A). The degree of stocking or weed control determines the proportion of potential productivity that is captured by trees. But, depending on a site's resistance and resilience, this potential can be altered through soil modification – either enhanced through such treatments as fertilization

(Ballard 1984; Chappell et al. 1992), or degraded through soil compaction or erosion (Childs et al. 1989; Powers et al. 1990). Many forest managers – particularly industrial foresters – are philosophically aligned with agronomists who are not satisfied with the natural productivity of the land. They take an aggressive policy of capturing the whole of inherent site productivity or enhancing it further (the right portion of Figure 4B) by correcting soil fertility, drainage, or tilth limitations (Fisher 1984). In contrast, public land managers usually take a more conservative tack by working within the limits of natural productivity. Because the National Forest Management Act of 1976 requires such forests to be managed in a way that protects their long-term productivity (U.S.D.A. Forest Service 1983), national forest managers are concerned particularly with the left portion of Figure 4B. Forest management takes a more defensive posture. Management shall not degrade site productivity. Assuming that productive forests will always be valued by society and that productivity can be enhanced by understanding the controlling factors, why is there controversy over the value of planted forests?

The debate

Despite more than a century of world success in tree planting, the productive stability of planted forests still is questioned on a variety of technical fronts. Concern stems partly from agricultural experiences where repetitive cropping of corn or cotton without replacing nutrients has led to yield declines (Mitchell et al. 1991). Concern also traces to historical misconceptions about the influence of plantations of conifers (usually monocultures, sometimes exotics) on soil and site processes (Maser 1988). The recurrent themes are complexity versus simplicity and sustainable productivity in plantations.

Complexity versus simplicity

The stability of monocultures

Many ecologists believe that diversity is the foundation of stability in natural forests. It follows, they reason, that planted forests are inherently less stable because they are structurally and biologically simpler and less resistant to natural agents of disturbance. Therefore, plantation productivity also must be unstable and – most likely – unsustainable (Maser 1988). Although less dramatic, this concept is illustrated somewhat by entomological research. In one study, Berisford and Kulman (1967) surveyed infestations of Nantucket pine tip moth (*Rhyacionia frustrana*) incidence in loblolly pine (*Pinus taeda*

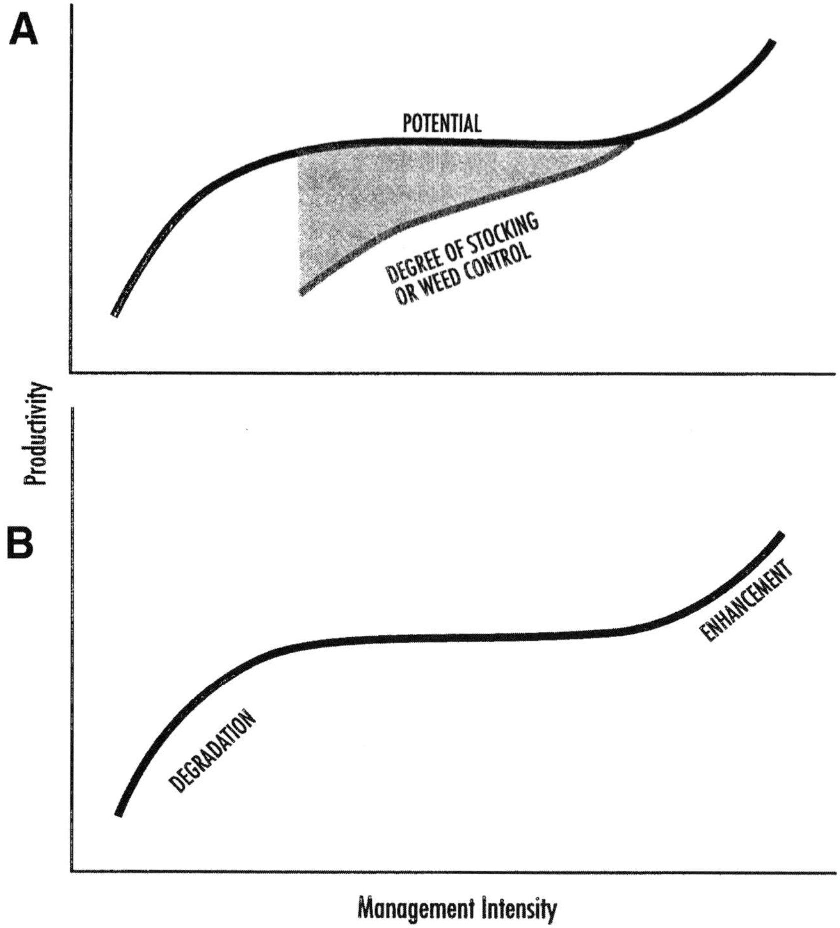

Figure 4. Relationship of management intensity to site productivity. (A) The degree to which the natural potential of a site is captured depends on stocking and weed control. (B) A site's potential productivity isn't static, but may be degraded by careless management (erosion, compaction), or enhanced by favorable soil treatment (fertilization, drainage, irrigation, subsoiling).

L.) stands in North Carolina. They concluded that infestations were greater in plantations than in old field stands, in widely spaced trees than at denser spacings, and in trees under lower competition from hardwoods. Although this seems to be an indictment against intensively managed pine plantations, subsequent studies show that weed control treatments also increased height growth, thus allowing infested trees to recover quickly (Ross et al. 1990). Infestations may be lower on water and nutrient stressed sites, but weakened trees also are less tolerant of damage (Meeker and Kulhavy 1992).

Thus, although intensively managed plantations may have higher rates of infestation, they also are more resilient.

Schowalter (1995) conducted a chronosequence survey of canopy arthropods in Douglas-fir (*Pseudotsuga menziesii* [Mirb.] Franco) forests in Oregon. He found that young plantations lacked many of the predator and detritivore taxa present in older stands, and concluded that herbivorous insects would not be regulated by their natural predators. The implication is that planted, intensively managed forests are more prone to pest damage than are natural forests. If this were true to any significant degree, insect problems would be common in intensively managed plantations. This is not the case.

The argument that plantations are inherently unstable is flawed for several reasons. First, it assumes that planted forests are biologically simple and that all are cast from the same "corn-row" mold of genetic, spatial, and structural uniformity. Actually, the principal distinction between planted and natural stands is in their intended use, for this governs the ways in which they are managed. Watershed protection plantations might be managed under extended rotations for continual ground cover and minimal disturbance beyond that of sanitation harvests to maintain plantation vigor. As outlined by Allen et al. (1996), plantations can be managed to enhance wildlife values through irregular stocking with canopy gap recruitment to ensure understory and overstory diversity in species and structure. Such plantations might be maintained longer than those meant for wood production. Finally, plantations meant for wood production are managed for profit. Rotations are shortened to capture a good return on the costs of investment. Hallmarks of such plantations are regular stocking and periodic entry to optimize merchantable wood production relative to cost of operation. But even these descriptions are too confining, because each general aim of plantation management contains a continuum of possibilities that lead to alternative futures. Flexibility in species mix and spatial arrangement makes plantations our best means for achieving a myriad of management options.

Under what conditions might monoculture simplification trigger disaster in planted forests? Perhaps the epitome of simplification exists in the extensive industrial plantings of radiata pine (*Pinus radiata* D. Don) introduced as an exotic monoculture to the geographic region of Australia known as the "Green Triangle." In 1961 the wood wasp *Sirex noctilio* was discovered in one large plantation in the state of Victoria. By 1972, *Sirex* had destroyed about 12 percent of the marketable volume despite the presence of both parasitoids and nematodes, and had killed two-thirds of the trees in severely infested stands (Madden 1988). Of this, about half of the mortality occurred in a single year. Damage was greatest in unthinned stands, and was most pronounced after a period of progressively lower summer rainfalls. Infestations declined after

1978. Earlier outbreaks reported in Tasmanian and New Zealand plantations of radiata pine reflected similar conditions of overstocking and a drought period interrupted by moderate summer rainfall or by thinning operations during the flight period of *Sirex*.

Short periods of tree stress, followed by respite, followed again by stress favor *Sirex* attack. Summer drought characterizes the radiata pine region of Victoria, and drought stress is accompanied by a shrinkage of phloem. Moderate summer rain rehydrates tissues, but if rehydration is followed by a second drying period, phloem shrinkage may release a pulse of water vapor and monoterpenes which attract *Sirex* during flight season (Madden 1988). Similar chemical pulses are emitted when thinning slash or pruning wounds are created during the flight season. The female wood wasp then bores into the xylem, breaks the capillary columns of high-tension water, and introduces mucus and fungi into xylem vessels. This intensifies water and nutrient stress, and leads to death. Although *Sirex* populations can be controlled by chemically attracting the wood wasp to trees inoculated with nematodes, careful timing of weeding and thinning should help maintain tree vigor and reduce risk – even in pure monocultures.

Powers and Ferrell (1996), in a designed stress experiment with pine plantations across an extreme range of sites in California, found no evidence of insect or disease problems, even at extreme intensities of weed control and fertilization. Outbreaks of forest pests generally trace to trees under stress from lack of management or poor timing of silvicultural operations, rather than intensive management per se. Collectively, trees stressed from overstocking because of a *lack* of management create the "forest health problem" that has captured the attention of politicians and the general public. Berryman (1988) presents an excellent review of forest insect outbreaks and management implications.

In the West, plantations managed for wood production often begin as monocultures, but they seldom remain so. California ponderosa pine planted as pure stands for timber production often contain varied understories of conifers and shrubs through crown closure and beyond. Although understory growth is reduced at high stocking densities, diversity persists (Oliver and Dolph 1992). Recently, Miller and Anderson (1995) compared six matched plantations and natural stands originating from wildfire in the coastal Douglas-fir region of Washington state. Although planted purely with Douglas-fir, many tree species had invaded the plantations. By 65 years, more than half of all living trees were other species. Nor are plantations necessarily less diverse than older, natural forests. Halpern and Spies (1995), studying forest chronosequences in the Oregon Cascades and Coast Range, found that changes in understory diversity were fairly short lived after harvesting and relatively few taxa were eliminated by soil disturbance or light burning. By crown closure, species richness on logged sites can recover to preharvest levels. At the stand or watershed scale, flora typical of old-growth understories were maintained.

Component linkages

A companion notion is that ecosystem components and linkages vital to forest health and stability are altered severely or eliminated entirely in planted forests, thus rendering these forests vulnerable to calamity because they are buffered poorly against natural disturbances. From this it follows that plantation productivity probably is unsustainable. Studies have not shown any clear linkages among species diversity, varied structure, and forest health or susceptibility to forest pests. For example, popular claims that large woody debris provides essential repositories of mycorrhizal fungal spores (Maser 1988) are not supported by any objective research. Furthermore, Moldenke (1992) found that young plantations entirely free of surface residues contained the same functional guilds of soil arthropods as occurred in adjacent mature, natural stands except for those of the forest floor (which had not yet developed in the plantations). In fact, arthropod density, diversity, and biomass all were higher on fertilized than on unfertilized plots in the plantations.

Studies on nutrient cycling have not found a scientific basis for the notion that cycles become more complex or efficient with succession (Vitousek and Reiners 1975). Certainly, linkages among components are more varied and structures are more complex in older natural forests than in plantations (Franklin 1993), but this pattern does not emerge until the second century after disturbance (Franklin and Spies 1991). Except for specialized niches, such as those for epiphytes characterizing old-growth forests and the animal communities dependent upon them, the multiple linkages regarded by some as vital to forest health and sustainability are seen merely as fortuitous redundancies by others. Despite heated proclamations on both sides, neither hypothesis has been tested broadly or with rigor.

Resistance and resilience

Ecosystem stability consists of two traits: resistance and resiliency (Webster et al. 1975). Resistance to change is an inertial force associated with large storage of organic matter and nutrients, long turnover times, redundancy, and a complex pathway in the cycling of matter favoring closed nutrient cycles. These are traits of fully stocked forests in more advanced stages of succession, and of soils with a high capacity to absorb stress and immobilize added chemicals either biotically or physically. Resiliency is a dissipative force that returns a forest back to near its original state following disturbance. Earmarks are rapid turnover and recycling of matter (the normal characteristic of planted forests). Although both traits contribute to stability, they are related somewhat inversely to each other (Webster et al. 1975). Those factors which increase resistance tend to decrease resilience, and vice versa. Compared to older natural stands, intensively managed plantations may have low accumulations of organic matter and less complexity (less resistance); however, they also have higher rates of nutrient turnover and the capacity to expand their crowns and increase their leaf areas when canopy gaps appear (more resilience). Therefore, planted forests are not necessarily less stable than natural forests.

Natural forests sometimes are said to be more resistant and resilient to severe natural disturbances. Wildfire illustrates a severe natural disturbance, but its effect for a given size of tree is no greater in plantations than in natural stands. In fact, given the historical policy of fire suppression in natural forests of the United States, damage from wildfire may be substantially less in well-tended plantations. Weatherspoon and Skinner (1995) studied effects of wildfire on 246 plantations and 874 other stands covering 14,000 ha in northern California. Damage was minimal in uncut natural stands, but was severe in partially cut stands lacking fuel treatment. Where slash had previously been burned in plantations, fire damage often decreased with distance from the edge of the burned natural forest. Plantation damage was spotty and varied where slash had been piled by tractor. Plantations lacking any fuel treatment burned uniformly and severely. Uneven-age management and residue retention in natural stands – a concept with its advocates (Franklin 1993) – may be a recipe for disaster in summer-dry climates without adequate fuel treatment.

That planted forests tend to have low resistance and high resilience is abundantly useful to management. In fact, treatments meant to boost plantation productivity are aimed deliberately at overriding resistance. For example, fertilizers are applied in doses great enough to override the buffering effect of weeds, microbes, and the nutrient-fixing capacity of soil. The aim is to achieve strong, positive, and predictable responses to treatment investments. Such silvicultural tools are not for the exclusive use of timber growers. The variety of intensive silvicultural practices available for regulating forest growth can be used to accentuate other values for those same plantations (Lippke and Oliver 1993). For example, treatments that create escape and hiding cover, improve both quality and quantity of forage for wildlife, and adjust the spatial arrangement of habitats can lend multiple values to plantations (Allen et al. 1996). Resiliency of planted forests also means that they respond in consistent and predictable ways to treatment. This allows managers to predict the magnitude and duration of treatment response, and thereby determine how treatments should be scheduled in an effective management regime.

Sustainable productivity in plantations

Although many arguments against forest simplification are not supported well by field experience, critics of plantation forestry may present more tangible arguments based on cases that suggest productivity decline. Can such cases stand the test of scientific objectivity? If real declines have occurred, can we infer the causes?

Spruce sickness

In the early nineteenth century, many of Central Europe's depleted areas and forests of hardwoods were converted to plantations of more profitable Norway spruce (*Picea abies* [L.] Karst.). By the second rotation of spruce, however, yields on some sites were lower than those in the first rotation. From this spread a belief that conifer monocultures degraded the soil (Wiedemann 1923; Ovington 1953; Grigor 1868), dogma that persists today (Sheppard 1986). More modern investigations show that "spruce sickness" was relegated to poorly drained lowland sites with heavy clay soils. There, old root channels from the original hardwood forest gradually plugged, and led to poor internal drainage. Waterlogging in the wet season confined spruce roots to shallow depths, thus leaving them perched and dry during drought (Krauss et al. 1939; Holmsgaard et al. 1961). On better-drained sites, spruce sickness either did not occur, or was relegated to lands already degraded by past practices. Clearly, the cause for spruce decline was quite explainable. Today, this would be known as "off-site planting." In his review of European research on growth changes attributed to plantings of spruce and pine, Stone (1975) concludes that declines often were more mythical than real. Where detailed records were kept, they indicated that growth generally was greater in second generations. Where declines did occur, they could be attributed to pre-existing differences in soils caused by preforestry land uses, to intermediate agricultural use, and to litter gathering which was practiced commonly for decades.

Radiata decline

These concerns resurfaced following a separate event in the southern hemisphere. Beginning about 1920, plantations of exotic conifers, principally radiata pine, were established in Australia on previously nonproductive scrublands of native species. By 1927, softwood plantings had been established in all states except Tasmania. Of these plantings, nearly 6,500 ha (41 percent of the total) were in South Australia which was considered the most climatically favorable region, and rotations were set roughly at 35 years (Gray 1935). Early success led to further planting with the aim of meeting all domestic wood needs and perhaps a surplus for export. By 1930 the planting area had doubled, and in the next 3 years it grew another 40 percent. By the 1950s the oldest plantations were being harvested and replanted. In some cases, thorough records were kept of first rotation performance, and that of the second rotation as well.

In 1966, Andrew Keeves published a landmark paper in which he compared changes in yield capability classes in first- and second-rotation stands of radiata pine on sandy soils of South Australia's Penola and Mount Burr Forest Reserves (Figure 5). Mapping units of the highest yield class in the first rotation had nearly disappeared in the second, and most mapped units had dropped by a yield class or more (mean annual increment declines of 30–60 percent). Concern spread quickly that pine monocultures somehow were poisoning the soil, negating a huge national investment in plantations. The yield decline between the first and second rotations was indeed real, and speculation abounded about the possible cause (Florence 1967). Similar concerns and possible causes of growth declines in second-rotation pine stands were expressed in New Zealand (Stone and Will 1965) and South Africa (Robinson 1973).

Using innovative methods of matched plots and stem analysis, Squire et al. (1985) showed that second-rotation decline probably could be eliminated merely by retaining logging slash and forest floor after harvest. In fact, slash retention produced greater early growth rates in second rotations than in first. Smethurst and Nambiar (1990) achieved similar results by weed control and N fertilization. Both studies show that the common practice of slash burning following logging led to weed development, soil drought, and reduced N availability on sandy soils. Organic matter retention produced soil moisture and temperature regimes favoring N mineralization. However, higher rates of N mineralization without weed control led to accelerated weed growth. This growth, coupled with declining rates of N mineralization as nutrient demand by trees increases (Figure 1, sector A), exacerbates nutrient deficiency and reduces growth in young stands – at least, on sandy soils (Smethurst and Nambiar 1990).

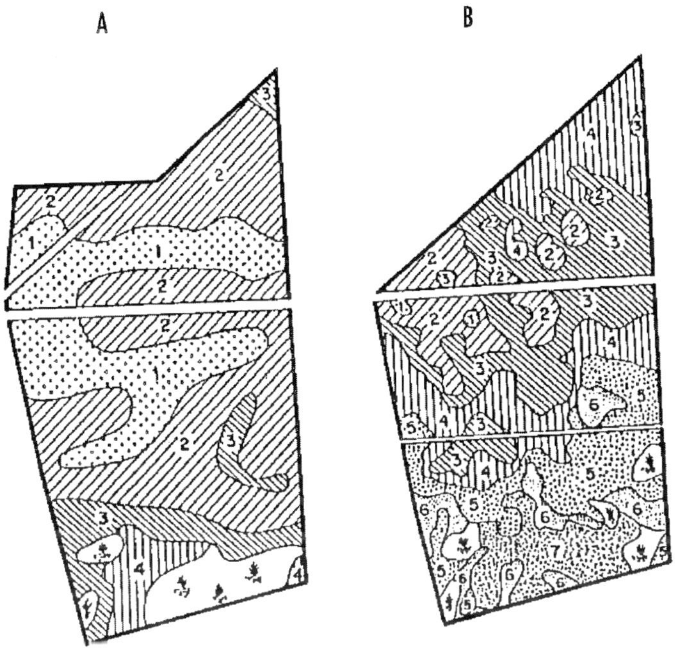

Figure 5. Site class mappings for first (A) and second (B) rotations of radiata pine within the Penola Forest Reserve, South Australia. Numbers 1–7 indicate site-quality classes between 1,500 (1) and 400 (7) m^3/ha over 50 years (Keeves 1966).

The southern pines

Opinions in the United States largely are based on chronosequence studies or retrospective analyses of current stands. For example, findings from repeated forest inventories in Georgia show a progressive decline in diameter growth of pine between 1956 and 1982 (Sheffield et al. 1985). However, declines seem to be restricted to nonindustrial private forest land where shrub and hardwood competition had increased from the absence of regular underburning. On more intensively managed industrial plantations, growth rates were stable or had increased (Sheffield and Cost 1986). Whether the latter is the result of improved genetic selection, better stocking, or weed control, or of maintenance or improvement in potential site productivity is unknown. In fact, one might question why plantation productivity is not greater in the southern United States, given decades of research in tree improvement, stocking control, and fertilization.

Repeated forest inventory may reveal trends, but it cannot establish cause and effect. A study in Louisiana suggests that declines in productivity of the next rotation can occur and may be caused by treatments meant to increase productivity in the present rotation. Tiarks and Haywood (1996) compared burning only, burning plus disking, and burning combined with disking and bedding before planting loblolly and slash (*Pinus elliottii* Engelm.) pines. The plantation was harvested after 22 years and replanted with the same species. Slash was burned, but mechanical site preparation was not repeated. After 10 years, standing volumes in the second rotation were 56 percent and 38 percent less than in the first rotation for loblolly and slash pine, respectively. Mechanical site preparation before the first rotation increased yields during that rotation, but apparently suppressed yields in the second.

Soil strength was measured in the second rotation, about 34 years after the original site preparation treatments. With few exceptions, strengths remained below 2 MPa throughout the upper 50 cm in the burned-only treatment. In the burned and disked plots, soil strengths exceeded 2 MPa in a continuous band at the 20–25 cm depth, and again at depths beyond 50 cm. Soil strength also was elevated in the burned and bedded plots, but the pattern was discontinuous. High soil strength is an element of compaction and reflects the force required by a small object (such as a root tip) to overcome particle-to-particle resistance as it moves through the soil. For many agricultural crops (Whalley et al. 1995) and forest trees (Sands et al. 1979), root activity declines dramatically as soil strength increases from 1 MPa to 3 MPa, a value at which root growth essentially ceases. In Tiarks and Haywood's (1996) study, increases in soil strength as a result of tillage may have exacerbated an incipient P deficiency, especially on the disked plots where the compacted pan was continuous.

How general is the second-rotation phenomenon noted in South Australia and Louisiana? Evans (1978), comparing first- and second-rotation growth rates on more than 100 matched plots of planted *Pinus patula* Schiede & Deppe in Swaziland, found few instances of statistically significant declines. Interestingly, earlier observations of the same plots suggested that a general decline had occurred (Evans 1975), which underscores the risk of hasty conclusions. In their recent review of the world experience, Morris and Miller (1994) conclude that evidence supporting the notion that long-term productivity generally declines in planted forests is scant.

The case for improvement

Agronomic studies have shown conclusively that yields of cereal and grain crops can be maintained or improved through fertilization, genetic improvement, and crop rotation (Mitchell et al. 1991). The same can be demonstrated in forestry. Plantation productivity – both current and potential – can be increased substantially through soil treatment. Classical examples include drainage and bedding that revolutionized pine planting on wet coastal sites of the southern United States (Pritchett 1979), fertilization (Ballard 1984),

N-fixation (Davey and Wollum 1984), and irrigation when combined with fertilization on dry, infertile sites (Snowdon and Benson 1992). If the change is permanent, it enhances site potential (the right of the curve in Figure 3B). The duration of such effects depends on treatment. On overly wet sites, improved soil aeration through drainage leads to greater growth and the effect remains permanent. However, growth gains resulting from irrigation last only as long as the treatment is sustained.

Productivity gains through fertilization are complex. Because of the general immobility of phosphate in the soil and its high rate of internal retranslocation, P fertilization effects may last for decades and extend perhaps from one rotation to the next. Fertilization rates are high (50–100 kg P/ha), relative to the quantities of P present in stand biomass (5–70 kg/ha; Ballard 1984). Because of its immobility in most soils, massive doses of fertilizer P favor extended uptake. Once absorbed, P shows high rates of internal retranslocation from needles at all stages of maturity (Nambiar and Fife 1991). After senescence, organic P, remaining in litterfall and root sloughage, concentrates in surface horizons in the vicinity of feeder roots where decomposition sustains P availability in a tight nutrient cycle.

In contrast, response to N fertilization seldom extends beyond a decade. As with P, a high proportion of N is retranslocated internally before leaf fall (Nambiar and Fife 1991). Unlike P, amounts typically applied (100–300 kg N/ha) are but a fraction of the mass stored in living vegetation and the forest floor (Ballard 1984). Some N may be volatilized within the first few days following fertilization with urea, particularly if temperatures are warm and granules have lodged in vegetation. Losses of 18–78 percent are common (Wollum and Davey 1975). Regardless of source, ammonia losses are exacerbated in neutral-alkaline soil. Also, surplus ammonium from fertilizer can be oxidized microbially to nitrate – an anion that can be leached through the soil profile beyond the influence of roots. Elevated nitrate concentrations in soil solutions extracted at 1 m depth can extend for up to 2 years following a single N fertilization (Miles and Powers 1988), and losses are greater on coarse-textured soils and on weakly developed soil profiles (Powers 1983). Biological denitrification to oxide gases also is possible if nitrate is abundant. However, competition for N by microbes and higher plants in aggrading plantations, relatively high C:N ratios, and the generally aerobic nature of the soil of conifer plantations suggest that denitrification rates will be acceptable (Davidson et al. 1990). Denitrification has not been studied extensively in plantations. It could be a significant factor if nitrate-forming fertilizers are applied during early stages of stand development when soil moisture is relatively high.

A high proportion of many nutrients absorbed following fertilization and translocated to tree foliage are retranslocated back into the phloem prior to leaf fall. The proportion retained is more a function of crown cover than fertilization per se (Figure 2). At senescence, foliar N concentrations may be somewhat higher in fertilized trees. California pine plantations fertilized repeatedly with N to a total of 1,074 kg N/ha had 11–35 percent higher N concentrations in needles at litterfall than did unfertilized plots in the same plantations. However, high rates of P, K, Ca, Mg, S, Zn, Cu, and B fertilization did not produce a similar effect. Thus, fertilization can result in N-enriched litterfall. Ultimately, some of this forms recalcitrant soil humus, essentially uncoupling a fraction of fertilizer N from the biological cycle. So, in contrast to the tight, closed cycle of fertilizer P, the fertilizer N cycle is comparatively leaky. Like a mechanical gear that wears a little as it turns, portions of N gradually abrade with each cycle.

Planting conifers with N-fixing species may enhance plantation growth under certain site conditions. On an N-deficient site at the Wind River Experimental Forest in Washington state, dominant Douglas-fir interplanted with (but a few years in advance of) red alder (*Alnus rubra* Bong.) was 20 percent taller after five decades than Douglas-fir planted in pure stands, and stand volumes were over 90 percent greater in the mixed planting when all species were considered (Miller and Murray 1978). Better sites show a lesser effect. On Hawaii's Big Island, stands of *Eucalyptus* (*Eucalyptus saligna*) interplanted with the N-fixing *Paraserianthes falcataria* near Hilo were equal to or larger than pure stands of *Eucalyptus* that had been fertilized repeatedly (DeBell et al. 1989), but mixed plantings led to no improvement on a drier site. Yet, strategies can be developed for drier sites. Nambiar and Nethercott (1987) demonstrate that annual lupine (*Lupinus*) seeded between rows of radiata pine on droughty, infertile sands serve the double purpose of excluding more persistent weeds while adding N and organic matter to the soil in South Australia. By the fourth year, pines so treated had twice the mass of pure pine controls. Thus, creative use of symbiotic N-fixation may be a way to improve site potential while avoiding chemical treatment.

Conclusions to date

Recent reviews (Powers et al. 1990; Morris and Miller 1994) conclude that direct evidence of productivity decline in managed forests is rare, whereas the converse seems common. Most records indicate sizable gains in actual productivity when planted forests replace natural forests, and suggest that potential productivity may be sustained as well. But most findings are short term and narrowly focused, and many are confounded by factors that add ambiguity. The rare instances in which true declines seem to have occurred can offer clues to help us adjust forest practices and guide our research programs. True declines point to changes in either of two site characteristics: biologically significant losses in soil porosity and in site organic matter (Powers et al. 1990). These two properties can be visualized as "gatevalves" that regulate more fundamental processes controlling site productivity.

Changes in soil porosity

There is no doubt that soil porosity has been reduced from compaction occurring during mechanized harvesting and site preparation (Greacen and Sands 1980; Froehlich and McNabb 1984). However, the consequences of this on potential productivity are largely unknown. The mechanism by which compaction can degrade productivity is the loss of soil macroporosity and the build-up of strength between soil particles to a level that restricts root growth. Natural soil porosity is a continuum of void sizes that varies with the mineral nature of the parent material and its degree of weathering, the tunneling activity of soil fauna and plant roots, and cycles of freezing and thawing. Reduced porosity lowers the exchange of water and gases between the atmosphere and the earth, the ease with which moisture and nutrients flow to plant roots and plant roots extend through the soil, and the biological activity of aerobic soil organisms. Lowered rainfall infiltration leads to surface runoff and soil erosion.

Few reports of soil compaction are based on controlled field experiments. Most are retrospective studies of operational trials where plot sizes are small (skid trails), true controls (noncompacted areas) are lacking, starting conditions are not clearly known, confounding factors such as weeds are present, and observational periods are short. On the basis largely of retrospective studies, Froehlich and McNabb (1984) conclude that every 10 percent increase in soil density spelled a 6 percent decrease in tree height growth. In contrast, Miller et al. (1996) describe a designed experiment involving planted Douglas-fir in coastal Washington state. Observations over 18 years led them to conclude that neither heights nor stem volumes differ among trees planted on tilled and nontilled compacted skid trails and in the bordering trafficked stands. Trees in all treatment classes were the same height (between 16 and 17 m at 18 years). Mean tree heights were more than four times the width of the skid trails (a small plot problem). Neither report demonstrates that trees in compacted and uncompacted plots were under similar degrees of weed competition, or that the "control plots" of logged-only areas had not been compacted to some degree during logging. Thus, reports from operation-level studies should be reviewed with some skepticism. Without demonstrable proof that compaction leads ultimately to lower yields, forest managers and equipment operators are skeptical about the worth of avoidance or mitigation (Miller et al. 1996).

The fundamental questions facing field compaction studies are these: (1) Did management activities affect soil porosity? (2) Did soil porosity changes affect actual and potential productivity? (3) Is the experiment rigorous enough to detect this? (4) Are these findings universal? (5) If not, why not?

Recent findings from California installations of the North American network of Long-Term Soil Productivity (LTSP) installations (Powers and Avers 1995) provide strong inferences into how soil compaction affects tree growth, and whether or not effects are universal. Two sites were examined: Challenge – a mesic site averaging 1,650 mm of precipitation annually, and Vista – a xeric site where precipitation is less than half that at Challenge. The Challenge soil is a deep, fine-textured clay-loam, weathered from metamorphosed volcanic rocks. That at Vista is shallower, sandy, and weathered from granodiorite. Following LTSP protocol, the mature natural forests were clearcut at each site and factorial combinations of organic residue removal and soil compaction treatments were applied to 0.4-ha plots. The LTSP treatments included complete removal of all organic surface residues and varying degrees of soil compaction applied as uniformly as possible across the entire treatment plot. Severe compaction increased soil bulk density by 28 percent at Challenge (Table 1), but only 8 percent at Vista (which had a higher bulk density initially). Planting holes then were augered to about 0.5 m, and several conifer species were planted, including ponderosa pine and white fir (*Abies concolor* [Gord. & Glend.] Lindl. ex Hildebr.) – two species that contrast strongly in their response to drought stress (Lopushinsky 1969). Competing vegetation was controlled on one-half of each treatment plot and allowed to develop on the other.

Measurements taken a few years after planting at Challenge indicated that, when weed competition was eliminated, the volumes of planted conifers averaged 42 percent less in the compacted treatment (Table 1), an outcome supporting the Froehlich and McNabb (1984) model. Pore size analysis and water retention studies showed that compaction had reduced surface soil macroporosity by 36 percent, and that the available water holding capacity (A.W.C.) had been reduced by 24 percent. This was substantiated by lower predawn plant water potentials for both pine and fir. Vista, with its coarser textured soil, responded differently. There, compaction reduced macroporosity by only 18 percent. Forcing sand-sized particles closer together reduced large voids and increased A.W.C. by 65 percent. The net result was a two-thirds improvement in seedling growth and a substantial improvement in predawn plant water potential in white fir – an effect contrary to the Froehlich and McNabb (1984) model.

Table 1. Ecological interactions of soil compaction and weed competition on soils of contrasting texture on two LTSP sites in California. Data are means of several measurements in August 1994 at the end of the fourth (Challenge) and second (Vista) growing seasons. Soil depth is 10–20 cm.

Characteristic species	Challenge (clayey texture)				Vista (sandy texture)			
	Not compacted		Compacted		Not compacted		Compacted	
	Weeds	No weeds	Weeds	No weeds	Weeds	No weeds	Weeds	No weeds
Seedling volume growth (1,000 cm)								
Pinus ponderosa	105	321	152	194	15	16	20	27
Abies concolor	4	18	6	10	2	3	3	5
Vegetative cover (%)	91	Trace	56	Trace	55	Trace	68	Trace
Soil bulk density (mg/m^3)	0.88	0.88	1.13	1.13	1.06	1.06	1.14	1.14
Total soil porosity (%)	67	67	57	57	60	60	57	57
Change in aeration (%)	0	0	−38	−34	0	0	−18	−18
Change in A.W.C. (%)[a]	0	0	−24	−24	0	0	+65	+65
Soil moisture 50 cm (%)	29	32[b]	30	33[b]	–[c]	–[c]	–[c]	–[c]
Predawn plant water potential (MPa)								
Pinus ponderosa	−0.88	−0.60	−0.87	−0.66	−1.61	−1.05	−2.05	−1.41
Abies concolor	−1.74	−0.54	−1.15	−0.63	−2.37	−1.13	−3.47	−0.93

[a] Available water holding capacity.
[b] Soil water potential exceeds −1.5 MPa.
[c] Not measured.

The LTSP experiment illustrates that, whereas soil compaction can reduce forest productivity, the effect is not universal – at least, in the short run. On droughty sites, any treatment improving the retention of available soil water without impairing aeration should improve forest productivity. As suggested by Miller et al. (1996), the biological significance of soil compaction must be tempered with regard to climate and soil textural differences.

Changes in site organic matter

Organic matter influences the interception and retention of solar heat by the soil. It dissipates the energy of falling water. It is the ultimate source of substances that bind soil particles together into stable aggregates that resist erosion. Through its carbon compounds, organic matter constitutes the energy source for soil fauna and microbes. Organic matter constitutes a concentrated reservoir of plant nutrients supplied to the soil through pulses in litterfall and root sloughage.

Without question, a sustained flow of organic matter from primary producers to the forest floor and into the soil is vital to sustained site productivity through its influence on the aggregation and physical stability of soil particles, soil water holding capacity, the activity of beneficial soil organisms, and nutrient supply (Jurgensen et al. 1990; Powers et al. 1990; Henderson 1995; Van Cleve and Powers 1995). However, virtually all findings from field experiments show that plantation survival and early growth are favored by removing surface materials during site preparation (Morris and Miller 1994). Therein lies the paradox. Why does practical experience often contradict theory? Is theory too simplistic, or are experiences too short sighted? The question should be examined from a first-principles position.

Temperature, moisture, and biotic activity in the surface soil are affected quickly by organic removal. Particularly, this is noticeable at high latitudes and elevations where surface organic residues insulate the soil. The resultant lowering of soil temperature means that water viscosity rises, soil faunal and microbial activity falls, and nutrients are less mobile.

Studies in boreal, interior British Columbia *Picea* forests (T. A. Black, unpublished; Fleming et al. 1994) showed that soil beneath scalped surfaces was as much as 4 °C warmer during the growing season (Table 2) and had fewer weeds than unscalped controls. In another study in interior British Columbia, both scalping the forest floor and mounding surface materials into raised planting beds improved the initial growth of planted *Picea engelmannii* Parry ex. Engelm. × *P. glauca* (Moench) Voss, but only the mounding treatment (which retained and incorporated surface organic matter) produced appreciably larger seedlings after 27 months (Bassman 1989). Soils remained warmer and better drained within mounds. Similar results were shown for

Table 2. Effect of presence (present) or absence (absent) of forest floor litter on the temperature and moisture in the surface 15–20 cm of soil in boreal and temperate second-year plantations (*Source*: T. A. Black, Department of Soil Science, University of British Columbia, Vancouver, unpublished; Fleming et al. 1994; and R.F. Powers, unpublished).

Month	Soil temperature (°C)				Soil moisture (%)			
	Boreal forest		Temperate forest		Boreal forest		Temperate forest	
	Present	Absent	Present	Absent	Present	Absent	Present	Absent
April	2.1	4.5	12.8	13.3	31	33	34	34
May	6.6	7.7	17.8	21.5	35	34	26	14
June	10.3	12.4	19.0	22.8	26	26	21	13
July	12.6	16.5	20.2	24.6	20	25	25	18
August	10.1	11.4	18.5	23.7	25	27	20	13
September	9.8	11.6	17.5	21.0	25	26	15	12

Pinus monticola Dougl. ex D. Don on cool, dry sites in northern Idaho (Jurgensen et al. 1990). After 3 years, soil N availability was ten times greater for mounded and control treatments than for scalped treatments, and seedlings growing on mounds were twice as large as in any other treatment.

The same insulating properties of surface residues that retard tree growing processes in cold forests produce a contrasting and beneficial effect in warm, dry regions. On a temperate LTSP site in California's Sierra Nevada, surface soils remained 3–4 degrees cooler throughout the growing season where logging slash had been retained, and the period of plant-available soil moisture was extended for several weeks (Table 2).

Temperate forest studies show that substantive or persistent removals of surface organic matter can trigger growth decline under certain conditions. On marginally fertile, pumice soil in New Zealand, Dyck and Skinner (1990) found that overall productivity in a windrowed *Pinus radiata* plantation with minimal displacement of topsoil was only two-thirds that of an adjacent plantation where slash had been left. Although soil compaction may have been a contributing factor to poor growth, foliar N, Mg, and B concentrations in trees between windrows were at marginal levels. Elsewhere in New Zealand, litter raking repeated annually for 26 years and unconfounded by machine compaction led to depressions both in total soil N and C and in extractable soil P and Ca, and increases in soil bulk density (Dyck and Skinner 1990). In Washington state on uncompacted sites of moderate to marginal fertility, planted Douglas-fir were nearly one-third shorter after 10 years where all surface residues had been removed than they were where residues had been retained (Compton and Cole 1991). Accelerated growth after fertilizing these plots at 5 years suggests that removing organic matter induced N deficiency in this temperate forest.

A progressive view of the value of surface organic residues is that value depends very much on climate. At higher latitudes, anything that reduces soil temperature reduces productivity. Surface residues accumulate and insulate the soil. There, soil temperature is lowered and shows little fluctuation, and biological processes in the rooting zone are slowed. Moist sites remain wet and aeration may be impaired. Such soils also warm slowly in the spring. On better drained sites, water stress may develop if the high viscosity of soil water (16 percent greater at 5 °C than at 10 °C) prevents soil supplies from replacing transpirational water losses. Barring significant disturbance from fire or mechanical operations, productivity will decline as surface residues accumulate. In contrast, surface residues on warm, xeric sites reduce evaporative losses of soil moisture in young, open stands. Residues also keep soil temperatures in a range more favorable for microbial activity and the release of organically bound nutrients. Obviously, slash must be modified if wildfire or insect risks are high. Alternatives include low-intensity burns, mechanical removal of some of the fuel load, and chipping residues either to provide a uniform mulch or to concentrate the chips into piles. On more mesic sites with less fertile soils, loss of surface residues will likely lead to deficiencies of N and P as canopies close and nutrient demand peaks (Figure 1B). However, special care must be taken to overcome problems of planting through slash and thick forest floors.

Although organic matter replenishment undoubtedly is crucial to sustained productivity in all ecosystems, its significance to important soil and site processes hinges on decomposition and on the climatic factors controlling it. Therefore, guidelines for organic matter retention during harvesting and site preparation operations must consider the overriding influence of climate.

Powers et al. (1990) created a conceptual model that indicates how soil porosity and site organic matter regulate net primary productivity within limits set by genetics and local climate (Figure 6). To some degree, all forest management activities affect one or both of these properties. The question, of course, is how much disturbance is too much? And how is this conditioned by climate and soil type? Although direct evidence of declining productivity in planted forests is rare, evidence of superior performance – at least, in the short run – is abundant. Forestry studies seldom are designed specifically to answer long-term questions, and short-term findings can be misleading (Evans 1975, 1978). Overall, the rarity of precise long-term records such as those for South Australia's Penola Forest (Figure 5) has hampered our understanding of the long-term effect of planted forests. The lack of a conclusive verdict merely

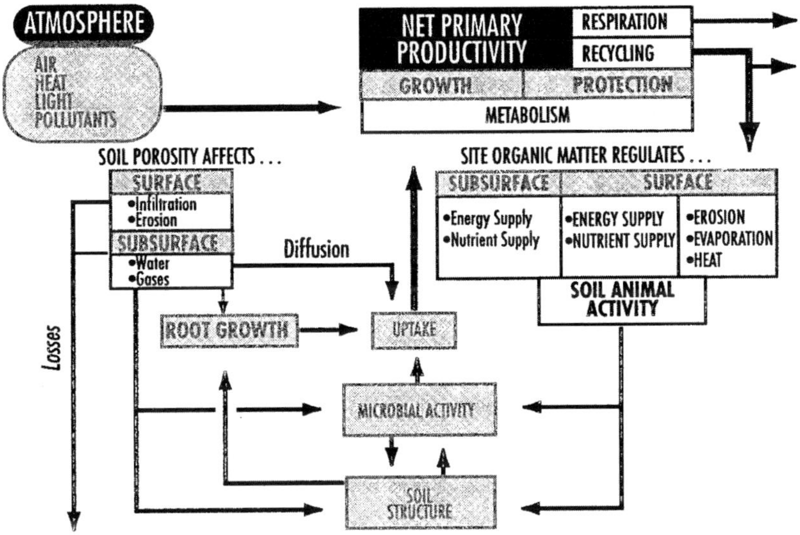

Figure 6. Conceptual model of the roles of soil porosity and site organic matter in regulating the processes controlling site productivity within the limits set by climate and genotype (Powers et al. 1990).

may mean that the hypothesis has not been tested rigorously. Until it is, critics of plantation forestry will continue to oppose modern forest practices.

Obtaining reputable evidence

Basically, there are three approaches to obtaining objective findings on the sustainable productivity of planted forests. The first is to compare cumulative patterns of plantation productivity in consecutive rotations. The second is to do the same with matched stands differing mainly in how they originated. The third is to conduct experiments to test critical hypotheses about planted forest productivity. Growth patterns that were superior, inferior, or equal to those for adjacent or previous stands would suggest improved, degraded, or stable site productivity (Figure 7), and causes might be inferred.

Paired stand comparisons and ambiguity

The direct way to detect productivity change is to compare growth patterns in an existing plantation with those for an adjacent natural stand or for previous stands growing on the same site. In western North America, plantations are first-generation stands that replaced natural stands or brush fields. Previous natural stands often are overmature, riddled by pests, or decadent. Trying

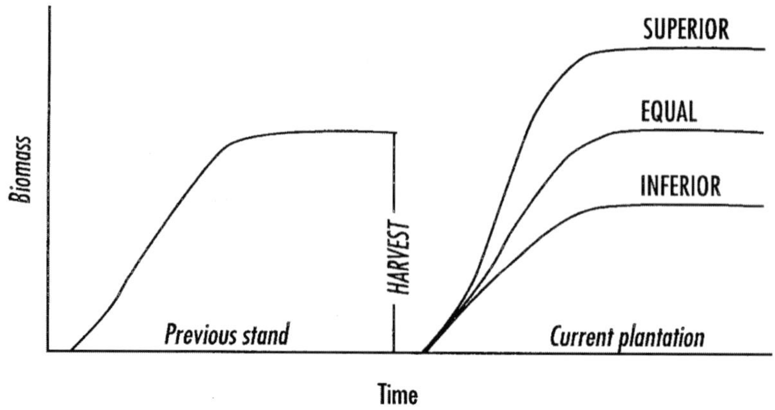

Figure 7. Growth patterns of current plantations can be compared to patterns from previous stands on the same site. Superior growth in the current plantation could indicate site improvement. Inferior growth could indicate site degradation.

to reconstruct past patterns of productivity is an exercise in extreme futility. Sites converted to plantations from grass or shrub communities commonly lack a historical record of tree growth, and site carrying capacity can be estimated only crudely from soil or environmental variables (MacLean and Bolsinger 1973). Therefore, successive rotation comparisons possible in the southern United States, Europe, Australia, and New Zealand are not yet feasible in the West. Furthermore, consecutive stand comparisons are inappropriate if stands differ greatly in structure, stocking, or genotype, or if climate differs appreciably between rotations. In the Pacific Northwest, the second alternative is to turn to productivity comparisons between adjacent stands of natural and planted origin.

This is not easily done. Natural stands vary immensely in age distribution, stand structure, and management history, and valid measures of potential productivity are difficult or impossible to obtain. The standing volume, biomass, or leaf area in irregularly structured natural stands preceding plantations rarely are practicable measures of a site's carrying capacity. Such data are physically difficult to collect in multilayered, heterogeneous forests. Also, stocking is irregular, stands may be senescing, and productivity may be declining. Even-aged natural stands offer mensurational advantages, but may be outside the period of relative stability in current productivity (Figure 1, sectors B-C). For example, they may be understocked at the time of harvest (Figure 3A) for reasons that have no bearing on the site's potential. Reconstructing growth patterns in natural stands via stem analysis has limited value because of uncertainties about stocking and crown conditions in the past. Moreover, genotypic differences between natural and planted stands may suggest increases or decreases in potential productivity that are unrelated to the site itself (the actual cause may be genotypic adaptation). In essence, comparing natural stands with plantations risks comparing apples with oranges. Only rarely do planted and natural stands of similar ages and management history exist side by side.

Two recent reports from the Douglas-fir region do seem to meet the requirements of sound comparisons. Miller and coworkers conclude that planted stands are at least as productive as those regenerating naturally when their disturbance and management histories are similar. Stand volumes at midrotation in the Cascades of western Washington and Oregon were 41 percent greater in plantations than in stands regenerated naturally (Miller et al. 1993). In older stands bordering Puget Sound, total volumes were essentially identical in planted and natural stands (Miller and Anderson 1995). The principal difference was that growth centered on Douglas-fir in the planted stands and on *Tsuga heterophylla* (Raf.) Sarg. in the natural stands. Although the possibility exists that planted sites have been degraded but that improved cultural treatments have masked the effect, the similarity of paired stand histories reported by Miller and colleagues argues strongly that this is not so. Such careful pairings are not common. Too often, such comparisons are marred by uncertainty about past conditions and by confounding factors that may have influenced stand development (Powers 1989; Powers et al. 1994). Generally, findings from paired stands studied retrospectively are ambiguous.

Eliminating ambiguity through controlled experiments

Two interrelated steps are proposed for eliminating ambiguity. The first is to recognize and control extraneous factors that may cloud experimental results. The second is to conduct standardized, powerful experiments designed to get to the root of the most critical questions.

Weeds as "demonic intruders"

The appearance of one or more unplanned, unwanted, and often unrecognized factors that can influence the outcome of a study is known as "demonic intrusion" (Hurlbert 1984). Generally, this traces not so much to demons from hell as to the experimenter's lack of foresight and to inadequate experimental control. A classic example of demonic intrusion is weed competition. Weeds often are ignored in Northwestern studies. However, their influence on early plantation growth can be profound – particularly in summer-dry climates.

Weeds have a surprisingly strong effect on nutrient availability in plantations. The presence of weeds can completely block tree response to fertilization, even when trees are under severe nutrient stress (Powers 1983).

Table 3. Effect of competing vegetation on elemental concentration in ponderosa pine needles and relative volume growth at 5 years (*Source*: Powers and Ferrell, 1996).

Site index (m)	Annual precipitation (mm)	Competing vegetation	Foliar concentration of N (g/kg)	P (g/kg)	K (g/kg)	S (mg/kg)	Al (mg/kg)	Relative volume growth (%)
17	1,015	Present	8.8a[a]	1.0a	6.7a	574a	159a	100a
		Absent	10.2b	1.2b	8.7b	659b	154a	307b
23	1,140	Present	9.5a	0.7a	4.4a	595a	161a	100a
		Absent	12.8b	0.8a	5.6b	748b	163a	240b
30	1,780	Present	11.3a	1.0a	6.3a	767a	176a	100a
		Absent	11.0a	1.0a	6.5a	773a	178a	152a

[a] Column means for a site quality class followed by the same letter do not differ significantly at $p = 0.05$.

Messier (1993) shows that removing ericaceous shrubs from young cutovers on Vancouver Island, British Columbia increased both N and P availability in the soil. In California, elimination of weeds from ponderosa pine plantations not only increased plant water potential during summer months, but also improved nutrient uptake (Powers and Ferrell, 1996). On the poorest and most droughty sites, weed control led to significant increases in foliar concentrations of all nutrients measured. In some cases, concentrations were raised above deficiency levels. On the best sites, foliar nutrient concentrations were unaffected by weeding (Table 3), but this does not mean that nutrient availability was unaffected. Rather, it suggests that availability and uptake kept pace with biomass increase (as indicated by 52 percent greater volume growth). As Nambiar and Sands (1993) point out, any treatment affecting water availability affects nutrient availability as well.

Weeds also can confound the effects of other treatments. This is illustrated by the LTSP experiment cited previously at Challenge and Vista (Table 1). Compaction increased soil bulk density from 0.88 to 1.13 mg/m^3(28 percent), which predicts about a 20 percent loss in height growth according to the model of Froehlich and McNabb (1984). When weeds were present, tree heights were as great or greater on compacted plots than on plots that had not been compacted. This contradiction to conventional wisdom can be explained by differential weed competition. Compacted plots also had one-third less weed cover (Table 1), meaning that trees there had less weed competition. Less weed competition implies greater moisture availability, which is verified by higher predawn water potentials in tree seedlings on compacted plots. Thus, the presence of weeds can mask the actual impact of soil compaction.

On plots free of weed competition, tree growth was substantially greater where soils were not compacted, predawn potentials were lower, and the true effect of soil compaction on this clay-loam was revealed.

Standardized experiments

Ironically, the ancient and noble practice of forestry has a dearth of long-term records concerning sustainable productivity. Until there is broad, convincing evidence that plantation productivity is sustainable, controversy and confrontation will persist. The subject is not academic. From an economic perspective, North America will rely increasingly on plantation growth to fill the supply and demand gap caused by harvest reductions in older, natural forests. From the political viewpoint, managers will face increasing challenges to prove that their practices are ecologically sound. Therefore, we need an objective means for measuring long-term changes in potential site productivity of managed forests. Morris and Miller (1994) propose three criteria:

1. Tree growth differences must be attributable to true changes in site conditions, and not merely the partitioning of site resources.
2. Substantive, long-term effects must be considered, because short-term trends can be misleading.
3. Adequate experimental control must be established.

In 1989 the U.S.D.A. Forest Service launched the Long-Term Soil Productivity program to tackle directly the question of sustained productivity (Powers and Avers 1995). LTSP is predicated on the principle that the fundamental processes controlling site productivity involve interactions between soil porosity and site organic matter – the conceptual model in Figure 6. The LTSP scientists concluded that these properties are the key variables affected by management. Accordingly, they agreed to a common experimental design that creates gradients in soil porosity and site organic matter after harvest. Realizing that no single answer will fit all situations, the design team developed guidelines that could be adapted to specific conditions of soil type and climatic regime. The experimental design for LTSP creates gradients in soil porosity and site organic matter following harvest. The result is a range of stress extending from minimal to extreme that is meant to encompass management disturbances now or in the future. The 3×3 factorial design is shown in Figure 8A. This simple but elegant design affords a mathematical response surface covering many possible combinations of disturbance. Furthermore, it meets all of the criteria of Morris and Miller (1994).

Major soil types meeting specific criteria are identified on public lands within major commercial forest types and climatic regions. Stands are harvested carefully, biomass and nutrient content are determined by major

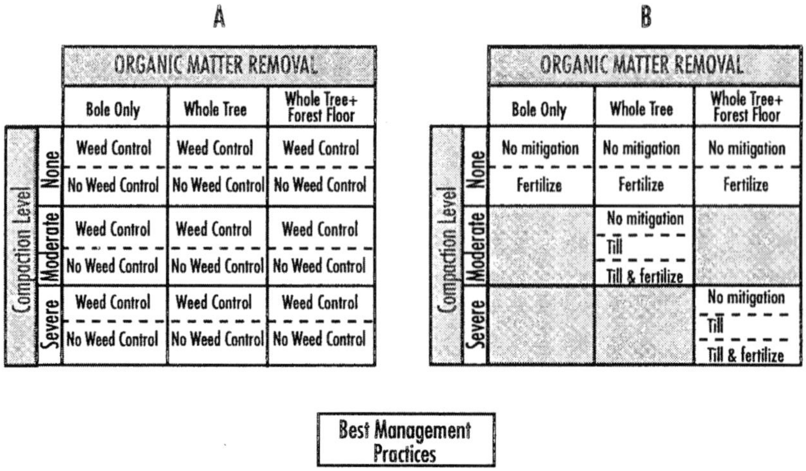

Figure 8. Two field designs for experiments on the long-term impacts of management practices on site productivity and the processes controlling it. (A) the standard LTSP design used by the U.S.D.A. Forest Service. Each plot is 0.4 ha with vegetation control/no control as a split plot. (B) An alternative design for satellite studies of amelioration and mitigation after harvest.

component, and the standard treatments (Figure 8A) are installed on plots measuring at least 0.4 ha. This plot size reduces edge effect problems inherent in growth studies on smaller plots maintained for many years. However, it does require a substantial investment in land, labor, and capital.

Initial soil conditions are measured on each plot. Then the site is regenerated with the tree species suited to the area. To avoid confounding weed competition with compaction effects, one-half of each treatment plot is kept weed-free. This split-plot design ensures that regional vegetation will develop naturally on the second half, thus yielding information on the long-term value of a diverse flora. Periodic measurements of vegetation on all plots provide direct comparisons of productivity as measured by volume, dry matter, and leaf area. Plots with "bole only removal, no compaction" (Figure 8A) serve as controls for testing the effects of all other treatments. Major soil properties (density, porosity, strength, organic matter and nutrient content, and moisture availability) also are measured at regular intervals, and continuous meteorological records are kept. Thus, both relative and absolute measures of productivity can be related to changes in soil properties as influenced by treatment and local climate. Each study site will be followed for a full rotation to overcome early trends that may change with time. Findings also will be compared to "best management practices" in operational plantations established nearby. To date, more than 60 installations exist across North America. In the West, nineteen are in the mixed-conifer forests of California, Oregon, and Idaho, and twelve others have been regenerated to *Picea glauca, Pinus contorta* Dougl. ex Loud., or *Populus tremuloides* Michx. by the Ministry of Forests in interior British Columbia.

The impetus for the LTSP study is the legal requirement established by the National Forest Management Act of 1976 that national forest lands be managed in ways that do not impair their long-term productivity (U.S.D.A. Forest Service 1983). For this reason, we have focused on defining the site degradation portion of Figure 4B, but site enhancement should be studied as well. In recent years, LTSP scientists have included mitigative and ameliorative treatments (generally tilling and/or fertilization of supplemental plots) at several locations, but prospects are dim for extending this costly design to new sites or "retrofitting" them to old ones.

A promising alternative would be to forge new partnerships with the forest industry by extending related experiments of joint interest to satellite installations on private lands. Some (but not all) treatments should be in common with LTSP to forge a direct link, and measurements should follow a similar protocol. New treatments involving mitigation and amelioration would be added in order to address issues of site enhancement and recovery from negative impacts. Conceptually, this could be done by modifying the LTSP design to somewhat smaller plots, by eliminating some of the factorial cells of a more academic nature, and by incorporating a standard suite of mitigative and ameliorative treatments (Figure 8B).

As proposed, treatments in common with LTSP are the three levels of organic removal without compaction, as well as intermediate and extreme levels of organic matter removal combined with compaction. Amelioration (site enhancement) could be addressed through fertilization of the uncompacted plots. Following compaction, mitigation (recovery from disturbance) could be tested through soil tilling both with and without fertilization. Cooperators likely will be tempted to test particular types of harvesting or site preparation equipment. However, the goal is not to test operational practices per se, because current operational or innovative practices will surely become *passe*. Obsolescence cannot be a part of long-term studies. Rather, we should commit to the goal of understanding the mechanistic pathways that lead to site productivity.

Other standardized experiments are possible and some are underway in the pine region of the southern United States and on degraded soils of the tropics (Powers et al., 1996). However, I believe that the experiments depicted in Figure 8 stand the best chance of meeting all the requirements of Morris and Miller (1994) and addressing the myriad questions facing plantation forestry in the future.

Epilogue

Planted forests are the world's best hope for meeting global wood requirements in the twenty-first century. In the West, intensive management practiced on lands already in plantations, or to be planted following stand-replacing wildfires or type conversions, will help supplant the reductions in annual wood harvest from natural forests. Thanks to advances in genetic selection, nursery practices, site preparation, planting techniques, stand tending, harvesting, utilization, and manufacturing efficiency, merchantable plantation yields stand at historical highs. But high yields may be short term if the soil support system is degraded through carelessness or ignorance. Plantation yields cannot be sustained, much less increased, unless the productivity capacity of the soil is maintained or enhanced. Understanding the impact of management practices on potential productivity is a responsibility of all land stewards and a necessity for sustainable forestry. Solutions will only be found through cooperative, integrated research programs which transcend agency and political boundaries to serve the greater good.

Central to the success of such a program is philosophical commitment by both scientists and administrators to make such programs work. Once treatments have been installed, commitment must be made that sites will receive at least the minimum maintenance needed to protect their integrity. Oversight must be provided by a cadre of scientists and administrators who believe in the worth of the effort and who will strive to ensure its success. In their review of forestry research programs around the world, Powers and Van Cleve (1991) conclude that all successful long-term programs are founded on two key elements. The first is core commitment to achieving the research goal, and the second is that the program address issues of continuing social relevance. Certainly, the issue of sustainable productivity of planted forests meets these criteria.

References

Allen, A.W., Bernal, Y.K. and Moulton, R.J. 1996. Pine Plantations and Wildlife in the Southeastern United States: An Assessment of Impacts and Opportunities. U.S.D.I. Information and Technology Report 3. National Biological Service, Washington, DC.

Ballard, R. 1984. Fertilization of plantations, pp. 327–360. In: Bowen, G.D. and Nambiar, E.K.S. (Eds) Nutrition of Plantation Forests. Academic Press, New York.

Barrett, L.I. 1949. Forest renewal, pp. 120–126. In: Stefferud, A. (Ed) Trees: The Yearbook of Agriculture. U.S. Government Printing Office, Washington, DC.

Bassman, J.H. 1989. Influence of two site preparation treatments on ecophysiology of planted *Picea engelmannii* × *glauca* seedlings. Can. J. For. Res. 19: 1359–1370.

Berisford, C.W. and Kulman, H.M. 1967. Infestation rate and damage by the Nantucket pine tip moth in six loblolly pine stand categories. For. Sci. 13: 428–438.

Berryman, A.A. (Ed) 1988. Dynamics of Forest Insect Populations: Patterns, Causes, Implications. Plenum Press, New York.

Binkley, D., Dunkin, K.A., DeBell, D. and Ryan, M.G. 1992. Production and nutrient cycling in mixed plantations of *Eucalyptus* and *Albizia* in Hawaii. For. Sci. 38: 393–408.

Cannell, M.G.R. 1989. Physiological basis of wood production: A review. Scand. J. For. Res. 4: 459–490.

Chappell, H.N., Weetman, G.F. and Miller, R.E. (Eds) 1992. Forest Fertilization: Sustaining and Improving Nutrition and Growth of Western Forests. Institute of Forest Resources Contribution No. 73. College of Forest Resources, University of Washington, Seattle.

Childs, S.W., Shade, S.P., Miles, D.W.R., Shepard, E. and Froehlich, H.A. 1989. Soil physical properties: Importance to long-term forest productivity, pp. 53–66. In: Perry, D.A., Meurisse, R., Thomas, B., Miller, R., Boyle, J., Means, J., Perry, C.R. and Powers, R.F. (Eds) Maintaining the Long-Term Productivity of Pacific Northwest Forest Ecosystems. Timber Press, Portland, OR.

Compton, J.E. and Cole, D.W. 1991. Impact of harvest intensity on growth and nutrition of successive rotations of Douglas-fir, pp. 151–161. In: Dyck, W.J. and Mees, C.A. (Eds) IEA/BE T6/A6 Report No. 5. FRI Bulletin No. 159. Forest Research Institute, Rotorua, New Zealand.

Cramer, H.H. 1984. On the predisposition to disorders of middle European forests. Pflanzenschutz-Nachr. 37: 98–207.

Davey, C.B. and Wollum II, A.G. 1984. Nitrogen fixation systems in forest plantations, pp. 361–377. In: Bowen, G.D. and Nambiar, E.K.S. (Eds) Nutrition of Plantation Forests. Academic Press, New York.

Davidson, E.A., Myrold, D.D. and Groffman, P.M. 1990. Denitrification in temperate forest ecosystems, pp. 196–220. In: Gessel, S.P., Lacate, D.S., Weetman, G.F. and Powers, R.F. (Eds) Sustained Productivity of Forest Soils, Proceedings of the Seventh North American Forest Soils Conference. Faculty of Forestry, University of British Columbia, Vancouver, British Columbia.

DeBell, D.S., Whitesell, C.D. and Schubert, T.H. 1989. Using N_2-fixing *Albizia* to increase growth of Eucalyptus plantations in Hawaii. For. Sci. 35: 64–65.

Della-Tea, F. and Jokela, E.J. 1991. Needlefall, canopy light interception, and productivity of young intensively managed slash and loblolly pine stands. For Sci. 37: 1298–1313.

Devlin, R.M. and Barker, A.V. 1971. Photosynthesis. Van Nostrand Reinhold Company, New York.

Dyck, W.J. and Skinner, M.F. 1990. Potential for productivity decline in New Zealand radiata pine forests, pp. 318–332. In: Gessel, S.P., Lacate, D.S., Weetman, G.F. and Powers, R.F. (Eds) Sustained Productivity of Forest Soils, Proceedings of the Seventh North American Forest Soils Conference. Faculty of Forestry, University of British Columbia, Vancouver, British Columbia.

Evans, J. 1975. Two rotations of *Pinus patula* in the Usutu forest, Swaziland. Commonw. For. Rev. 54: 64–81.

Evans, J. 1978. A further report on second rotation productivity in the Usutu Forest, Swaziland – results of the 1977 reassessment. Commonw. For. Rev. 57: 253–261.

Fernow, B.E. 1911. A Brief History of Forestry in Europe, the United States, and other Countries. University of Toronto Press, Toronto.

Fisher, R.F. 1984. Predicting tree and stand response to cultural practices, pp. 53–65. In: Stone, E.L. (Ed) Forest Soils and Treatment Impacts, Proceedings of the Sixth North American Forest Soils Conference. Department of Forestry, Wildlife and Fisheries, University of Tennessee, Knoxville, TN.

Fleming, R.L., Black, T.A. and Eldridge, N.R. 1994. Effects of site preparation on root zone soil water regimes in high-elevation forest clearcuts. For. Ecol. Manage. 68: 173–188.

Florence, R.G. 1967. Factors that may have a bearing upon the decline of productivity under forest monoculture. Aust. For. 31: 50–71.

Franklin, J.F. 1993. Lessons from old-growth. Fueling controversy and providing direction. J. For. 12: 10–13.

Franklin, J.F. and Spies, T.A. 1991. Composition, function, and structure of old-growth Douglas-fir forests in Oregon and Washington, pp. 71–80. In: Ruggiero, L.F., Aubrey, K.B., Carey, A.B. and Huff, M.H. (Eds) Wildlife and Vegetation of Unmanaged Douglas-fir Forests. General Technical Report PNW-GTR-285. Pacific Northwest Research Station, USDA Forest Service, Portland, OR.

Froehlich, H.A. and McNabb, D.H. 1984. Minimizing soil compaction in Pacific Northwest forests, pp. 159–192. In: Stone, E.L. (Ed) Forest Soils and Treatment Impacts, Proceedings of the Sixth North American Forest Soils Conference. Department of Forestry, Wildlife and Fisheries, University of Tennessee, Knoxville, TN.

Gholz, H.L., Ewel, K.C. and Teskey, R.O. 1990. Water and forest productivity. For. Ecol. Manage. 30: 1–18.

Gray, H.R. 1935. Coniferous Planting in Australia. Commonwealth Forestry Bureau Bulletin Number 15. Commonwealth Government Printer, Canberra, Australia.

Greacen, E.L. and Sands, R. 1980. Compaction of forest soils: A review. Aust. J. Soil Res. 18: 163–189.

Grier, C.C., Lee, K.M., Nadkarni, N.M., Klock, G.O. and Edgerton, P.J. 1989. Productivity of Forests of the United States and Its Relation to Soil and Site Factors and Management Practices: A Review. General Technical Report PNW-GTR-222. Pacific Northwest Research Station, USDA Forest Service, Portland, OR.

Grigor, J. 1868. Arboriculture. Enmonston and Douglas, Edinburgh.

Halpern, C.B. and Spies, T.A. 1995. Plant species diversity in natural and managed forests of the Pacific Northwest. Ecol. Appl. 5: 913–934.

Henderson, G.S. 1995. Soil organic matter: A link between forest management and productivity, pp. 419–435. In: McFee, W.W. and Kelly, J.M. (Eds) Carbon Forms and Functions in Forest Soils, Proceedings of the Eighth North American Forest Soils Conference. Soil Science Society of America, Madison, WI.

Hermann, R.K. 1976. Man and forests – a prodigal relation, pp. 29–51. In: Forests and Future Resource Conflicts. Northwest Area Foundation Series. School of Forestry, Oregon State University, Corvallis, OR.

Holmsgaard, E., Holstener-Jorgensen, H. and Yde-Anderson, A. 1961. Bodenbildung, Zuwachs und Gesundheitszustand von Fichtenbestanden erster und zweiter Generation. 1. NordSeeland. Forstl. Forsogv. Danm. 27: 1–167.

Hoskins, W.G. 1955. The Making of the English Landscape. Penguin Books, Middlesex, England.

Hurlbert, S.H. 1984. Pseudoreplication and the design of ecological experiments. Ecol. Monogr. 54: 187–211.

Jurgensen, M.F., Harvey, A.E., Graham, R.T., Larsen, M.J., Tonn, J.R. and Page-Dumroese, D.S. 1990. Soil organic matter, timber harvesting, and forest productivity in the inland Northwest, pp. 392–415. In: Gessel, S.P., Lacate, D.S., Weetman, G.F. and Powers, R.F. (Eds) Sustained Productivity of Forest Soils, Proceedings of the Seventh North American Forest Soils Conference. Faculty of Forestry, University of British Columbia, Vancouver, British Columbia.

Keeves, A. 1966. Some evidence of loss of productivity with successive rotations of *Pinus radiata* in the south-east of South Australia. Aust. For. 30: 51–63.

Krauss, G., Muller, K., Gartner, G., Hartel, F., Schanz, H. and Blanckmeister, H. 1939. Standortsgemasse Durchführung der Abkehr von der Fichtenwirtschaft im nordwestsachsischen Niederland. Tharandter Forstliches Jahrb. 90: 481–715.

Lieth, H. 1975. Primary production of the major vegetation units of the world, pp. 203–215. In: Lieth, H. and Whittaker, R.H. (Eds) Primary Productivity of the Biosphere. Springer-Verlag, New York.

Lippke, B. and Oliver, C.D. 1993. Managing for multiple values. A proposal for the Pacific Northwest. J. For. 12: 14–18.

Lopushinsky, W. 1969. Stomatal control in conifer seedlings in response to leaf moisture stress. Botanical Gazette 134: 258–263.

Lugo, A.E., Brown, S. and Chapman, J. 1988. An analytical review of production rates and stemwood biomass of tropical forest plantations. For. Ecol. Manage. 23: 179–200.

MacLean, C.D. and Bolsinger, C.L. 1973. Estimating Dunning's Site Index from Plant Indicators. Research Note PNW-197. Pacific Northwest Forest and Range Experiment Station, USDA Forest Service, Portland, OR.

Madden, J. 1988. Sirex in Australasia (chapter 20), pp. 407–429. In: Berryman, A.A. (Ed) Dynamics of Forest Insect Populations: Patterns, Causes, Implications. Plenum Press, New York.

Maser, C. 1988. The Redesigned Forest. R.&E. Miles, San Pedro, CA.

Meeker, J.R. and Kulhavy, D.L. 1992. Pine tip moth (Lepidoptera: Tortricidae) infestation rates influenced by site and stand characteristics in loblolly pine plantations in east Texas. Envirn. Entomol. 21: 534–541.

Messier, C. 1993. Factors limiting early growth of western red cedar, western hemlock and Sitka spruce seedlings on ericaceous-dominated clear-cut sites in coastal British Columbia. For. Ecol. Manage. 60: 181–206.

Miles, S.R. and Powers, R.F. 1988. Ten-Year Results of Forest Fertilization in California. Earth Resources Monograph 15. Region 5, USDA Forest Service, San Francisco, CA.

Miller, H.G. 1984. Dynamics of nutrient cycling in plantation ecosystems, pp. 53–78. In: Bowen, G.D. and Nambiar, E.K.S. (Eds) Nutrition of Plantation Forests. Academic Press, New York.

Miller, R.E. and Anderson, H.W. 1995. Stand Characteristics of 65-Year-Old Planted and Naturally Regenerated Stands Near Sequim, Washington. Research Paper PNW-RP-482. Pacific Northwest Research Station, USDA Forest Service, Portland, OR.

Miller, R.E. and Murray, M.D. 1978. The effects of red alder on growth of Douglas-fir, pp. 283–306. In: Briggs, D.G., DeBell, D.S. and Atkinson, W.A. (Eds) Utilization and Management of Alder. General Technical Report PNW-70. Pacific Northwest Forest and Range Experiment Station, USDA Forest Service, Portland, OR.

Miller, R.E., Bigley, R.E. and Webster, S. 1993. Early development of matched planted and naturally regenerated Douglas-fir stands after slash burning in the Cascade Range. West. J. Appl. For. 8: 5–10.

Miller, R.E., Scott, W. and Hazard, J.W. 1996. Soil compaction and conifer growth after tractor yarding at three coastal Washington locations. Can. J. For. Res. 26: 225–236.

Mitchell, C.C., Westerman, R.L., Brown, J.R. and Peck, T.R. 1991. Overview of long-term agronomic research. Agron. J. 83: 24–29.

Moldenke, A.R. 1992. Non-target impacts of management practices on the soil arthropod community of ponderosa pine plantations, pp. 78–103. In: Habitat, Fiber, Society:

Balance or Bias? Proceedings of the Thirteenth Annual Forest Vegetation Management Conference. Forest Vegetation Management Conference, Redding, CA.

Morris, L.A. and Miller, R.E. 1994. Evidence for long-term productivity change as provided by field trials, pp. 41–80. In: Dyck, W.J., Cole, D.W. and Comerford, N.B. (Eds) Impacts of Forest Harvesting on Long-Term Site Productivity. Chapman and Hall, London.

Moulton, R.J., Lockhart, F. and Snellgrove, J.D. 1996. Tree Planting in the United States – 1995. State and Private Forestry, Cooperative Forestry, USDA Forest Service, Washington, DC.

Nambiar, E.K.S. and Fife, D.N. 1991. Nutrient retranslocation in temperate conifers. Tree Physiol. 9: 185–207.

Nambiar, E.K.S. and Nethercott, K.H. 1987. Nutrient and water availability to and growth of young radiata pine plantations intercropped with lupins. New For. 1: 117–134.

Nambiar, E.K.S. and Sands, R. 1993. Competition for water and nutrients in forests. Can. J. For. Res. 23: 1955–1968.

Nef, J.U. 1952. Mining and metallurgy in medieval civilisation (chapter 6). In: The Cambridge Economic History of Europe (volume 2). Harvard University Press, Cambridge.

Oliver, W.W. and Dolph, K.L. 1992. Mixed-conifer seedling growth varies in response to overstory release. For. Ecol. Manage. 48: 179–183.

Ovington, J.D. 1953. Studies of the development of woodland conditions under different trees. J. Ecol. 3: 12–34.

Postel, S. and Heise, L. 1988. Reforesting the Earth. Worldwatch Paper 83. Worldwatch Institute, Washington, DC.

Powers, R.F. 1983. Forest fertilization research in California, pp. 388–397. In: Ballard, R. and Gessel, S.P. (Eds) Forest Site and Continuous Productivity. General Technical Report PNW-163. Pacific Northwest Forest and Range Experiment Station, USDA Forest Service, Portland, OR.

Powers, R.F. 1989. Retrospective studies in perspective: Strengths and weaknesses, pp. 47–62. In: Dyck, W.J. and Mees, C.A. (Eds) Research Strategies for Long-Term Site Productivity. IEA/BE A3 Report Number 8, FRI Bulletin 152. Forest Research Institute, Rotorua, New Zealand.

Powers, R.F. and Avers, P.E. 1995. Sustaining forest productivity through soil quality standards: A coordinated U.S. effort, pp. 147–190.
In: Powter, C.B., Abboud, S.A. and McGill, W.B. (Eds) Environmental Soil Science: Anthropogenic Chemicals and Soil Quality Criteria. Canadian Society of Soil Science, Brandon, Manitoba.

Powers, R.F. and Ferrell, G.T. 1996. Plantation development as constrained by moisture, nutrition and insects: The "Garden of Eden" study. N. Z. J. For. Sci. 26: 126–144.

Powers, R.F. and Van Cleve, K. 1991. Long-term ecological research in temperate and boreal perennial ecosystems. Agron. J. 83: 11–24.

Powers, R.F., Alban, D.H., Miller, R.E., Tiarks, A.E., Wells, C.G., Avers, P.E., Cline, R.G., Fitzgerald, R.O. and Loftus, Jr, N.S. 1990. Sustaining site productivity in North American forests: problems and prospects, pp. 49–79. In: Gessel, S.P., Lacate, D.S., Weetman, G.F. and Powers, R.F. (Eds) Sustained Productivity of Forest Soils, Proceedings of the Seventh North American Forest Soils Conference. Faculty of Forestry, University of British Columbia, Vancouver, British Columbia.

Powers, R.F., Mead, D.J., Burger, J.A. and Ritchie, M.W. 1994. Designing long-term productivity experiments, pp. 247–286. In: Dyck, W.J., Cole, D.W. and Comerford, N.B. (Eds) Impacts of Forest Harvesting on Long-Term Site Productivity. Chapman and Hall, London.

Powers, R.F., Tiarks, A.E., Burger, J.A. and Carter, M.C. 1996. Sustaining the productivity of planted forests, pp. 97–134. In: Carter, M.C. (Ed) Growing Trees in a Greener World: Industrial Forestry in the 21st Century. School of Forestry, Wildlife and Fisheries, Louisiana State University, Baton Rouge, LA.

Pritchett, W.L. 1979. Properties and Management of Forest Soils. John Wiley and Sons, New York.

Robinson, R.K. 1973. Mycorrhizae and "second rotation decline" of *Pinus patula* in Swaziland. S. Afr. For. J. 84: 16–19.

Ross, D.W., Berisford, C.W. and Godbee, Jr., J.F. 1990. Pine tip moth, *Rhyacionia* spp., response to herbaceous vegetation control in an intensively site-prepared loblolly pine plantation. For. Sci. 36: 1105–1118.

Ryan, M.G. 1991. The effect of climate change on plant respiration. Ecol. Appl. 1: 157–167.

Ryan, M.G., Binkley, D. and Fownes, J.H. 1997. Age-related decline in forest productivity: Pattern and process. Adv. Ecol. Res. 27: 214–262.

Sands, R., Greacen, E.L. and Gerard, C.J. 1979. Compaction of sandy soils in radiata pine forests. I. A penetrometer study. Aust. J. Soil Res. 17: 101–113.

Schowalter, T.D. 1995. Canopy arthropod communities in relation to forest age and alternative harvest practices in western Oregon. For. Ecol. Manage. 78: 115–125.

Sereni, E. 1974. Storia del Paesaggio Agrario Italiano. Gius Laterza & Figli Spa, Rome.

Sheffield, R.M. and Cost, N.D. 1986. Evidence of pine growth loss in Forest Service inventory data, pp. 74–85. In: Atmospheric Deposition and Forest Productivity, Proceedings of the Fourth Regional Technical Conference. Appalachian Society of American Foresters, Raleigh, NC.

Sheffield, R.M., Cost, N.D., Bechtold, W.A. and McClure, J.P. 1985. Pine Growth Reductions in the Southeast. Resource Bulletin SE-83. Southeastern Forest Experiment Station, USDA Forest Service, Asheville, NC.

Sheppard, K.R. 1986. Plantation Silviculture. Martinus Nijhoff, Dordrecht.

Simmons, G.L. and Pope, P.E. 1987. Influence of soil compaction and vesicular-arbuscular mycorrhizae on root growth of yellow poplar and sweet gum seedlings. Can. J. For. Res. 17: 970–975.

Smethurst, P.J. and Nambiar, E.K.S. 1990. Effects of contrasting silvicultural practices on nitrogen supply to young radiata pine, pp. 85–96. In: Dyck, W.J. and Mees, C.A. (Eds) Impact of Intensive Harvesting on Forest Site Productivity. IEA/BE T6/A6 Report Number 2, FRI Bulletin Number 159. Forest Research Institute, Rotorua, New Zealand.

Snowdon, P. and Benson, M.L. 1992. Effects of combinations of irrigation and fertilisation on the growth and above-ground biomass production of *Pinus radiata*. For. Ecol. Manage. 52: 87–116.

Squire, R.O., Farrell, P.W., Flinn, D.W. and Aeberli, B.C. 1985. Productivity of first and second rotation stands of radiata pine on sandy soils. Aust. For. 48: 127–137.

Stoeckeler, J.H. and Williams, R.A. 1949. Windbreaks and shelterbelts, pp. 191–199. In: Trees: The Yearbook of Agriculture 1949. U.S. Government Printing Office, Washington, DC.

Stone, E.L. 1975. Effects of species on nutrient cycles and soil change. Philos. Trans. Royal Soc. London B. 271: 149–162.

Stone, R.L. and Will, G.M. 1965. Nitrogen deficiency of second generation radiata pine in New Zealand, pp. 117–139. In: Youngberg, C.T. (Ed) Forest Soil Relationships in North America, Proceedings of the Second North American Forest Soils Conference. Oregon State University Press, Corvallis, OR.

Switzer, G.L. and Nelson, L.E. 1972. Nutrient accumulation and cycling in loblolly pine (*Pinus taeda* L.) plantation ecosystems: The first twenty years. Soil Sci. Soc. Am. Proc. 36: 143–147.

Tiarks, A.E. and Haywood, J.D. 1996. Effects of site preparation and fertilization on growth of slash pine over two rotations. Soil Sci. Soc. Am. J. 60: 1654–1663.

Toumey, J.W. and Korstian, C.F. 1942. Seeding and Planting in the Practice of Forestry. John Wiley and Sons, Inc., New York.

USDA (United States Department of Agriculture) 1949. The Yearbook of Agriculture. U.S. Government Printing Office, Washington, DC.

USDA (United States Department of Agriculture) Forest Service 1983. The Principal Laws Relating to Forest Service Activities. Agriculture Handbook 453. USDA Forest Service, Washington, DC.

Van Cleve, K. and Powers, R.F. 1995. Soil carbon, soil formation, and ecosystem development, pp. 155–200. In: McFee, W.W. and Kelly, J.M. (Eds) Carbon Forms and Functions in Forest Soils, Proceedings of the Eighth North American Forest Soils Conference. Soil Science Society of America, Madison, WI.

Vitousek, P.M. and Reiners, W.A. 1975. Ecosystem succession and nutrient retention: A hypothesis. BioSci. 25: 376–381.

Waring, R.H. and Schlesinger, W.H. 1985. Forest Ecosystems: Concepts and Management. Academic Press, New York.

Weatherspoon, C.P. and Skinner, C.N. 1995. An assessment of factors associated with damage to tree crowns from the 1987 wildfires in northern California. For. Sci. 41: 430–450.

Webster, J.R., Waide, J.B. and Patten, B.C. 1975. Nutrient recycling and the stability of ecosystems, pp. 1–27. In: Howell, F.G., Gentry, J.B. and Smith, M.H. (Eds) Mineral Cycling in Southeastern Ecosystems. National Technical Information Service, U.S. Department of Commerce, Springfield, VA.

Whalley, W.R., Dumitru, E. and Dexter, A.R. 1995. Biological effects of soil compaction. Soil Tillage Res. 35: 53–68.

Wiedemann, E. 1923. Zuwachstruckgang und Wuchstockungen der Fichte in den mittleren und den unteren Hohenlagen der Sachsischen Staatsforsten. Tharandt. (U.S. Department of Agriculture Translation Number 302, 1936.)

Winters, R.K. 1974. The Forest and Man. Vantage Press, New York.

Wollum, A.G., II and Davey, C.B. 1975. Nitrogen accumulation, transformation, and transport in forest soils, pp. 67–106. In: Bernier, B. and Winget, C.H. (Eds) Forest Soils and Forest Land Management, Proceedings of the Fourth North American Forest Soils Conference. Laval University Press, Quebec.

New Forests **17:** 307–327, 1999.

Private forest investment and long-run sustainable harvest volumes

RALPH J. ALIG[1], DARIUS M. ADAMS[2], JOHN T. CHMELIK[3] and PETE BETTINGER[4]
[1]*USDA Forest Service, Pacific Northwest Research Station, Corvallis, OR, USA;* [2]*Department of Forest Resources, College of Forestry, Oregon State University, Corvallis, OR, USA;* [3]*USDA Forest Service, Pacific Northwest Research Station, Portland, OR USA;* [4]*Department of Forest Resources, Oregon State University, Corvallis, OR, USA*

Key words: non-industrial forests, planted forests, projection models, timber markets, timber supply

Abstract. Private timberlands in the United States have the biological potential to provide larger quantities of timber on a sustainable basis than they do today. Most opportunities for increasing growth and harvest lie on nonindustrial private lands in the South. Past studies, based on fixed scenarios of future prices, also suggest that many of these opportunities for intensified management can be undertaken with positive economic returns. Translation of these physical and apparent economic potentials into projections of future management and harvest requires a model of private timber management investment behavior. This study examines the dynamics of investment in private forest management according to a model of timber markets and timber supply in which intertemporal levels of private investment, harvest, and timber prices are all endogenous. The results of this model are used to examine the extent and types of possible future private management investments and how these will affect timber supply. In addition, the sensitivity of these projections to variations in key market and behavioral determinants is examined through simulation of alternative scenarios involving reduced public timber harvest and constraints on planting investment of nonindustrial private owners.

The base case illustrates the substantial potential of timberlands for increased growth and harvest. This requires, however, investments in planting well beyond those observed in recent years. Given this, the area in planted forests would almost triple within the next 30 years. Expanded investment would allow immediate increases in timber harvest and sustained increases in timber inventory, with virtually no trend in softwood log prices. Projected increases in plantation area would concentrate timber production on fewer hectares, with more hectares managed passively. Naturally regenerated forests in the future would cover at least three-quarters of the private timberland area, with hardwoods continuing to dominate. Restricting nonindustrial private plantation investment to levels observed in the recent past markedly alters projections for softwoods, thus raising prices and reducing timber harvest relative to the base case across the full projection period. In contrast, reductions in public timber harvest alone result in increased prices and reduced total cut in the near term, but have limited impact on the outlook three-five decades hence, because private investment effectively compensates for public timber harvest reductions.

Introduction

Private timberland in the United States comprises 145 million ha and supplies the largest part of the country's wood requirements (Powell et al. 1993). Despite increased substitution and slower economic and population growth, the long-term demand for wood products is not expected to decline (Haynes et al. 1995). At the same time, timber harvest reductions on public lands are likely to continue, as will the conversion of forest land in all ownerships to nonforest or non-timber uses. As a consequence, private timberlands will become increasingly important in the nation's timber supply, and there will be some incentive to expand long-term growth and harvest from these private lands.

Previous studies have identified large areas of private timberland with biological potential for increased net growth through more intensive management. On the basis of static investment analyses with fixed prices, these studies suggest as well that substantial portions of these investments could provide an attractive financial return (USDA Forest Service 1990; Alig and Wear 1992). Translation of these physical and apparent economic potentials into projections of future management and harvest requires some model of private timber management investment behavior. With few exceptions, however, past analyses of private timber supply have simply used one or more sets of fixed assumptions for future investment activity in which planting and other investment actions occur in a preset and invariant sequence over time (see, for example, USDA Forest Service 1990). Although this approach provides a general picture of the impacts of various investment levels, it may significantly misrepresent future time paths of timber management and harvests by ignoring any interaction or feedback between investment decisions and current or future timber market conditions.

This study examines the dynamics of investment in private forest management by using a model of timber markets and timber supply in which intertemporal levels of private investment, harvest, and timber prices are all endogenous. We developed projections of the extent and types of possible future private management investments and how these will affect timber supply. Could private investment significantly modify the future timber supply outlook? Would this conclusion change if the public cut continues to fall, and are sustainable nonfederal harvest volumes possible without large timber price changes? How important are the contributions of forest industry and nonindustrial private lands in this process? For each scenario, we report projections of plantation area for forest industry (FI) and nonindustrial private forest (NIPF) owners, timber management intensity, timber inventory levels, and log prices.

Extent and management of planted forests

Changing forest product market and policy conditions in recent years have been associated with marked intensification of timber management on some private lands and with an increasing share of total harvest from private lands. Total private timber inventory volume has increased by 37 percent, although the private U.S. timberland[1] base currently is 3 percent smaller than in 1962. Rising timber prices induced investment to enhance timber yields, primarily through the use of planting stock.

Changes in planted forest area

Prior to the 1930s, only a few thousand hectares of NIPF land were planted annually. The area of NIPF tree planting increased sharply to over 400,000 ha/yr during the peak of the Soil Bank Program in the late 1950s (Alig et al. 1990). Spikes in NIPF planting have typically been associated with government programs, including those that followed the Soil Bank Program: the Forestry Incentives Program begun in the 1970s and the Conservation Reserve Program (CRP) in the late 1980s and early 1990s (Figure 1). The CRP was part of the 1985 and 1990 USDA Farm Bills, and subsidized the afforestation of about 1 million ha of highly erodible cropland, mostly in the South. Annual planting rates in the United States have fluctuated notably since the early 1960s (USDA Forest Service 1995). The area of FI tree planting increased steadily between the early 1960s and the mid-1980s, and peaked in about the same year as NIPF planting during the CRP period.

The area of plantations in the South and Pacific Northwest Westside (PNWW) regions[2] of the United States has grown over the last several decades. Expansion of plantation area is consistent with broad trends in other key timber-growing regions of the world, where plantations increasingly are the source of industrial wood. Plantations in many cases offer timber supply advantages in terms of location, accessibility, operability, wood type, and wood quality. The vast majority of tree planting on private timberland consists of softwood species, mainly because they produce larger volumes of higher value sawtimber in less time, relative to hardwoods.

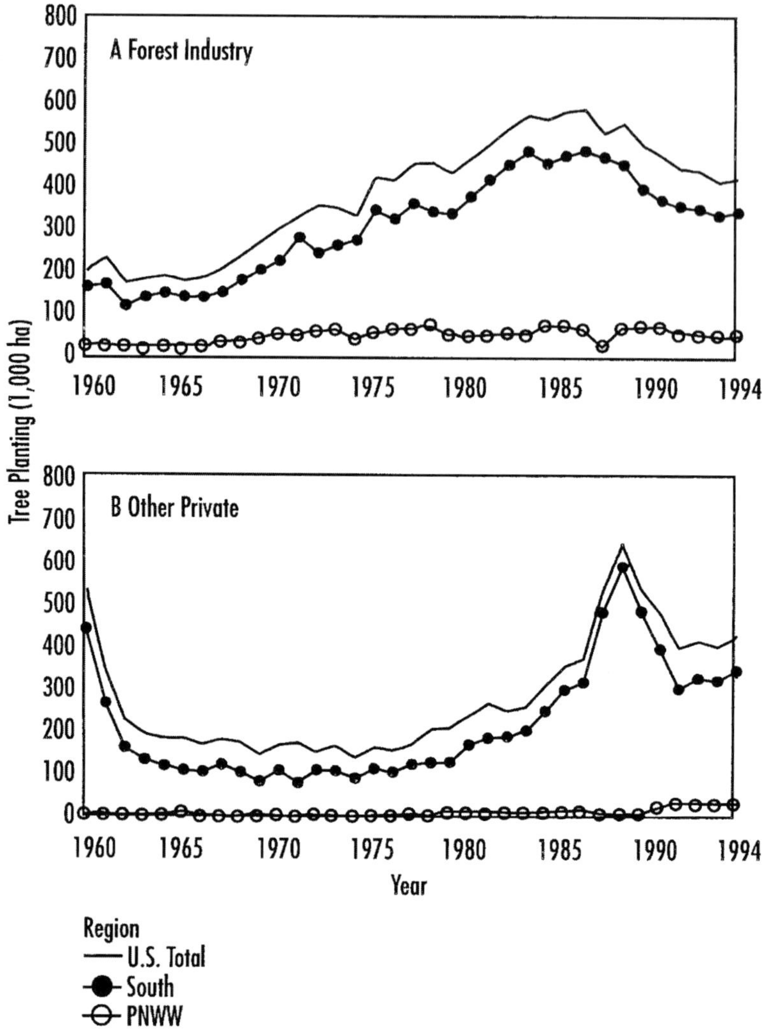

Figure 1. Historical tree planting levels for forest industry and other private landowners, 1960–1994.

The South has about two-thirds of the fast-growing coniferous plantations in the world (Sedjo 1995), equal to about 12 million ha of southern pine plantations. The South contains two-thirds of the plantation[3] area of the United States (Brooks 1993), with planted pine area in the South increasing more than 10-fold since 1952 (USDA Forest Service 1988). In 1994, more than 80 percent of the area planted to trees in the United States was in the South (USDA Forest Service 1995). Within the South, a proportionately larger amount of FI timberland (32 percent) is planted compared to NIPF timberland (6 percent). In 1994, the FI planted slightly more than half of new private plantations in the United States, on a timberland base about one-quarter the size of that of the NIPF. The next highest level of private tree planting in 1994 was in the PNWW region, equal to about one-tenth of private plantings. In contrast to the South, most of the PNWW private tree planting was on FI lands. The area of private tree planting in the PNWW region has also increased since 1962 (Figure 1), but at a slower rate than in the South.

Several studies have found a positive relationship between available timber growing stock and timber harvested, thereby indicating that tree planting may have an effect on short-term as well as long-term timber supply (Alig et al. 1990). Research examining NIPF tree planting decisions has indicated a weak response to stumpage prices (Newman and Wear 1993). Reforestation decisions by NIPF owners were correlated with reforestation costs and availability of cost-sharing (e.g., Brooks 1985; Royer 1987). Forest industry planting decisions have not been researched as much as their NIPF counterparts, but available evidence suggests that FI planting decisions in the South are positively influenced by increases in stumpage prices and land prices (where land is a substitute for plantations), and negatively influenced by increases in interest rates (Lee et al. 1992)

Changes in intensity of plantation management

Estimates of timber management are based on expert opinion in most cases, but indicate that the large NIPF timberland base is generally managed in a custodial or low-intensity mode (e.g., USDA Forest Service 1988; Adams et al. 1992). In contrast, the FI base has a much higher percentage of plantations and receives a higher proportion of intermediate treatments (e.g., thinning). Activities associated with intensive management (e.g., fertilization, weeding, and thinning) can effectively reduce time to reach a minimum merchantable standard by increasing incremental volume growth on each stem. However, even short rotations (20–30 years in the South) represent a long time horizon filled with considerable uncertainty, so that noncorporate forest owners are sometimes reluctant to undertake such investments.

Past studies of investment opportunities and behavior

A number of regional and national studies have examined and inventoried the area of private timberland on which timber management could be intensified (see, as examples, USDA Forest Service 1973, 1988, 1990; Alig and Adams 1995). These studies have consistently shown a large potential for intensification on NIPF lands and, to a lesser extent, on forest industry ownerships. For example, in the 1989 RPA Timber Assessment by the USDA Forest Service (1990), more than 24 million ha of NIPF land were identified as having potential for increased growth through intensified management, while returning a real rate of return of at least 4 percent. These opportunities are concentrated in the South and include more than 8 million ha of timberland that offer attractive rates of return if regenerated to plantations. If these investments were undertaken and sustained, they would result in a timber volume increment of some 19.8 million m^3 at harvest.

Past studies of private investment behavior, which have treated investment as endogenous, can be grouped according to their modeling methodology. Vaux (1954, 1970), USDA Forest Service (1963), and Hyde (1980) use a relatively simple model in which levels of management and steady-state timber harvest are linked to stumpage price for various strata of the inventory in a schedule that might be interpreted as a long-term timber supply function. Optimal long-term supply, price, and timber management intensity are found where this function intersects long-run demand. In a short-term, dynamic context, Adams et al. (1982) employ TAMM, the Timber Assessment Market Model (Adams and Haynes 1996), together with a price-linked ranking of investment opportunities, to identify the equilibrium level of management investment. Assuming a fixed temporal pattern of investment implementation, they make price projections with TAMM at various investment levels in an iterative fashion until solutions stabilize. Newman and Wear's (1993) econometric model of long-term timber production decisions by private owners in the South includes timber management investment in the form of a planting input. Although the model is not used to make projections, it does provide estimates of the sensitivity of planting to changes in timber and other input prices. Sedjo and Lyon (1990), in a global study, and Adams et al. (1996a), in an analysis of the United States, employ intertemporal, spatial equilibrium market models of the forest sector in which investment decisions are determined together with harvest timing. Although their respective approaches are conceptually similar, Sedjo and Lyon use a continuous, real-dollar measure of management input, whereas Adams et al. use a range of discrete classes or groups of management practices.

Forest sector model

This study employs the forest sector portion of the Forest and Agricultural Sector Optimization Model (FASOM) to project future levels of private management investment and timber harvest in the United States. The Forest and Agricultural Sector Optimization Model is a multiperiod, price-endogenous, spatial equilibrium market model of the U.S. forest and agricultural sectors (for a detailed discussion, see Adams et al. 1996a,b). The model operates on a decadal time step, with projections made for ten decades to accommodate treatment of terminal inventories. Policy analysis is limited to results for the 50 years from 1990 to 2040. Exogenous model elements in the forest sector component are held constant after the fifth decade. Terminal inventories (at the end of the finite projection period) are valued, with the assumption of perpetual, steady-state timber management following the last year of the time horizon.

The Forest and Agricultural Sector Optimization Model treats only the log market portion of the forest sector. Log demand is derived from the markets for processed products such as lumber, plywood, and paper. Logs are differentiated by six product classes: hardwood and softwood sawlogs, pulpwood, and fuelwood. A single national demand region for forest products is used, and markets for logs are assumed to be competitive. The demand functions for softwood and hardwood sawtimber and pulpwood are derived from solutions of the TAMM (Adams and Haynes 1996) and NAPAP (Ince 1994) models by summing derived regional demands for logs from manufacturing at higher market levels (sawlogs from TAMM, pulpwood from NAPAP). Sawlog and pulpwood processing facilities have some maximum capacity to produce output in any given period, and, hence, log demand has some upper bound. Decisions to purchase additional capacity in each period to augment current and future log consumption are endogenous. "Downward" substitution is permitted in the model from sawlogs to pulpwood to fuelwood. Log trade with regions outside the United States was treated by including price-sensitive, product-specific demand (export) or supply (import) functions for each region as appropriate, based on historical or anticipated offshore trading patterns. Public timber harvests, differentiated by national forest and other public lands, are treated as exogenous. Land use changes (shifts from forest to other uses) are also exogenous. For the present study, these projections were drawn from Alig and Wear (1992), and portray a continued net loss of NIPF timberland area, mostly to urban and developed uses.

The Forest and Agricultural Sector Optimization Model describes private timberland in terms of strata differentiated by region (nine), class of ownership (forest industry and nonindustrial), forest type (four classes describing species composition, either softwoods or hardwoods, in the current and immediately preceding rotation), site productivity (three levels of potential for wood volume growth), management intensity (four discrete types of timber management regime applied to the area), and 10-year age class (ten). Central to examining potential changes in private forest management are the four timber management intensity classes, from which FASOM endogenously selects the optimal form for both softwood and hardwood hectares: passive – no management intervention of any kind between harvests of naturally regenerated aggregates; low – custodial management of naturally regenerated aggregates; medium – minimal management in planted aggregates; and high – genetically improved stock, fertilization, or other intermediate stand treatments in planted aggregates. Specific practices differ by region, site quality, and forest type.

Growth of existing and regenerated stands is simulated by means of timber yield tables that give the net wood (growing stock) volume/ha in unharvested stands by age class for each stratum (a similar approach is described by Mills and Kincaid 1992). Harvest of a hectare of timberland involves, in general, the simultaneous production of some mix of softwood and hardwood timber volume.

Forestry budgets for private timberland include costs for establishment, growing (e.g., fertilization), timber harvest, and delivery to a mill. Costs differ by region, species, and timber management practice. A constraint reflects limits on the extent of hardwood area that is biologically suitable for conversion to softwood timber types.

Base projections and scenarios

As a datum for comparison, we develop a base case (BASE) that assumes a continuation of current levels of public harvest and no limitations on the extent of private management investment. Base case assumptions for demand and other market conditions derive from the USDA Forest Service's 1993 RPA Update (Haynes et al. 1995). To partially examine the sensitivity of the BASE results to changes in conditions, we modified assumptions in three additional scenarios: reduced public timber harvest but no private investment limits (RPH), constrained private investment with current public timber cut (CPI), and a combination of reduced public harvest and constrained private investment (R + C). We discuss five aspects of the BASE projections in some detail – plantation area, forest type changes, intensity of plantation management, timber prices, and harvest levels – and then contrast the alternative scenarios.

Base case

Plantation area

Table 1 shows the FASOM BASE projections of private plantation areas from 1990 to 2040. We restrict our attention to the South and Pacific Northwest

Table 1. Forest and Agricultural Sector Optimization Model projections of plantation areas on private timberland in the U.S. south (South) and the Pacific northwest westside (PNWW) regions, for the base case and three scenarios, 1990–2040.

Decade	Projected plantation area (1,000 ha)							
	BASE		RPH[a]		CPI[b]		R + C[c]	
	South	PNWW	South	PNWW	South	PNWW	South	PNWW
	Forest industry							
1990	5.267	1.191	5.267	1.191	5.267	1.191	5.267	1.191
2000	6.794	1.827	7.437	1.802	12.375	1.860	12.412	1.774
2010	10.820	2.091	11.638	2.110	13.551	2.126	13.565	2.126
2020	11.350	2.282	12.198	2.282	14.224	2.317	14.224	2.316
2030	11.302	2.446	10.797	2.446	14.224	2.468	14.224	2.468
2040	11.311	2.579	10.797	2.579	14.224	2.579	14.224	2.579
	Nonindustrial private forest							
1990	4.051	0.277	4.051	0.277	4.051	0.277	4.051	0.277
2000	14.290	0.926	16.954	0.869	5.696	0.469	5.306	0.473
2010	22.115	1.149	27.370	1.149	6.803	0.637	6.389	0.641
2020	23.381	1.250	29.296	1.250	6.433	0.800	6.818	0.800
2030	23.684	1.364	29.707	1.364	5.412	0.949	5.332	0.949
2040	23.650	1.432	29.633	1.432	5.428	0.981	5.102	0.981

[a]Reduced public timber harvest, but no private investment limits.
[b]Constrained private investment with current public timber cut.
[c]A combination of reduced public harvest and constrained private investment.

Westside (PNWW) regions, as the two regions with greatest forest investment activity. In the BASE, about 28 million ha of forest plantations are added in those two regions, with about half of that area added during the first decade.

The area of FI plantations is projected to more than double by 2040 in both regions. Nonindustrial private forest plantation area is projected to increase more than 5-fold in both regions, growing from 4.3 million ha in 1990 to 25.1 million ha in 2040. The amount of overall plantation area increases rapidly in the initial decades, and then stabilizes toward the end of the projection period.

In the South, where timber rotations are shorter, plantation area peaks in 2020 for FI lands and in 2030 for NIPF lands. The area of FI plantations in the South expands the most during the 2000–2009 period, whereas NIPF owners in that region add the largest proportion of their plantations (over 40 percent) during the 1990–1999 period. In the PNWW, additions to both FI and NIPF plantation areas are projected throughout the period 1990–2040, but are largest during the 1990–1999 period.

Potential increments in private plantation area are consistent with investment opportunities identified in earlier studies (e.g., Alig and Wear 1992; Alig et al. 1992). When FASOM projections of planted areas were compared to TAMM projections prepared for the 1993 RPA Assessment Update (Haynes et al. 1995), base case results for FASOM generally differed more in degree than in form. This may reflect, in part, the consideration of data on "economic opportunities" for silvicultural investments by Forest Service analysts as part of the assumptions made regarding intensity shifting for the Update. The tendency would be to assume that more area moves to higher management intensity classes (MICs) in regions and ownerships where such opportunities are abundant (see, for example, USDA Forest Service 1990, Chapter 9). Our BASE showed similar movements, but, lacking any limits on investable funds or other constraints, the FASOM model selects much higher levels (larger areas) of plantation investment.

Forest type changes

Projected plantation activity includes conversion of substantial areas of hardwood types to softwoods, especially in the South and less so in the PNWW region (Figure 2). Type conversion in the first several decades is fueled by a relative shortage of softwoods. This contributes in turn to declining hardwood harvest volumes and rising hardwood prices in later periods (beginning about 2020).

Although plantation areas expand markedly in the BASE relative to current levels, future private forests would still be predominantly of natural origin. By 2040, U.S. private timberland would include, at a maximum, about one-quarter softwood plantations, with somewhat higher proportions in the South and PNWW regions. About four-fifths of the NIPF timberland area in the United States would still be concentrated in the lowest two management intensity classes that involve naturally regenerated stands, in contrast to about one-half for the forest industry. The growing area of plantations effectively reduces pressure on naturally regenerated forests as a source of industrial wood (Sedjo 1995).

Intensity of plantation management

Figure 3 shows the projected distribution of planted FI and NIPF softwood timberland area by medium and high MIC classes. Industry owners are projected to apply a higher proportion of relatively intensive plantation management (e.g., precommercial thinning, fertilization, and commercial

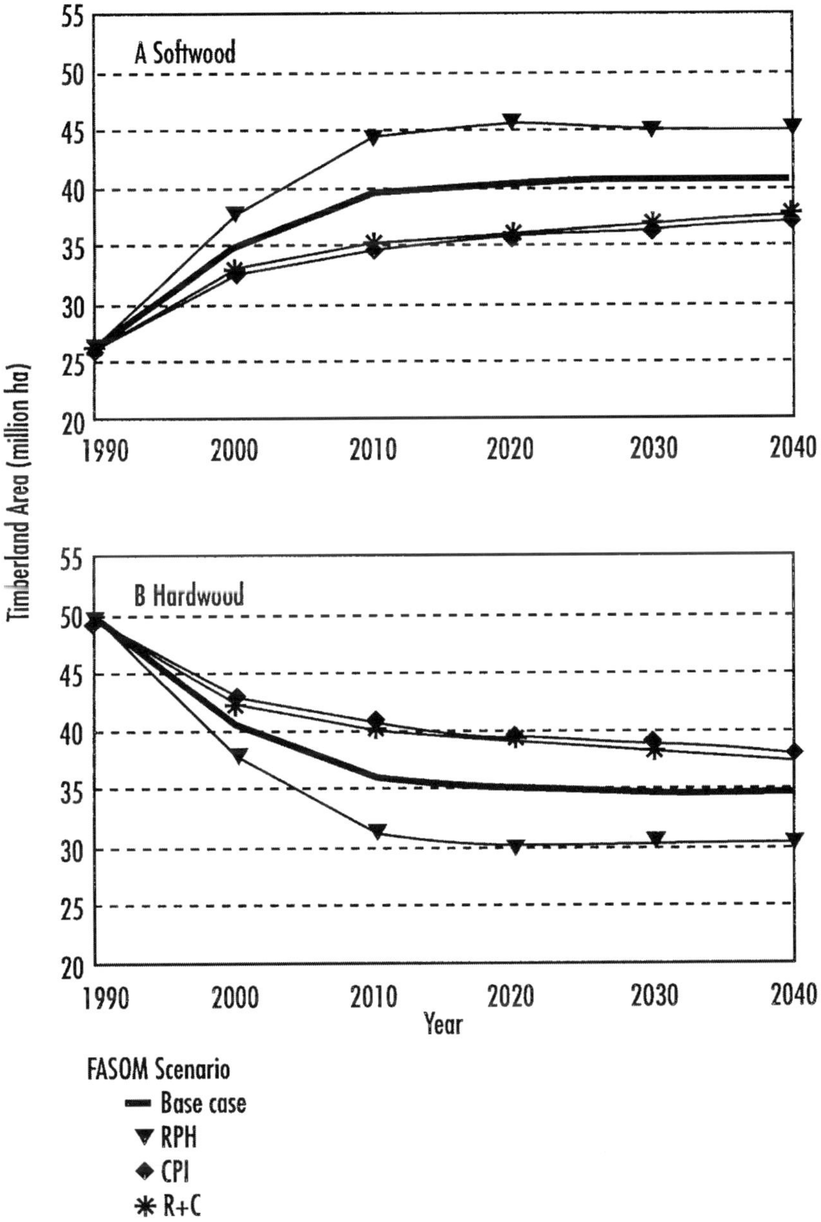

Figure 2. Total private softwood and hardwood timberland area for the base case and three alternative scenarios (RPH – Reduced Public Harvest; CPI – Constrained Private Investment; R + C – Reduced Public Harvest and Constrained Private Investment) for the U.S. south and Pacific northwest westside, 1990–2039.

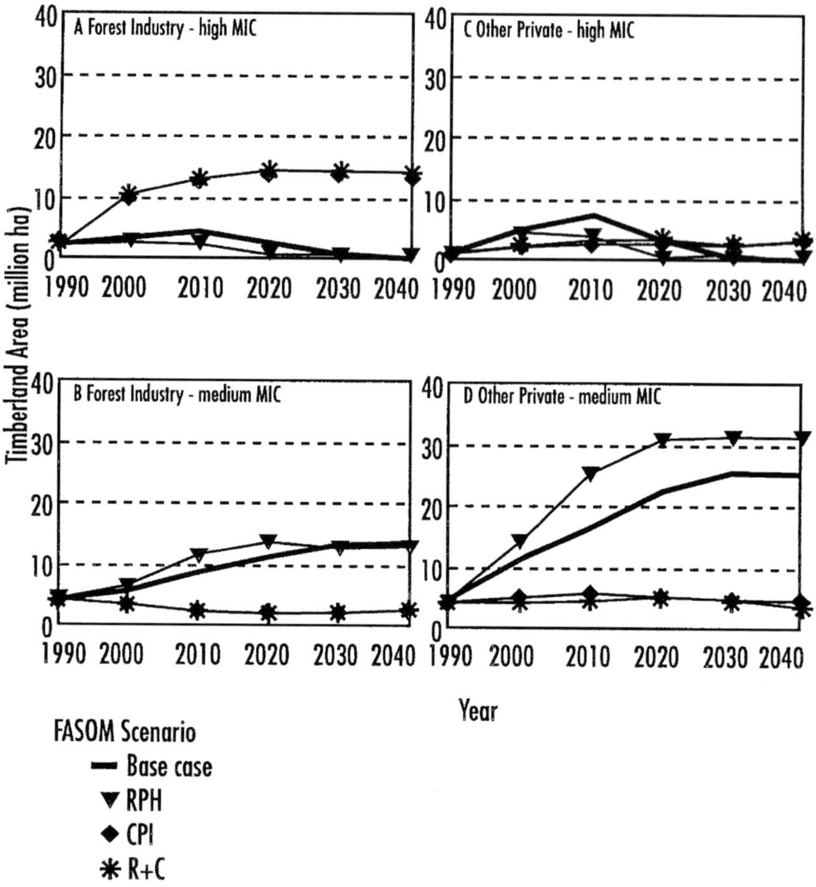

Figure 3. Projected forest industry timberland area for (A) high and (B) medium management intensity classes, and projected other private timberland area for (C) high and (D) medium management intensity classes for the base case and three alternative scenarios (RPH – Reduced Public Harvest; CPI – Constrained Private Investment; R + C – Reduced Public Harvest and Constrained Private Investment) for the U.S. south and Pacific northwest westside, 1990–2039.

thinning) than are NIPF owners. This does not mean that the bulk of private lands will be more intensively managed than at present. Indeed, averaged over all private lands, most plantations would receive less intensive treatments. Scarce investment funds are allocated to the most productive lands that receive a significant increase in management intensity. Other areas receive less investment. For example, some lands in the low management intensity class (not shown in the figures) are shifted into the lowest (passive) class.

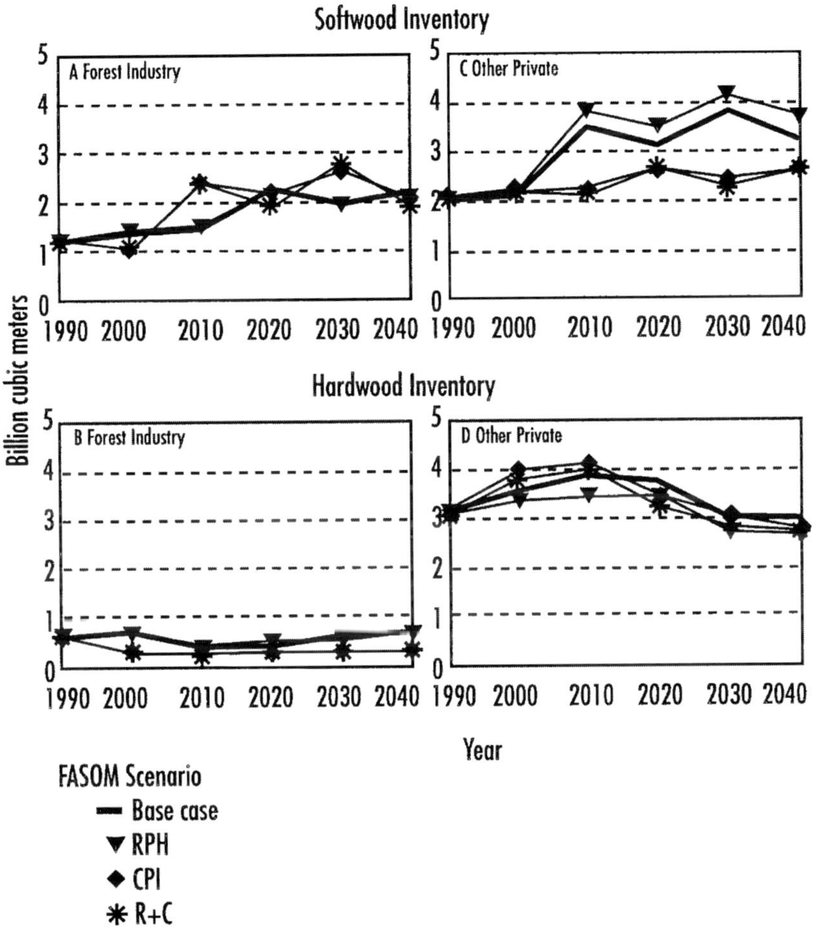

Figure 4. Projected forest industry (A) softwood and (B) hardwood inventory and other private (C) softwood and (D) hardwood inventory for the base case and three alternative scenarios (RPH – Reduced Public Harvest; CPI – Constrained Private Investment; R + C – Reduced Public Harvest and Constrained Private Investment) for the U.S. south and Pacific northwest Westside, 1990–2039.

Timber volumes

Projected softwood and hardwood inventory levels, combined for the South and PNWW regions, are shown in Figure 4. Softwood growing stock inventories on both ownerships are projected to increase through the 2000–2009 decade, and then to roughly level off at amounts at least 50 percent higher than initial 1990 levels. Projected levels of hardwood growing stock rise slightly in the 1990s for both owner groups. The FI level drops after 2000, as a large area of hardwoods is converted to softwood plantations, then rises later in the projection. Nonindustrial private forest hardwood inventories follow a similar but somewhat delayed pattern, declining after 2010 and then turning up slightly after 2030. The 2040 FI hardwood inventory level is back up above the initial 1990 level, whereas the 2040 NIPF level is slightly lower than the 1990 level.

Price and harvest projections

Given the investment activity described above, harvests can expand readily to meet growing demand and softwood product prices differ little over the course of the projection (Figure 5). In contrast to previous studies that used fixed investment schedules and market models with only a single period's scope, the BASE softwood price projections are far less volatile and exhibit little growth. Within FASOM's highly flexible structure, private timber producers can anticipate future price movements and plan current investment and harvest accordingly.

The BASE timber harvest rises for both softwoods and hardwoods over the full projection. As indicated in Figure 5, softwood harvest is larger in all decades than the BASE projection in Haynes et al. (1995), whereas softwood sawtimber prices are lower. This same general relationship holds for hardwood projections as well. In the present model, unrestricted management investment allows larger harvests in all periods, including the 1990 and 2000 decades, in anticipation of future growth increases.

Alternative scenarios

Our model of private investment decision presumes that owners pursue, in effect, an objective of land value maximization, limited only by conditions of the market and of their initial inventories of land and timber. There are, of course, many conditions that act to restrict investment behavior. We examined one class of these constraints – availability of investable funds – in scenarios limiting nonindustrial investment expenditures to levels observed in the recent past. Also, because changes in volumes of public harvest represent shifts in demand for private timber, and because these volumes are highly uncertain, we examined both "current" and "reduced timber cut" scenarios for public ownerships. The nature of the alternative scenarios and expectations regarding results are as follows:

Reduced public timber harvest (RPH). Harvests from most classes of public lands in the United States are established by mechanisms independent of

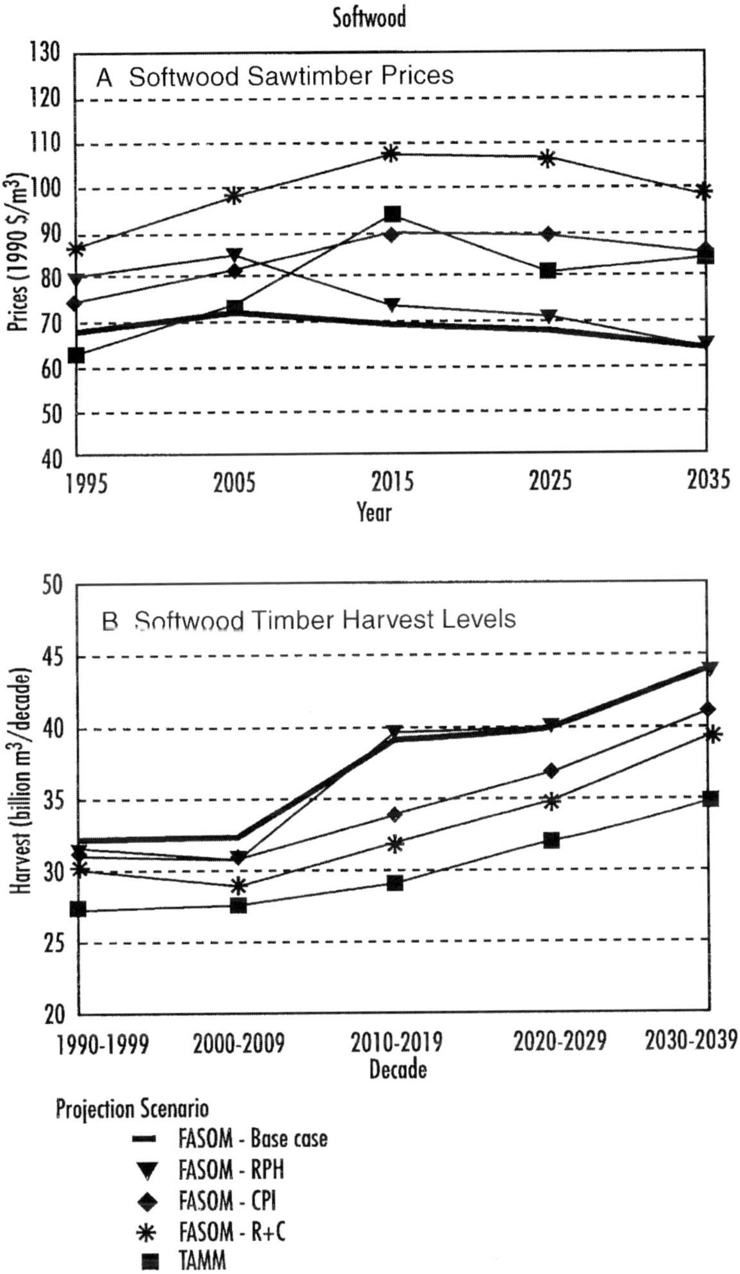

Figure 5. Projected softwood prices for the FASOM base case and three alternative scenarios (RPH – Reduced Public Harvest; CPI – Constrained Private Investment; R + C – Reduced Public Harvest and Constrained Private Investment), and projected U.S. softwood (sawtimber and pulpwood) timber harvest levels for the FASOM base case, three alternative scenarios, and TAMM, 1990–2039.

markets and prices for forest products. Thus, public cut is treated as an exogenous input to FASOM. Changes in public harvests have the partial effects of shifting the demand for private timber and, because forest owners in FASOM have perfect foresight, of changing the extent and timing of current and future private timber harvests and forest management investments. In this modification of the BASE scenario, we eliminated national forest timber harvests, but left those on other public lands unchanged (Adams et al. 1996a). These restrictions on public timber production should lead to higher timber prices, at least in the near term, and more incentives for private owners to intensify timber management.

Constrained private investment (CPI). In FASOM any change in future conditions is optimally anticipated (from a net social welfare viewpoint) and investment is freely flexible to range over time. A representation of "real world" behavior would doubtless be somewhat less adaptable, in recognition of the limitations of the decision maker. In this modification of the BASE scenario, we limited the future annual expenditures for NIPF tree planting investments to the level observed in 1993 (USDA Forest Service 1995). The constraints were imposed by region. We would expect that restricted NIPF capability to invest will lead to higher log prices, less area in softwoods on NIPF lands, and increased forest investment by forest industry owners, although their capability to respond is limited by the smaller size of their land base.

Combined: reduced public harvest and constrained private investment (R + C). Under this scenario, we combined the assumptions of less public timber harvest and the restricted capability of NIPF owners to invest in management.

Plantation area

Projected areas of planted private timberland differed widely among the three alternative scenarios and in comparison to the BASE (Table 1). In all scenarios, planted area increases at least through 2009. The largest increase is for NIPF planted timberland in the South under the RPH scenario, which is 6 million ha larger than the BASE by 2040. In contrast, NIPF planted timberland area in the higher-cost PNWW region changes little between the BASE and the RPH scenario.

Constraining the level of plantation investment by NIPF owners in the CPI scenario substantially lowered NIPF plantation area. This is especially important in the South, where the 2040 area is less than one-quarter of the BASE level. In response, FI planting in the South is one-quarter larger by

2029. Any further increases in southern FI softwood plantations in the CPI scenario are constrained by a limit on the area of convertible hardwood.

Projected areas of planted private timberland were similar between the CPI scenario and the combined scenario (R + C), where the reduced public harvest condition is imposed as well. The investment constraint is clearly the controlling condition for the projection, and is binding in all periods. The effects of the public cut restriction are to further raise demand for private timber, this leading to markedly higher log prices than in the CPI scenario (Figure 5).

Adjustments across policy scenarios, especially over the next 10–20 years, reflect the strong influence of the age class distribution and limited merchantable timber volumes available for harvest in the current timber inventory. For example, there is an interaction between projected increases in private plantation area under the RPH scenario and the lower near-term supply of timber and lower total U.S. harvest. The lower short-term harvest causes prices to rise. This exacerbates conditions of limited merchantable private inventory during the first three decades of the projection, which amplifies the very modest price increases observed in the BASE in this period. In the RPH scenario, higher prices elicit more private investment, and gradually force prices back toward BASE levels.

Forest type changes

All scenarios involved significant conversion of hardwood areas to softwood types (Figure 2). Higher softwood log prices elicit more investment in conversion of hardwood types to softwoods. Reduced public cut stimulates this process in the first two decades, and restrictions on NIPF investment reduce it by nearly equal amounts. After 2020 there is little incentive to continue conversions in any of the scenarios. Type conversion constraints were binding after 2020 for the amount of hardwood land available for conversion to softwoods on FI lands in the South under the CPI scenario. The FI steps in to plant more area when the level of NIPF plantation investment is constrained to recent historical levels. The hardwood conversion constraint also is binding in that period for the combined R + C scenario.

Intensity of plantation management

Timber management intensity on planted lands appears to be somewhat more sensitive to scenario changes than to type conversion, particularly in the long term. Limiting NIPF investment raises log prices across the full projection span (Figure 5). As a result, more land could be enrolled in the higher (and more costly) plantation intensity class (high MIC class in Figure 4). A significantly larger proportion of these limited regeneration expenditures by NIPF owners are for planting high site lands. The range of adjustment is insufficient, however, to bring log prices back to BASE levels. Restriction of NIPF investment clearly dominates behavior in the combined R + C scenario.

Reduced public cut without an investment constraint, in contrast, stimulates substantially more hardwood site conversion. Although this additional land remains primarily in the lower plantation intensity category (medium MIC), private investment increases enough to return prices to BASE levels. Under the RPH scenario, inventory is managed more intensively than in the BASE because of higher product prices and timberland values. With lower supplies of stumpage available from the public sector, product and timberland prices increase and private landowners expand their inventory holdings in line with product demands.

Price and harvest projections

All scenarios acted to raise softwood sawtimber prices above the BASE (Figure 5a). Increases under the RPH scenario were limited to the first two decades for reasons noted. Limiting private investment prevents owners from fully exploiting future gains, and prices remain higher than in the BASE. Reducing public cut, in the face of an NIPF investment restriction, ratchets prices still further upward, because there are few additional investment options accessible (FI lands are already at hardwood conversion limits).

Softwood harvests in all scenarios are below BASE levels (with the exception of the 2010 decade for the reduced public cut case), and, similar to the BASE, show steady growth after 2000. Restricting either public cut or NIPF investment forces some decline in total softwood cut between the first and second decades of the projections. This reflects more rapid depletion of initial stocks in the first decade of these scenarios, and the lag between investment and harvest. The addition of reduced public timber cut to investment limits leads to a nearly constant downward shift in softwood harvest throughout the projection (Figure 5b), because private owners are left with very limited means of response in the face of the investment restriction.

Discussion and conclusions

Our base case suggested that U.S. private timberlands have considerable potential for sustainable wood production under intensified management. Indeed, investment at the levels envisioned here could lead to substantially greater harvest volumes and lower prices than those projected in the 1993 RPA Timber Assessment (Haynes et al. 1995). Under assumptions of perfect foresight and perfect capital markets and no constraints on investable funds in our model, harvest could potentially expand to meet growing demand such that softwood log prices would vary little over the projection. The requisite levels of aggregate private investment were, however, well beyond those observed in recent years. The area in planted forests would almost triple within the next 30 years, with the bulk of this increment in the South where plantation investments are less costly. On industrial ownerships, projected plantation rates would differ little from recent historical activity. For NIPF lands, however, our projections involved a substantial increase over current levels, despite the rising trend of investment in recent decades. This reflected in part the small portion of that ownership currently devoted to plantations, the site quality and growth potential of its lands, and its overall size.

Private harvest in the United States over the next two decades will be strongly influenced by current timber inventory characteristics, particularly the limited areas and timber volumes in older merchantable age classes in virtually all regions. Despite these conditions, our projections indicate that expanded investment would allow some immediate increments in timber harvest, sustained increases in timber inventory, and virtually no long-term trend in softwood log prices. In the longer term, continued conversion of rural land to urban and developed uses will act to reduce the timberland base, in some cases removing the most productive lands. Our results suggest, however, that great potential for expanding sustained production levels would still exist.

At the same time, these results do not necessarily portend a future forest comprising solely planted stands. Projected increases in plantation area would concentrate timber production on fewer hectares, with more timberland passively managed and with less harvest pressure on naturally regenerated forests. Naturally regenerated forests would cover three-quarters of the future private timberland base, with hardwoods continuing to dominate.

Under the assumptions of perfect foresight and perfect capital markets in our model, changes in prices and profitability stimulated changes in investment that acted to counter the effects of the various policies examined, and to limit the movement of prices and quantities away from the base case results. In the reduced public harvest case, for example, investment rose in response to higher prices and largely offset public losses within two decades. Because the source of this flexibility was management intensification, however, any restrictions on private investment have significant impacts. Thus, the scenario limiting NIPF investment to recent historical levels yielded higher log prices and reduced aggregate cut. Combining any other supply restriction or demand expansion (such as the reduction in public harvest) with the investment limit only served to exacerbate the market impacts.

Notes

1. Timberland is forest land that is not reserved for other uses and is capable of producing 1.4 m^3 of industrial wood/ha/yr.
2. The U.S. South consists of thirteen states, from Virginia to Texas; the PNWW region consists of the portions of Oregon and Washington that are west of the crest of the Cascade Mountains (Haynes et al. 1995).
3. Planted area and plantation are synonymous in this paper.

References

Adams, D. and Haynes, R. 1996. The 1993 timber assessment market model: Structure, projections, and policy simulations. General Technical Report, PNW-GTR-368. Pacific Northwest Research Station, USDA Forest Service, Portland, OR.

Adams, D., Haynes, R., Dutrow, G., Vasievich, J. and Barber, R. 1982. Private investment in forest management and the long-term supply of timber. Am. J. Agric. Econ. 64(2): 232–241.

Adams, D., Alig, R., Anderson, D., Chmelik, J. and Stevens, J. 1992. Future prospects for western Washington's timber supply. Institute of Forest Resources Contribution Number 74. College of Forest Resources, University of Washington, Seattle.

Adams, D., Alig, R., McCarl, B.A., Callaway, J. and Winnett, S. 1996a. An analysis of the impacts of public timber harvest policies on private forest management in the U.S. For. Sci. 42(3): 343–358.

Adams, D., Alig, R., Callaway, J., McCarl, B. and Winnett, S. 1996b. The forest and agricultural sector optimization model (FASOM): Model structure and policy applications. Research Paper PNW-RP-495. Pacific Northwest Research Station, USDA Forest Service, Portland, OR.

Alig, R. and Adams, D. 1995. Productivity of nonindustrial private forests in western Washington: Alternative futures. West. J. Appl. For. 10(1): 29–35.

Alig, R. and Wear, D. 1992. Changes in private timberlands: Statistics and projections for 1952 to 2040. J. For. 90(5): 31–37.

Alig, R., Lee, K. and Moulton, R. 1990. Likelihood of timber management on nonindustrial private forests: Evidence from research studies. General Technical Report SE-60. Southeastern Forest Experiment Station, USDA Forest Service, Asheville, NC.

Alig, R., Vasievich, J. and Lee, K. 1992. Economic opportunities to increase timber growth on timberland, pp. 115–125. In: Qureshi, A. (Ed.) Forests in a Changing Climate. Climate Institute, Washington, DC.

Brooks, D. 1985. Public policy and long-term timber supply in the South. For. Sci. 30(3): 697–704.

Brooks, D. 1993. U.S. forests in a global context: An issue paper for the resource planning act assessment. General Technical Report RM-GTR-228. Rocky Mountain Forest and Range Experiment Station, USDA Forest Service, Ft. Collins, CO.

Haynes, R., Adams, D. and Mills, J. 1995. The 1993 RPA timber assessment update. General Technical Report RM-GTR-259. Rocky Mountain Forest and Range Experiment Station, USDA Forest Service, Ft. Collins, CO.

Hyde, W. 1980. Timber Supply, Land Allocation, and Economic Efficiency. Johns Hopkins University Press for Resources for the Future, Baltimore, MD.

Ince, P. 1994. Recycling and long-range timber outlook. General Technical Report RM-GTR-242. Rocky Mountain Forest and Range Experiment Station, USDA Forest Service, Ft. Collins, CO.

Lee, K., Kaiser, F. and Alig, R. 1992. Substitution of public for private funding in planting southern pine. South. J. Appl. For. 16(4): 204–208.

Mills, J. and Kincaid, J. 1992. The aggregate timberland assessment system – ATLAS: A comprehensive timber resource projection model. General Technical Report PNW-281. Pacific Northwest Research Station, USDA Forest Service, Portland, OR.

Newman, D. and Wear, D. 1993. Production economics of private forestry: A comparison of industrial and nonindustrial private forest owners. Am. J. Agric. Econ. 75(3): 674–684.

Powell, D., Faulkner, J., Darr, D., Zhu, Z. and MacCleery, D. 1993. Forest resources of the United States, 1992. General Technical Report RM-GTR-234. Rocky Mountain Forest and Range Experiment Station, USDA Forest Service, Ft. Collins, CO.

Royer, J. 1987. Determinants of reforestation behavior among southern landowners. For. Sci. 33(3): 654–667.

Sedjo, R.A. 1995. The potential of high-yield plantation forestry for meeting timber needs: Recent performance and future potentials. Discussion Paper 95-08. Resources for the Future, Washington, DC.

Sedjo, R.A. and Lyon, K.S. 1990. The Long-term Adequacy of World Timber Supply. Resources for the Future, Washington, DC.

USDA (United States Department of Agriculture) Forest Service. 1963. Timber trends in western Oregon and western Washington. Research Paper PNW-5. Pacific Northwest Forest and Range Experiment Station, USDA Forest Service, Portland, OR.

USDA (United States Department of Agriculture) Forest Service. 1973. The outlook for timber in the United States. Resource Report 20. USDA Forest Service, Washington, DC.

USDA (United States Department of Agriculture) Forest Service. 1988. The South's fourth forest: Alternatives for the future. Resource Report 24. USDA Forest Service, Washington, DC.

USDA (United States Department of Agriculture) Forest Service. 1990. An analysis of the timber situation in the United States: 1989–2040. General Technical Report RM-199. Rocky Mountain Forest and Range Experiment Station, USDA Forest Service, Ft. Collins, CO.

USDA (United States Department of Agriculture) Forest Service. 1995. Tree planting in the United States. State and private forestry. Cooperative Forestry, Washington, DC. (Various issues, 1973–1994.)

Vaux, H. 1954. Economics of Young-growth Sugar Pine Resources. Bulletin 78. Division of Agricultural Sciences, University of California, Berkeley.

Vaux, H. 1970. Public Timber Supply Alternatives in the Douglas-fir Region. Hill Family Foundation Series. College of Forestry, Oregon State University, Corvallis.

New Forests **17:** 329–338, 1999.

International expectations for sustainable forestry: a view from the U.S. forest industry

JOHN P. MCMAHON
Timberlands External and Regulatory Affairs, Weyerhaeuser Company, Tacoma, WA 98477, USA

Key words: customers, certification, forest products, multilateral governments

Abstract. Today several global forces are raising expectations that forestry must be practiced on a sustainable basis. This is true for forestry as practiced in tropical, temperate, and boreal regions, as well as in developed and developing countries. Several institutional processes are in place to develop criteria for and indicators of sustainable forestry. As customers and consumers have become more knowledgeable, they are becoming more demanding about the environmental attributes of the products they buy. Forest products companies and forest landowners who wish to maintain access to these more demanding markets will need to provide assurances of their commitment to sustainable forestry practices.

Forest landowners and forest products manufacturers in Washington, Oregon, and California are currently better positioned to meet these expectations than are producers in other regions. In these western states, state forest practices laws and regulations are generally consistent with emerging international criteria and indicators of sustainable management.

However, U.S. competitors in such countries as Canada and Sweden are acting rapidly to establish sustainable forestry practices, and to communicate their commitments to their stakeholders and customers. The assurance of sustainable forestry practices is now the latest means by which environmentally aware and demanding customers and consumers differentiate among forest products producers.

Introduction

Release of the 1987 report by the Bruntland Commission, *Our common future* (WCED 1987), can be viewed as a landmark event in advancing concepts related to the sustainable development of forest resources. In this light, sustainable forestry practices are a key subset of the concepts under review.

The report initiated the first serious discussions of sustainability issues among governments in both developed and developing countries, among nongovernmental organizations, and among professional foresters around the world. Further, recognition of the need for sustainable forestry was heightened during the United Nations Conference on Environment and Development (UNCED) in Rio de Janeiro in June 1992.

Some analysts maintain that the Rio conference did not fulfill its potential for producing an international agreement on forestry. However, the resulting statement of non-binding principles for forestry had a very definite and positive influence on the way that governments, forestry professionals, the forest industry, and many non-governmental organizations now view the practice of forestry in an international context (UNCED 1992).

First, UNCED stimulated several governmental efforts to describe common criteria for sustainable forestry. The criteria have to be applicable on a worldwide basis.

Second, because many people from nongovernmental organizations were dissatisfied with the Rio conference, momentum was established to develop additional sets of principles and criteria for sustainable forestry. This momentum resulted in the formation of such groups as the Forest Stewardship Council, the Rainforest Alliance, and the Rogue Institute.

Third, purchasers of forest products in the United Kingdom, northern Europe, and to some extent within the U.S. have begun to question whether or not the wood and paper products are being supplied from forests that can be considered to be sustainable. That is, are these forests managed in a way that will allow them to maintain their biodiversity, productivity, and regenerative capacity are maintained in perpetuity?

The initial questions were directed at those who produce tropical forest products. The same questions are now being asked of forest products companies in the Northern Hemisphere. For example, in the U.S. questions come from such groups as the Environmental Defense Fund Paper Task Force, the Magazine Publishers Association, and some well-known large customers for paper and wood products. They are asking suppliers in the forest industry for detailed information about industrial forest management practices.

In view of these developments, this paper reviews the response of forest land managers, particularly from an industrial forestry standpoint, to evolving concepts of sustainable forestry. In this regard, planted forests have played, and will continue to play, a vital role in this regard.

Multilateral government efforts since 1992

Important multilateral governmental efforts regarding sustainable forestry have been initiated followed UNCED. These include several international working group processes, all of which have developed criteria for sustainable forestry that are applicable in their own situations. Examples are the International Tropical Timber Organization, the Helsinki Process among the countries of the European Community, and the Montreal Process, which involved the U.S., Canada, Australia, New Zealand, Chile, and five other countries representing about 90 percent of the world's temperate and boreal forests (Canadian Forest Service 1995). The U.S. was represented in the Montreal Process by the State Department and the USDA Forest Service. The most recent results of the Montreal Process were endorsed by the participating countries in February 1995. The endorsement was in the form of support for what is known as the "Santiago Declaration." This declaration was published as a list of nonlegally binding criteria and indicators for the conservation and sustainable management of temperate and boreal forests. The Santiago Declaration recognizes two basic values. The first is "the value of having an internationally accepted understanding of what constitutes sustainable management of temperate and boreal forests. The second value is an agreement on the criteria and indicators for sustainable forest management in advancing such an understanding" (Canadian Forest Service 1995).

Two other noteworthy events in 1995 focused on sustainability at the international level and promoted sustainable forestry. They are focusing on sustainability at the international level. First, in March, the Food and Agriculture Organization of the United Nations (FAO) convened a ministerial meeting on forestry. The meeting produced The Rome Statement on Forestry, which reaffirms the importance of working to achieve the forestry goals established at the 1992 Rio conference. Further, it identifies several areas for action to achieve those goals. These goals include, including continuing the development of criteria and indicators for sustainable forestry, and assessing the role of voluntary certification schemes in promoting sustainable forest management (FAO 1995).

The Rome Statement on Forestry represented timely advice to the United Nations Commission on Sustainable Development (UNCSD). An Intergovernmental Panel on Forests was established during the Commission's April 1995 meeting in New York. Its purpose is to "pursue consensus on and formulation of coordinated proposals for action" on several issues. Key examples include: the encouragement of national implementation of criteria for and indicators of sustainable forestry, and the examination of the issue of voluntary certification and labelling of forest products. The Intergovernmental Panel is expected to provide its conclusions, recommendations, and proposals for action to the UNCSD's fifth session in 1997 (UNCSD 1995). Thus, the momentum to increase international attention onto sustainable forestry, which began in Rio de Janeiro in 1992, continues today under the auspices of UNCSD, albeit with the support of the ministers of forestry of the major forested nations.

Forest industry response to international sustainable forestry: North America and Western Europe

Not all of the recent focus on sustainable forestry has occurred at the governmental level. Companies who manufacture and sell forest products, and their trade associations, have taken note of the increased interest in forest sustainability among their customers. This is particularly true in Europe. Companies have begun to examine their own preparedness to address the kinds of questions that are now being asked.

Subsequent to the Rio conference in 1992, the American Forest and Paper Association (AF&PA) conducted an opinion poll of the American public. Results showed that the concept of sustainable forestry had strong appeal. In several focus groups conducted around the U.S., consumers wanted to know whether or not the products they were buying were resulting in destruction of tropical rain forests, depletion of old-growth timber, or loss of endangered species' habitats.

Given the results of this opinion poll, the senior executives of the forest industry, represented by AF&PA, to approved the development of new sustainable forestry principles. The principles would be applicable to AF&PA member companies engaged in private forest management and raw material procurement throughout the U.S. (AF&PA 1994). The AF&PA Sustainable Forestry Initiative was approved in 1994, and is now being implemented by member companies plus state forestry and contract loggers' associations in all of the major timber-producing states.

The Canadian forest industry recognized a similar need has been engaged in an intensive effort with the Council of Forest Ministers and the Canadian Standards Association to develop sustainable forestry principles that will be applicable across the Canada (Canadian Standards Association 1995). The forest industry in the Nordic countries, led by Sweden, have recently committed themselves to the joint development of criteria and indicators for sustainable forestry. This follows an extensive dialogue in Sweden over the past several months among the forest industry, the World Wide Fund for Nature, and, more recently, the Swedish Nature Federation. Although this effort is continuing, there is general agreement on the overall goal. Part of the goal places considerable importance on the re-establishment of biodiversity in forests that have had a history of harvest and regeneration for over 500 years.

The role of non-governmental organizations

Several international nongovernmental organization have been actively engaged in developing their own independent principles and criteria for sustainable forestry. Such organizations include the World Wide Fund for Nature, Greenpeace, and the Rainforest Alliance. These organizations generally remain somewhat at "arm's length" from the efforts of the forest industry and governments. This is because the nongovernmental organizations believe that industry and governments have too much of a vested interest to develop or implement objective principles and criteria.

Perhaps best known of the efforts of non-governmental organizations is that of the Forest Stewardship Council (FSC), led by the World Wide Fund for Nature and the World Resources Institute. The strategy of the FSC is to convince wholesalers and retailers of forest products, and their consumer customers, that they should buy building materials and paper products that have been certified as being produced from sustainable forests. Such forests must be managed according to sustainable forestry practices. Evaluations are made by independent firms who re approved by the FSC for granting certification (World Wildlife Fund 1995).

Support for the FSC approach has gained a following in Europe, particularly among some large retailers in the United Kingdom. In the U.S., several groups, among them small landowners, at least three relatively larger nonindustrial private landowners, and one Native American forest group, have been certified through the FSC process. These organizations believe that they will gain an advantage in the marketplace as a result (Forest Stewardship Council 1995).

The forest industry in the U.S., Canada, the Nordic countries, and elsewhere has several reservations about the FSC approach. Their reservations stem from: (1) the reluctance of the FSC to include forest landowners as key stakeholders in their process; (2) the ambiguity in FSC's Natural Forest and Plantation Principles; (3) inadequate recognition of existing state and provincial forest practices requirements; and (4) their expectation of holding forest products manufacturers are accountable for the "chain of custody" of all raw material sources, including logs or stumpage purchased from nonindustrial private landowners, through the manufacturing process to the marketplace (Berg and Olszewski 1995).

The ISO 14000 process

Some members of the forest industry in Canada, the U.S., and elsewhere have been examining alternatives to the FSC approach. One option is that of using the recently developed International Standards Organization's (ISO) 14000 Environmental Management Systems (EMS) process. The Process allows the industry to demonstrate its commitment to managing the environmental aspects of the business, including forestry. Manufacturers of pulp and paper, in particular, have gained both experience and confidence in using the ISO 9000 series process to certify their compliance with the product quality expectations of international customers.

The ISO 14000 process offers several advantages to companies or forest landowners who wish to obtain international recognition for their efforts in environmental management. Advantages include the following:

- Building on their internal experience with the ISO 9000 process, and using one system to manage the quality and environmental aspect of both their manufacturing and forestry operations.
- Having a well-documented, internationally recognized process for incorporating their legal and regulatory obligations, and their own internal environmental management commitments in one system.
- Avoiding the need to participate in multiple, potentially conflicting, certification systems, which would create confusion for customers, as well as additional management expense and complexity.

The ISO 14000 EMS process would provide several other important advantages to subscribing organizations. Members could specify the environmental goals and targets to which it expects to be held accountable; document the management processes that they have in place to manage against those goals; and obtain third-party certification, or provide self-certification, that it is doing what it claims to be doing (CEEM 1995).

Pressures for certification of sustainable forestry

With this growing interest in certification of sustainable forestry, the question could reasonably be asked, "why should forest landowners have to concern themselves with engaging a third-party certifying group at considerable additional cost?" For example, in states like Oregon, Washington, or California, with their comprehensive state forest practices acts, why would landowners need to satisfy someone in the United Kingdom, or Germany, or Phoenix, Arizona, that their products are, in fact, being produced from a forest that is being managed on a sustainable basis?

The reasons are varied and debatable. Primarily they are:

- Public concern continues about the worldwide rate of deforestation, particularly of tropical forests, because of serious consequences in the loss of biodiversity and the potential for adverse effects on global climate.

- Consumers who have concerns about the global environment are seeking reassurance that the forest and paper products they buy are coming from sustainable sources.
- Governments are being pressured by nongovernmental organizations to examine their own domestic forest policies, as well as the influence that government policies have on international trade in forest products.
- Large retailers want to be able to assure consumers that their suppliers are providing wood and paper products from sustainable forests.
- Most customers and consumers have little knowledge of the specifics of the Oregon, (or Washington or California) forest practices acts, and can't be expected to know whether or not they address sustainability to the extent necessary. Therefore, having some type of evaluation of forest practices by an independent, third party can be viewed as a means of addressing this concern.
- Some consumers and nongovernmental organizations, particularly in Europe, don't trust government to address the question of resource sustainability in an objective or adequate manner, or with a sufficient sense of urgency. They believe that some form of independent, third-party involvement is necessary to keep sufficient pressure on governments, forest landowners, and manufacturers.
- Forest products companies who wish to protect their position in international markets, particularly in Europe, must be willing to consider the involvement of third parties in quality commitment to meeting customer expectations. The larger the company, and the more committed it is to international markets, the more likely it is to consider the merits of going beyond self-certification of its practices.
- Forestry, as it is now being practiced worldwide, is highly visible. Modern remote sensing and communications technologies can quickly make what once would have been a local issue the subject of front page news in the international media.

To cite a few examples, consider the Clayoquot Sound controversy in British Columbia over old-growth harvesting, the recent international concern about new forest concessions in Surinam, or the international non-governmental organization attention to forest development in Siberia. There is literally nowhere in the world today where a forest landowner or operator is not likely to be subject to some degree of oversight, if not by government, then by non-governmental organizations or local citizens groups.

Conclusions

These realities suggest a few conclusions. It is clear that, in coming years, forest practices in North America and elsewhere will be evaluated by international criteria. It will no longer be sufficient for manufacturers and suppliers of forest products to view forest practices as only a local or state-level matter of no concern to their customers.

The governments of the major timber-producing countries, their forest industries, and many nongovernmental organizations will reach agreement on the basic criteria and indicators of sustainable forestry. These will be used to evaluate forest policies and practices within their own particular areas of interest.

By the year 2000, there will likely be a convergence of the various international efforts to develop such criteria. Failure to do so will result in unnecessary confusion for customers and consumers, a less-than-level "playing field" for trade in the international forest products market, and lack of acceptance by forest landowners.

Large, industrial forestry organizations, whether they are managing their own private land in the U.S. or operating under long-term tenure arrangements in other countries, will be expected by their major customers to provide assurance of their commitment to sustainable forestry. Therefore, these organizations must be willing to subject their practices themselves to some level of independent review.

In order to provide such assurances to their stakeholders and customers, many forest products companies will implement formalized environmental management systems and auditing processes in their forestry and manufacturing operations. The approach could resemble those being developed by the ISO.

Forest products companies that wish to serve the European market and environmentally aware customers in the U.S. will position themselves to meet these new expectations. Competitors who fail to respond to these expectations will risk loss of market share.

In the U.S., forest landowners and manufacturers in the western states, that have comprehensive state forest practices acts will have a competitive advantage over producers in states who are slow to adopt similar requirements. While other regions struggle to address the chain-of-custody question, for example, forest landowners in the Pacific Coast states can provide assurances today that they are meeting the basic requirements of sustainable forestry.

Summary

Today, there are several global forces that are raising expectations that forestry must be practiced on a sustainable basis. This is true for forestry whether in tropical, temperate, or boreal regions, and in both developed and developing countries. Several institutional processes are in place to develop criteria and indicators of sustainable forestry. Customers and consumers are becoming more knowledgeable and demanding about the environmental attributes of the products they buy. Forest products companies and forest landowners who wish to maintain their access to these more demanding markets will need to provide assurances of their commitment to sustainable forestry practices.

Forest landowners and forest products manufacturers in Washington, Oregon, and California are currently better positioned to meet these expectations than are producers in other regions because our state forest practice laws and regulations are generally consistent with the emerging international criteria and indicators of sustainable

However, our competitors in countries like Canada and Sweden are acting rapidly to establish sustainable forestry practices, and communicate their commitments to their stakeholders and customers. Being willing and able to provide assurance of sustainable forestry practices is now becoming the latest means of differentiating forest products producers to environmentally aware and demanding customers and consumers.

References

AF&PA (American Forest and Paper Association). 1994. Sustainable Forestry Principles and Implementation Guidelines. AF&PA, Washington, DC.

Berg, S. and Olszewski, R. 1995. Certification and labeling: A forest industry perspective. Journal of Forestry 93(4): 30–32.

Canadian Forest Service. 1995. Criteria and Indicators for the Conservation and Sustainable Management of Temperate and Boreal Forests. Natural Resources Canada, Hull, Quebec.

Canadian Standards Association. 1995. Guidance Document Z808: Sustainable Forest Management (working draft 008).

CEEM. 1995. What is ISO 14000? Questions & Answers. CEEM Information Services, Fairfax, VA.

FAO (Food and Agriculture Organization of the United Nations). 1995. The Rome Statement on Forestry. UN Rome: FAO, Rome, Rome. March 17, 1995.

Forest Stewardship Council. 1995. Principles and criteria for natural forest management. Oaxaca, Mexico: Forest Stewardship Council, Oaxaca, Mexico. Oaxaca, Mexico

International Tropical Timber Organization. 1994. Criteria for the Measurement of Sustainable Tropical Forest Management. Policy Development Series No. 3. International Tropical Timber Organization; Policy Development Series No. 3. Yokohama, Japan: International Tropical Timber Organization, Yokohama, Japan.

UNCED (United Nations Conference on Environment and Development). 1992. Non-legally binding authoritative statement of principles for a global consensus on the management, conservation, and sustainable development of all types of forests. U.N. Conference on Environment and Development, Rio de Janeiro, 3–14 June 1992.

UNCSD (United Nations Commission on Sustainable Development). 1995. UNCSD Open-ended Inter-governmental Panel on forests. Annex I to Chapter 11, Combatting deforestation and the non-legally binding authoritative statement of principles for a global consensus on the management, conservation, and sustainable development of all types of forests. UNCSD, New York.

WCED (World Commission on Environment and Development). 1987. Our common future. Oxford University Press, Oxford, 400 pp.

World Wildlife Fund. 1995. A WWF Guide to Forest Certification. World Wide Fund for Nature, Surrey, United Kingdom.

New Forests **17:** 339–359, 1999.

The potential of high-yield plantation forestry for meeting timber needs

Recent performance, future potentials, and environmental implications

ROGER A. SEDJO
Forest Economics and Policy Program, Resources for the Future, Washington D.C., USA

Key words: demand, production, financial returns, pulp wood, supply

Abstract. This study examines the performance and potential of intensively managed plantation forests as a source of industrial wood, and their environmental implications. The perspective of the study is global. Although it includes the United States and parts of Europe, much of the focus is on what are called the "emerging" plantation regions – countries largely in the semitropical areas of the southern hemisphere – which have not historically been important wood producers, but are growing in importance as a result of the productivity of their planted forests. The first section of this paper documents the growing importance of plantations as a source of industrial wood since the late 1970s. The study finds that plantations from nontraditional (new) regions have been growing rapidly in size and economic importance, and, thus, have been playing an increasing role as a source of the world industrial wood. Furthermore, experience seems to suggest that plantations are playing an environmentally beneficial role in (1) reducing pressure on greater areas of natural forests and (2) generating positive environmental effects as they replace degraded marginal agricultural lands. The second section of the paper examines the likely role of plantation forests in the future, and includes an assessment of financial, political and environmental considerations. This section pays particular attention to the concerns frequently expressed by environmentalists regarding plantations. Many of the objections directed at forest plantations on environmental grounds appear to ignore the substantial beneficial role of plantations on the environment. Plantations, which are financially very attractive in many locations, offer the potential of meeting large portions of the world industrial wood needs even while reducing substantially the disturbances on large areas of natural forests. This is possible because the very high productivity of plantation forests requires less area to produce industrial wood.

Introduction

Over a decade ago I wrote a book entitled the *Comparative economics of plantation forestry* (Sedjo 1983). In that book, I examined the potential financial returns to plantation forests in a number of promising regions around the globe. The economics of intensively managed forest plantations were examined in Australia, Brazil, Chile, Indonesia, New Zealand, South Africa,

Figure 1. Major regions of actual or potential premise for industrial forest plantations (Sedjo 1983).

and west Africa, as well as in Nordic Europe, the Pacific Northwest, and the U.S. South. In this paper I revisit the question of the potential of plantation forestry. I provide an assessment of plantation forestry's performance over the past 15–20 years, and reexamine its prospects and likely problems over the next couple of decades. Specifically, in the first section of this paper I (1) review the performance of plantation forestry in meeting the world's timber demand requirements over the past 15 years, and (2) examine the potential of high-yield plantation forestry to meet an increasing proportion of the world's industrial wood demand in the future. In the second section I examine the political and environmental issues and risks that are associated with plantation projects in various locations, especially in the context of large-scale plantation projects which involve nonindigenous species.

Plantation performance

Background

A decade ago I examined eleven major regions that had actual or potential promise for industrial forest plantations. These regions were (see Figure 1): North America (the Pacific Northwest, the U.S. South), South America (Brazil: Amazonia, Southern Brazil, Chile), Oceania (Australia, New Zealand), Africa (South Africa, Gambia-Senegal), Europe (Nordic – Sweden, Finland), and Asia (Borneo, Indonesia).

In the early work I examined the economics of plantation forestry for these regions with a representative prototype plantation for each region. Further,

I examined the economic returns given conditions circa 1980 from both a pulpwood operation and an integrated sawtimber/pulpwood operation. For the Nordic region and the United States, the plantations involved domestic native species that were established by reforestation activities after the initial forest was cleared and often after an intervening period during which the land was used for other purposes, e.g., cotton and tobacco in the South.[1] For the nontemperate northern hemisphere regions, however, industrial plantations almost always involved the introduction of a non-native "exotic" species. Commonly, these were either from North America, as with a southern pine species, Caribbean or radiata (Monterey), *Pinus radiata* D. Don, or from Australia, as with one of the *Eucalyptus* species. These regions have been called the "emerging" forest plantation regions.

The financial returns to all the operations, including those of the temperate region, were "acceptable." However, in general the financial returns to the plantations in the semitropics appeared to be potentially the most favorable. The returns to plantations in New Zealand, Chile, and Brazil appeared very promising.

I should note that plantation activity today is important in all of these areas, with the possible exception of Gambia-Senegal in west Africa. One area that was neglected in the earlier study is Iberia (Spain and Portugal), where *Eucalyptus* plantations have become of significant importance. In addition, some other countries, e.g., Uruguay, have shown promise of becoming important sources of industrial wood from exotic plantations.

Recent performance: 1977–1992

Table 1 brings us to the present and provides the official Food and Agriculture Organization of the United Nations estimates of industrial wood product for 1977 and 1992, a 15-year period during which industrial forest plantation wood production would have been expected to experience growth. Regions given in boldface type have relied extensively on industrial forest plantations for their industrial wood production. These regions have all experienced rapid growth in their industrial wood production, growth largely provided by exotic plantation forests, and much more rapid than the world average. An analysis of the data reported in Table 1 reveals that, in 1977, timber production from the plantation regions accounted for 6.11 percent of the world's total. Fifteen years later, in 1992, exotic industrial plantations accounted for 9.80 percent of the world's total. It should be noted that plantations in the U.S. South, Europe, and elsewhere are not included in this figure.

Table 1 also provides data that show total world industrial wood production increasing only 16 percent over the past one and a half decades, an annual compounded growth of less than 1 percent. The South America plantation

Table 1. Industrial wood production, 1977–1992 (*Source:* FAO Forest Products Yearbook, 1983 and 1993).

Region	Production (1,000 m^3)		Proportional change
	1977	1992	
World	1,381,203	1,603,376	1.16
South America	56,894	116,669	2.05
Brazil[a]	*37,319*	*77,714*	*2.08*
Chile	*6,022*	*17,774*	*2.95*
Uruguay	*347*	*1,043*	*3.01*
Argentina	*4,029*	*7,577*	*1.88*
Europe	262,973	278,451	1.06
Sweden	44,260	49,150	1.11
Finland	28,900	35,745	1.24
Spain	*10,596*	*15,112*	*1.43*
Portugal	*7,420*	*10,309*	*1.39*
Oceania			
New Zealand	*9,492*	*14,992*	*1.58*
Africa			
South Africa	*9,181*	*12,601*	*1.37*
North America	476,002	592,900	1.25
United States	320,441	402,500	1.26
Canada	143,365	179,215	1.25

[a] Italic type denotes regions that have relied extensively on industrial forest plantations with exotic species.

countries of Brazil, Chile, Uruguay, and Argentina all have approximately doubled industrial wood production in the past one and a half decades, whereas Chile and Uruguay have tripled production in this period. Spain, Portugal, South Africa, and New Zealand have all substantially increased production, primarily because of the role of industrial plantation forests.

By contrast to the emerging plantation countries, European production has hardly increased. Of course, many of the forests of Europe represent conscious tree planting and reforestation. However, for the most part, their production has increased only modestly. For example, Sweden and Finland, countries that rely on extensive reforestation and intensive management, show only modest production increases.

On the other hand, North America has increased production well above the world average, with both the United States and Canada increasing about 25 percent. In Canada, much of the increased harvest has come from essen-

Table 2. Pulp exports: Selected countries (*Source:* FAO Forest Products Yearbook, 1983 and 1993).

Region	Export (1,000 t)		Proportional change
	1977	1992	
World	17,072	24,520	1.43
South America	433	2,665	6.15
Brazil[a]	*95*	*1,630*	*17.16*
Chile	*339*	*939*	*2.77*
Europe	6,130	7,752	1.26
Sweden	3,220	2,756	0.86
Finland	1,181	1,287	1.09
Spain	*83*	*569*	*6.86*
Portugal	*318*	*1,041*	*3.27*
Oceania			
New Zealand	*477*	*592*	*1.24*
North America	8,509	15,416	1.81
United States	2,415	6,552	2.71
Canada	6,094	886	41.45

Note: Emerging plantation sources represented 7.69% and 19.46% of total world export in 1977 and 1992, respectively.

[a] Italic type denotes regions that have relied extensively on industrial forest plantations with exotic species.

tially wild "old-growth" forests that have been opened to logging by new access, e.g., roads. However, this increase is unlikely to be repeated over the next decade because of limitations in sustainable harvests and environmental concerns. Recently, the British Columbia (B.C.) government has committed itself to reducing harvests on public (Crown) lands in that province. Some of the reductions in B.C. harvests are likely to be offset by increased harvests in Alberta and perhaps other provinces. In 1994 and 1995, partly in response to favorable prices, Canadian exports to the United States have increased. Nevertheless, the steady increases of the past are not likely to continue indefinitely.

Woodpulp exports

Many of the industrial plantations that have been established have been oriented toward the production of woodpulp, often for the international "market pulp" market. Table 2 compares world pulp exports in 1992 with the level in 1977. A number of points should be noted. First, total world pulp

Table 3. Pulp and paper trade balance – Brazil (*Source:* Suchek 1991; preliminary data for 1994 from the Government of Brazil).

Year	Paper (million U.S.$)			Pulp (million U.S.$)			Total
	Imports	Exports	Balance	Imports	Exports	Balance	
1970	6	5	–1	48	1	–47	–48
1975	32	30	–2	116	9	–107	–109
1980	35	364	329	165	156	–9	320
1985	15	279	264	98	263	174	438
1990	42	598	556	244	603	359	915
1994	326	934	608	61	819	758	1,366

exports increased by 43 percent, and most major producers experienced some increases. However, the really large increases in pulp exports were experienced by a handful of countries that have been actively establishing industrial forest plantations; these countries I have titled the "newly emerging plantation region." For example, in addition to Brazil, countries like Spain, Portugal, New Zealand, and Chile have all increased pulp exports by multiple factors. The portion of pulp production exported from the newly emerging plantation region increased from 7.69 percent in 1977 to just under 20 percent in 1992.

It is instructive to examine the growth of pulp exports for selected countries. In the decade and a half from 1977 to 1992, Brazilian pulp exports increased from 95,000 t in 1977 to 1,630,000 t, an increase of over seventeen times the one-and-a-half-decade period to 1992, the last year for which I have worldwide data (FAO Yearbook 1993). Moreover, Brazil's experience is not an isolated phenomenon within South America. The 1977 total pulp exports from South America, which were 433,000 t in 1977, had increased to 2,665,000 t by 1992. Brazil and Chile were the leading exporters.

Additionally, much of the increased pulp from other regions has come from countries where forest plantations are playing an increasing role in supplying wood. The United States, for example, almost tripled pulp exports. Much of this increase is related to plantation production, especially in the South. In addition, pulpwood exports from forest plantations are widely found in the Nordic countries and elsewhere in Europe.

Case study: Brazil

A slightly different perspective can be gained by examining the experience of a specific country, Brazil, with respect to its pulp and paper trade balance

Table 4. Countries providing a large portion of their industrial wood from plantation sources (*Source*: Pandey 1992).

Country	Forest area (1,000 ha)		Share of plantations (%)	
	Natural	Plantation	Total area	Total production
New Zealand	6,270	1,240	19.0	93
Brazil	396,000	6,500	2.0	60
Chile	6,300	1,400	22.0	95
Argentina	36,000	800	2.0	60
Zimbabwe	28,800	117	0.4	50
Zambia	12,900	60	0.4	50

timber trade of recent years. Table 3 shows Brazil's pulp and paper trading experience over the 1970–1994 period. The trade balance fell from –$48 million in 1970 to –$109 million in 1975. However, as a result of the plantation policy it initiated in the 1960s, by 1980 Brazil began to be a major net exporter of paper and pulp, and by 1994 Brazil showed a positive trade balance of $1,366 million in paper and pulp.

Plantations as a cause of deforestation of natural forests?

These data do not allow us to differentiate readily between harvests of plantation wood and that of native forests in these countries. However, a separate study has shown (Table 4) that most of the industrial harvest from the emerging forest plantation countries comes from their plantation forests.

For all of the countries for which data are presented, a disproportionate high fraction of industrial wood is from plantation resources. This finding reflects some inherent advantages of plantations in terms of location, accessibility, wood type, and wood quality.

Additionally, Table 4 suggests that plantations are likely to deflect harvests away from natural forests, thus contributing to their protection. Also, it should be noted that, although a common view is that plantations displace native forests, most forest plantations are actually established on former agricultural lands.[2] Thus, plantations are rarely an important cause of the deforestation of natural forests. Furthermore, in some countries, e.g., Brazil, almost 100 percent of the wood fiber for pulp and paper comes from industrial plantations.

The ability of the fairly modest, fast-growing, industrial plantation forests to make such a significant impact on the world's wood supply, to say nothing of their own domestic wood requirements, is surprising until it is recognized

Table 5. Worldwide timber yields (*Source:* adapted from Clapp 1993).

Site	Yield (m^3/ha/yr)	Rotation (yr)
Temperate and boreal softwood forests		
Canada average 1	—	
British Columbia	1.5–5.3	—
Sweden average	3.3	—
Finland	2.5	60–100
Russia	1–2.9	—
Siberia	1–1.4	70–200
Softwood plantations		
Britain (Sitka spruce)	14	40
South Africa (*Pinus* spp.)	10–25	20–35
New Zealand (Monterey pine)	18–30	20–40
East Africa (*Pinus* spp.)	25–45	20–30
Brazil (*Pinus* spp.)	15–35	15–35
Chile (Monterey pine)	20–30	15–35
Eucalyptus plantations		
Portugal and Spain	10–15	8–12
South Africa	15–20	10
Congo	30–40	7–20
Brazil	30–70	5–20
Chile	20–30	8–20

how rapidly plantation forests can grow. Worldwide timber yields, presented in Table 5, provide an idea of the rapid growth rates possible from these forests.

Trends in industrial forest plantation establishment

Historically, the source of industrial wood has been natural forests. Increasingly, however, the source is tree-growing plantations. Table 6 provides a recent estimate of the area of industrial forest plantations worldwide. The large number of plantations in Asia is attributable in large part to China, which has experienced fairly high seedling mortality; thus, this figure may be a bit high. Nevertheless, plantation forestry is active elsewhere in Asia, including Japan, Korea, Indonesia, and so forth. Furthermore, many of the

Table 6. Industrial plantations worldwide (*Source:* drawn from Bazett 1993).

Region	Plantation area (million ha)		Total
	Coniferous	Nonconiferous	
North America	12.0	0.5	12.5
Latin America	3.5	2.9	6.4
Asia	31.6	8.2	39.8
Africa	1.3	1.2	2.5
Europe	15.7	3.3	19.0
CIS	16.9	—	16.9
Oceania	2.1	0.1	2.2
Total	*83.1*	*16.2*	*99.3*

temperate plantations grow slowly, as, for example, in the Commonwealth of Independent States (CIS).

Table 7 presents data on what are called fast-growing plantations. These are found largely in the tropics and semitropics, and include the U.S. South. Fast-growing industrial plantations constituted about 25 percent of total industrial plantation area, or about 25 million ha, in the late 1980s. Although commercial harvests are commonly viewed as a major source of deforestation, most knowledgeable observers now recognize that commercial timber harvests generally do not involve the permanent conversion of land to another use. Rather, forest lands involved in commercial harvests typically remain as forest lands. Reforestation occurs either through tree planting, or, less commonly today, through natural regeneration.

Recent experience

A quick reconnaissance of some of the experiences of plantation countries might be of interest. Worldwide, we estimate that something over 4 million ha of forest plantations are being established annually,[3] with a significant portion for commercial purposes.[4]

The United States in recent years has planted between 1.0 and 1.2 million ha (about 2.5–3.0 million ac) annually. This amounts to the planting of 4–6 million seedlings *each day* over a 365-day year. In 1992, 70 percent of the total planting was in the South, and 84 percent of the total was on private lands (Moulton et al. 1993).

Table 7. Fast-growing industrial plantations, circa the late 1980s (*Source:* drawn from Bazett 1993).

Region	Plantation area (1,000 ha)		
Country	Coniferous	Nonconiferous	Total
North America			
U.S. South	12,000[a]	500	12,500
Latin America			
Brazil	1,600	2,300	3,900
Chile	1,140	60	1,200
Argentina	460	180	640
Venezuela	180	20	200
Mexico	60	20	80
Other	80	350	430
Total	3,520	2,930	6,450
Europe-Iberia			
Spain	—[b]	450	450
Portugal	—[b]	400	400
Total	3,520	2,930	6,450
Africa			
Republic of South Africa	500	800	1,300
Angola	20	50	70
Congo	—	40	40
Kenya	160	10	170
Zimbabwe	70	10	80
Other	550	330	880
Total	1,300	1,240	2,540
Oceania			
New Zealand	1,180	20	1,200
Australia	900	60	960
Other	50	30	80
Total	2,130	110	2,240
Asia			
Indonesia	—[c]	100	100
China	—	400	400
Other		170	170
Total		670	670
World total	18,950	6,300	25,250

[a] The conifer plantations of the southern United States are "borderline" fast-growing.
[b] Iberia also has up to 4 million ha of slow-growing conifers.
[c] Indonesia has about 700,000 ha of slow-growing conifers.

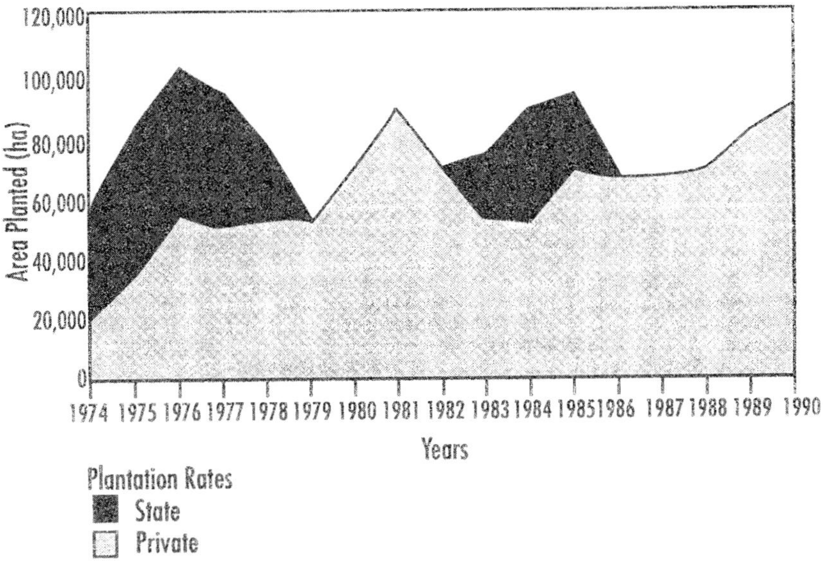

Figure 2. Plantation establishment rates in Chile.

New Zealand has about 1.3 million ha of industrial plantation forests. Although the establishment of new forests was reduced to 10–20 thousand ha in the early 1990s, planting for 1994 is estimated at a record 130,000 ha, the result, in part, of very strong international prices in 1993 and 1994. It is also estimated that another 3–4 million ha of land are potentially available for industrial plantations. Most of this land is from sheep pasture, which has been in decline for decades because of weak market prices.

Essentially all of New Zealand's industrial wood production is from plantations. Production is expected to increase from current production levels of about 15.5 million m^3 annually to more than 25 million m^3 after 2000, as trees already planted reach rotation age.

Chile has become a major focus of plantation establishment and production. In the past one-and-a-half decades, plantation establishment has fluctuated between 50,000 and 100,000 ha annually, and reached 107,500 ha in 1991 (Figure 2). Associated with this has been the development of both a pulping and a solid wood industry.

Indonesia has yet to become a major producer of industrial wood from plantations. However, the country has made very ambitious plans centered around the establishment of plantations and pulpmills. Planning documents call for the establishment of as many as thirty-two pulpmills over the next three decades. Although this may be overly ambitious, Indonesia recently put three new world class mills into operation to supplement the two that already exist. Furthermore, two more mills are expected during the decade of the 1990s, and three additional mills by 2005. Initially, much of the feedstock for the pulpmills is to come from mixed tropical hardwood sources. However, this source will gradually be replaced by plantation wood.

Argentina has about 800,000 ha of industrial forest plantations, and is adding about 25,000 ha/yr of new plantation forests. It has long-term plans for the planting of about 400,000 additional ha. Argentina's growing conditions are quite good and large amounts of low cost land are potentially available for industrial forest plantations. A major drawback with some of the sites in Argentina is their distance from major markets and a weak transportation infrastructure.

Australia has been establishing plantations at a modest rate throughout the post-World War II period. Since the late 1960s that annual rate of establishment has been in excess of 20,000 ha, and reached as high as 35,000 ha in 1978. The vast majority of the early plantings were softwood species, mainly radiata pine. Beginning in the late 1980s there has been a shift to hardwood species, predominantly indigenous. Nevertheless, softwoods still constitute well over half of the new plantings (NAFI News 1995).

In Brazil large areas of mostly degraded agricultural lands have been converted to plantation forests over the past several decades, and especially since the late 1960s. During the 1970s, annual plantings often exceeded 500,000 ha; Brazil has an estimated 6.5 million ha in plantation forest currently. The potential for additional plantations is great, both in terms of land area available and also given the high growth experienced in many Brazilian forests. Although the Jari project in the Amazon has received much notoriety, the great potential for plantations lays largely outside of the Amazon region in the tropical and semitropical region in the south of the country. In recent years, a number of new pulpmills have come into production with much of the output being exported.

Plantations into the future

The past 15 years have demonstrated that forest plantations can be viable financial investments in many parts of the world. Although it was not that long ago that many foresters argued for special criteria for forestry investments (e.g., see Leslie 1977), in recent years the financial returns to plantation forestry in many regions are adequate to promote forestry as an investment capable of competing with alternative uses of capital.

Forest plantations have provided the base for the development of new and vital industries, especially pulp and paper. They have allowed countries to utilize lands that have become largely degraded for agricultural purposes, and to redirect those lands into productive activities that also generate environmental benefits.

Financial returns

These results are consistent with the study results of over a decade ago that found that the financial returns to investments in forest plantations are likely to be favorable in many regions (Sedjo 1983). Signals from the market suggest that the financial returns to plantations have been more than adequate to generate continued and increased investments in planting in many regions. In addition to large and continuing investments in tree-growing and intensive management by many major multinational forest products firms, we now observe a variety of investment instruments being utilized to collect capital for tree growing. The instruments include various types of partnerships, limited partnerships, investments designed for pension programs and the like. These are not confined to the United States, but are also active in places like New Zealand and Argentina.

Additionally, plantation forestry began largely as a government activity in many countries, among them the United States, Brazil, Chile, and New Zealand. Recently, the trend in tree planting has been away from public sector direct planting and subsidies to private sector planting, with diminishing or no subsidies.

For example, in the early post-World War II period, most of the plantings in the United States were by public authorities, usually on public lands. Today, over 80 percent of the area planted is done by the private sector on private lands. Similarly, in New Zealand there has been a dramatic shift in tree planting away from the public sector to the private sector. Much of this has been driven by the broad privatization program of the New Zealand government, which privatized most of the plantation forests and largely eliminated planting and forest management subsidies. Although these actions were initially followed by a sharp decline in planting, recent plantings in New Zealand have been at record levels (MacLaren 1993). Furthermore, most of these plantings are reported as being undertaken by small landowners. In Brazil the initial plantings were propelled by very generous government tax write-offs, and were available almost exclusively to large companies that could use the write-offs. In recent years, however, these write-offs are reported to have been discontinued. Nevertheless, plantings are reported to be high despite the reduction in official government support. Similarly, in Chile the subsidy that once covered about 70 percent of the establishment costs has essentially been eliminated; nevertheless, tree planting is at record levels. Also, in Argentina much of the subsidy on tree planting has recently been reduced.

This is not to suggest that the world no longer contains tree planting subsidies. Rather, the point is that these subsidies are declining and, generally, are less critical to continued activity.

Political and environmental considerations

Plantations require long-term investments, and long-term investments require a degree of political stability. Even the fastest growing plantations require at least 7 years for a harvest rotation. Furthermore, if pulpmills are viewed as part of the investment package, investments can be well in excess of $1 billion and have production lives of perhaps 30 years.

The experience of the 1980s suggests that these types of investments are unlikely to be forthcoming in the absence of a degree of stability. However, in some cases, the political situation may be viewed differently by foreign investors than by locals. Thus, one of the lessons of the past 15 years is that domestic investments may be forthcoming even if international investments are not. After all, Jari in the Amazon continues production, although foreign investors have long since liquidated their interests.

Perhaps the most delicate current issue related to forest plantations is the environmental effects. This may also be one of the most troublesome for the future. Although plantation forests are not likely to ever be viewed as environmentally preferable to native forests, they are widely viewed as superior to most other land uses. Nevertheless, the difficulties associated with the acceptance of forest plantations can be substantial.

The environmental movement appears to have a high degree of ambivalence toward plantation forestry. On the one hand, some individuals with incontestable environmental credentials have agreed that plantation forests offer a viable alternative to management and logging natural forests in the tropics. In essence, plantation forests can serve to deflect logging away from native forests to renewable forest plantations. As shown in Table 6, countries with active industrial plantations tend to rely heavily on their plantation forests for industrial wood.

On the other hand, many environmentalists are suspicious of plantations. Forest plantations are often viewed as displacing either native forests or local agriculture and native peoples. Where the plantations are large, the discussion is often couched in terms of the huge multinational firms displacing workers' livelihoods in the pursuit of export dollars. Furthermore, plantation forests are often viewed as creating a sterile landscape, even when they are established on lands that have long been in marginal agriculture.[5] A related issue involves whether the species planted are native or exotics. Native species, perhaps not surprisingly, appear to face much less opposition than do imported exotics. However, in many parts of the world the timber exotics have a clear advantage in growth and overall financial viability, just as do many introduced species in agriculture (e.g., wheat has been introduced to the Americas, and potatoes to regions outside of the Americas). Especially despised is the *Eucalyptus* genus, which has been accused of all types of foul behavior, including absorbing excessive water, maintaining a sterile understory, and even political crimes.[6]

It is clear that if the plantations are viewed as leading to the destruction of native forests, they will face stiff opposition from environmentalists. On the other hand, if they are viewed as an alternative land use to marginal cropping or pasture, they are likely to be perceived more sympathetically. Finally, if plantations are viewed as an alternative source of industrial wood, which reduces the pressures on native forests, they are likely to have appeal both to environmentalists and to society more broadly.

Aracruz: A Brazilian case study

The political acceptability of plantations appears to be closely related to the degree of environmental hostility that they may create, and also to the extent that they do, or do not, disrupt the existing social order. Aracruz in Brazil provides an interesting case study of the demands and pressures that a plantation enterprise may face[7] (see Campinhos, this volume).

Aracruz Celulose S. A. is a holding company of a group that includes 5 subsidiaries. The operation, located along the coast north of Rio in Espirito Santo, began in the 1960s and includes a plantation area of 132,000 ha, as well as a pulpmill and other facilities. It employs about 3,300 people in its various forestry and processing activities. The mill has a capacity to produce 1,025,000 t/yr of bleached *Eucalyptus* pulp. It exports roughly 90 percent of its pulp, most to Europe and the United States, with the rest remaining in Brazilian markets. The company has roughly $2.5 billion in assets and its shares are traded on the New York stock exchange. Aracruz is famous for its work on improving its *Eucalyptus* genetic stock, and has produced some of the most rapid tree growth ever recorded. Commonly, trees in the Aracruz plantations reach 35 m in height and are ready for harvest in 7 years. The company also produces hybrid seeds, both for its own use and for export. Aracruz has received numerous awards, from the Brazilian government and internationally, for its accomplishments in forestry and exceptional work in developing cloning techniques. Aracruz has also been involved in social and environmental projects in its region, as well as in a program that provides local farmers with seedlings, assistance, and a market for their timber production.

Despite its accomplishments, Aracruz has been under heavy attack from some environmental organizations, including Greenpeace. The opposition to

Aracruz appears to have a number of bases. First, there is a general distrust of forest products companies based upon perceived "past harmful environmental activities." For example, Greenpeace claims that Aracruz destroyed much of the native forest and productive farmland in the process of establishing the forest plantation. Aracruz claims that the tree plantations were established on degraded agricultural lands.

Second, environmentalists express concern over the impact of forest plantations upon the land. Plantation forests are often referred to by environmentalists as "green deserts", which inhibit other vegetation from inhabiting the area. Although Lugo et al. (1993) have published scientific research results that seriously challenge this view of green deserts, this perception continues to be common. Environmentalists often view the *Eucalyptus* genus as particularly negative.

Third, concern is expressed over the effects of pulp production wastes on local water, fishery resources, and human health. Clearly, pulpmill discharges can be environmentally harmful if not treated properly.

Fourth, criticisms of claimed environmental damages are supplemented by criticisms of the impacts for plantations on local peoples. These include charges that agricultural land displaced by the forest plantations was reasonably productive, and conversion to plantations involved the expulsion of local peoples.

Finally, there is recited a litany of grievances against Aracruz. These grievances include poor working conditions, promises not kept, and so forth.

Some lessons

The above criticisms appear to be representative of the types of grievances that are often directed against plantation forests. Without trying to determine the precise facts in this situation, the nature of the charges suggests that there are some lessons as to the types of issues that are likely to arise and that need to be addressed if a large-scale plantation is to be successful. A provisional list of such issues includes the following:

1. *There is a need to try to avoid destroying native forests in the process of establishing plantations.* This rule should hold, if for no other reason than that the destruction of native forests, and their replacement by industrial plantations made up of non-native species, is likely to generate opposition.
2. *Land tenure is important. Private land is generally recognized as the domain of the owner.* Where private land is converted to plantations, there is likely to be a greater acceptance of that prerogative of ownership. Where public lands are involved, interest groups are more likely to believe that they are justified in vying for alternative land uses. Addition-

ally, local peoples who have occupied lands for some period are likely to be viewed as having legitimate interests.

3. *External impacts, or externalities, on other regional resources need to be considered.* To the extent that negative external effects, such as water quality issues, are associated with plantations and/or processing facilities, these larger concerns can become a basis for opposition.
4. *Opposition to plantations is likely to be more modest if native species are used.* For example, there is relatively little opposition to plantations in the U.S. South or the Nordic countries, especially where species native to the region are used. However, the introduction of non-native species is often viewed with hostility and can become the rallying point of opposition. However, because non-native species often have far superior growing characteristics, this dilemma is not easily addressed.

To address briefly each of these points, first, it is almost universally true that investors in forest plantations in the tropics prefer to establish plantations on lands cleared earlier, rather than to clear existing native forests. Areas cleared earlier for agriculture generally provide for much greater ease of establishment; this is because of the absence of competing vegetation, flatness of terrain, location with roads and other infrastructure, and access to markets. However, these areas of cleared agricultural lands are often adjacent to patches of forested area that have never been cleared. This generates a conflict between maintaining pockets of native forest and the establishment of efficiently designed plantations.

Second, where lands are privately owned with clear title, the transfer from agriculture to forestry will usually involve the sale of land, and with it the associated expectation that local peoples relinquish rights to inhabit the area. However, where land is in public ownership and occupied, government transfer of use rights will not necessarily be communicated to or accepted by peoples occupying the land. Commonly, compensation will not be forthcoming (e.g., see Anderson and Huber 1988; Sedjo 1990), and, indeed, the results may highly disadvantage local groups.

Third, external environmental effects on other resources in the area, such as impacts on water availability and quality, wild and aquatic life, and so forth, are important and need to be considered. Furthermore, to the extent that industrial plantations are viewed as crops, and land in plantations is viewed as an alternative to agriculture, forest plantations in general and those utilizing non-native species may receive greater tolerance. This acceptance may be enhanced if local peoples also recognize that many other crops are non-native, and that even non-native tree plantations provide habitat for many native plants and animals.

Finally, hostility to plantations may be mitigated if it is widely understood that plantations take pressure off of native forests for timber harvesting. However, since most deforestation is generated by the desire for agricultural lands, not commercial logging, industrial plantation forests can provide only part of an answer to deforestation.

From where will the world's future industrial wood come?

The exploration of the question of the likely sources of the world's wood supply is not a moot exercise. In the United States, much of the harvest increase over the past 15 years has come from "old-growth" forests, largely in the Pacific Northwest. However, recent concerns about the spotted-owl, the Endangered Species Act, and ecosystem management of public lands have, in effect, taken much of the U.S. public land out of the nation's timber base. Also, in British Columbia, the source of about 40 percent of Canada's industrial wood, the provincial government has committed itself to reducing harvests on public (Crown) lands.

A recent study (Sedjo et al. 1994) uses the Timber Supply Model (Sedjo and Lyon 1990) to examine the impact of reduced timber harvests in the western United States and in British Columbia. This study's projections indicate that reductions in harvests from U.S. public lands are likely to be offset by increased harvests on private lands in the United States, as well as by increased harvesting outside the United States. In Canada, some of the declines in B.C.'s harvests can be offset by increased harvests in Alberta, and the projections show the potential for increased production in eastern Canada. Also, the projections indicate that increasing wood production can be expected from the Nordic countries of Europe. Finally, the study indicates a growing role for the emerging plantation region of the southern hemisphere and the semitropics.

Overall, the study suggests a decreased role for the western United States and British Columbia, because accessible timberland in these areas is declining as land is being withdrawn from the timber base. Although western North America has recently accounted for about 12 percent of the world's industrial wood, this fraction can be expected to decline.

These projections are generated by the base case of the Timber Supply Model, which assumes that there are only modest increases in the establishment of new high-yield plantations worldwide. The base case used for the projections assumed that new high-yield plantations would be established at a rate of about 200,000 ha/yr for 30 years, thereby adding an additional 6 million ha of plantation to the timber base over that period (Sedjo and Lyon 1990). However, the original study examined a high plantation scenario in which the rate of plantation establishment was raised to 600,000 ha annually for 30 years, for a total new plantation area of 18 million ha. This scenario generated higher regional and total harvest levels and associated timber prices that were lower than they would have been in the absence of the greater plantings.

Evidence suggests that current levels of industrial plantation establishment are considerably above 200,000 ha annually. In fact, as noted, just two countries, New Zealand and Chile, are currently planting a total of well in excess of 200,000 ha of new plantation forests annually. Should these high rates of plantation establishment persist, one would expect plantations to play a greater role in providing future timber supplies than anticipated in the base case of the model.[8]

Another possible source of additional industrial wood is from the former Soviet Union and particularly Russia. It is well known that the forest resources in the region are vast, but that the infrastructure is weak. Currently, the industrial wood production of Russia has dropped considerably from levels of the late 1980s, apparently as a result of disruptions related to recent political changes. Nevertheless, despite the disruptions, log exports to Japan in recent years have risen in response to the financial incentives available in that market.[9] In the near term it appears unlikely that Russia could expand industrial production sufficiently to offset decreases elsewhere and increasing demand. In future decades, however, the potential for substantial increases in industrial wood harvest will exist.

Conclusions

This paper has shown how dramatically industrial wood from plantations has increased over the past 15 years and its likely rising role in the future. Based upon existing plantation inventories and planting rates, it appears likely that plantation source wood will grow even more rapidly in the next 15 years and beyond.

In recent years, there have been two impediments to plantations. The first relates to concerns over political stability and the unwillingness to make long-term financial commitments in an unstable political environment. This concern has generally been on the decline in much of an increasingly internationalized world, especially with the debt crises of developing countries largely behind us. However, the recent financial crisis in Mexico may have revived these concerns. The second impediment is found in the objections to plantations that are being made by some environmental groups. To overcome these objections it must be demonstrated that plantation forestry can serve a protective function for native forests, generate positive environmental benefits, and mitigate any associated environmental damages and social disruptions.

Despite impediments, substantial volumes of trees planted over the past 15 years will soon be ready for harvest in such countries as New Zealand and Chile. Also, the high wood prices of the last 2 years appear to have triggered an upsurge in planting, not only in the United States, but worldwide. Furthermore, mills are being constructed to utilize the wood that will be available. The rapid expansion of pulpmills associated with fast-growing plantations in the tropics and semitropics is likely to continue into the foreseeable future, and, indeed, it will likely become the major engine of worldwide pulp production expansion.

Notes

1. By some definitions these might be characterized as intensive forest management, rather than as plantations.
2. Despite some notable exceptions, e.g., the Jari plantations in the Amazon, there are distinct advantages to locating plantations on degraded agricultural lands. These advantages include availability of roads, modest slopes, favorable location *vis a vis* markets, and so forth.
3. Estimates of annual rates of plantation establishment vary. The WRI (1986) estimated that in 1980 there were 14.5 million ha of forest land is reforested or renewed annually. However, the area actually planted is probably substantially less. Mather (1990, p. 86) estimates plantings in the 1980s to be about 10 million ha annually. That estimate may also be too high as it assumes 4.8 million ha annual plantation establishment by China and 1.3 million ha annually by the former U.S.S.R. By the mid-1990s, given the political changes in both countries, these rates have probably decreased considerably. Hence, our estimate of about 4 million ha annually for the mid-1990s seems reasonable.
4. We estimate new exotic industrial plantation establishment in the tropics and semitropics to be about 600,000 ha annually. South America is establishing about 250,000 ha annually; New Zealand and Australia, 150,000 ha annually; and Southeast Asia-Pacific, about 200,000 ha annually. This does not include China.
5. This view has been challenged by the research of Lugo, Parrotta, and Brown (1993), which indicates that, contrary to the conventional wisdom, tree plantations located in the tropics provide habitat that allows for the flourishing of a large number of native plant species. Their work shows that number of native plant species increase in the understory of tree plantations in about the same numbers as in secondary native forests.
6. An FAO forester relays the experience of discussions with a Spanish environmentalist who was vehemently opposed to the establishment of *Eucalyptus* plantations. Frustrated that his objections were refuted with technical responses, the exasperated environmentalist finally blurred out, "But *Eucalyptus* is a fascist tree."
7. This section draws heavily from "Aracruz Celulose S. A. A Brazilian Kraft Pulp Company Confronts Sustainable Development", May 1994, a senior thesis by Meredith Mallis Persily submitted to the Center for Comparative Study of Development, Department of Portuguese and Brazilian Studies, Brown University.

8. The form that "foreign" wood might take is an open question. Although it is possible that it would be imported as logs or chips and processed in the United States, it is more likely that much or all of the processing would take place outside the United States.
9. The Japanese Wood Supply and Demand Information Service reported increases in Russian log exports to Japan beginning in 1993.

References

Anderson, R.S. and Huber, W. 1988. The Hour of the Fox: Tropical Forests, the World Bank and Indigenous People in Central India. University of Washington Press, Seattle.

Bazett, M. 1993. Industrial Wood. Shell/WWF Tree Plantation Review Study No. 3. WWF (United Kingdom), Panda House, Weyside Park, Godalming, Surrey.

Clapp, R.A.F. 1993. The forest at the end of the world: The transition form old-growth to plantation forestry in Chile. Ph.D. dissertation, Department of Geography. University of California, Berkeley, CA.

FAO Yearbook of Forest Products. 1982–93. Food and Agricultural Organizations of the United Nations, Rome, 1994.

Leslie, A.J. 1977. Where contradictory theory and practice co-exist. Unasylva *29* 115: 2–17.

Lugo, A.E., Parrotta, J.A. and Brown, S. 1993. Loss in species caused by tropical deforestation and their recovery through management. Ambio 22: 106–109.

MacLaren, J.P. 1993. Radiata Pine Growers' Manual. Forest Research Institute Bulletin No. 184. Forest Research Institute, Rotorua, New Zealand.

Mather, A. 1990. Global Forest Resources. Timber Press, Portland, OR.

Moulton, R.J., Mangold, R.D. and Snellgrove, J.D. 1993. Tree Planting in the United States 1992. USDA Forest Service, Washington, DC.

NAFI (National Association of Forest Industries) News. 1995. NAFI, Canberra.

Pandey, D. 1992. Assessment of tropical forest plantation resources. Department of Forest Survey, Swedish University of Agricultural Sciences. Unpublished report to Forestry Department, Food and Agriculture Organization of the United Nations, Rome.

Persily, M.M. 1994. Aracruz Celulose S. A.: A Brazilian kraft pulp company confronts sustainable development. Senior thesis, Center for Comparative Study of Development, Department of Portuguese and Brazilian Studies. Brown University.

Sedjo, R.A. 1983. The Comparative Economics of Plantation Forestry: A Global Assessment. Johns Hopkins Press for Resources for the Future, Baltimore, MD.

Sedjo, R.A. 1990. Review of "The Hour of the Fox." Econ. Dev. Cult. Chang. 39(1): 215–220.

Sedjo, R.A. and Lyon, K.S. 1990. The Long-term Adequacy of World Timber Supply. Resources for the Future, Washington, DC.

Sedjo, R.A., Wiseman, A.C., Brooks, D. and Lyon, K.S. 1994. Global forest products trade: The consequences of domestic forest land-use policy. Discussion Paper 94–13. Resources for the Future, Washington, DC.

Suchek, V.I. 1991. The role of the planted forest in the pulp and paper industry of Brazil. For. Chron. 67(6).

WRI (World Resource Institute, Washington D.C.) 1986. World Resources 1986. Basic Books, New York.

Case Studies of Planted Forests

New Forests **18:** 3–4, 1999.

Planted forests, agriculture, and the environment – a perspective

ROBERT E. BUCKMAN
Professor (retired), College of Forestry, Oregon State University, Corvallis, Oregon; and Past President, International Union of Forestry Research Organizations

This volume provides the opportunity to include in our view of forestry a focus on an agronomic perspective as well as a naturalistic one. Planted forests have many parallels and much complementarity with the practices and concerns of agriculture. For example, after more than a half century of spectacular increases in food production, world agriculture today searches for the next generation of productivity gains in plant and animal crops.

Likewise, forests the world over are required to produce increasing quantities of commodities – fuelwood, pulp and paper, solid wood products, extractives such as latex and resin, edible fruits and nuts, forage for livestock, and food and shelter for wildlife. Planted forests play a vital and growing role in satisfying many, if not most, of those needs.

After several millennia of only modest gains, spectacular increases in agricultural productivity began in the 1920s, driven primarily by science and the educational and outreach programs associated with it. Starting with hybrid corn, many technologies followed. The upshot is that today, for example, the United States feeds 260 million people and enjoys a substantial export market on an agricultural land base about the same size as existed with 100 million people in the 1920s.

Productivity gains in forestry, many of them associated with planted trees, have also been substantial, but not nearly as well documented as those in agriculture. Here in the Pacific Northwest, U.S.A., a vigorous Douglas-fir genetics program began with Thornton Munger's 1912 seed source study on the nearby Wind River Experimental Forest. Today an extensive cooperative Douglas-fir tree improvement program exists with generational gains in volume growth in the range of 8–12 percent. As documented in papers here, gains from other species have been far more spectacular. Nor are productivity gains confined to volume growth – similar improvements have been made in extractives such as latex, gum naval stores, maple syrup and many other forest-derived products, all of which depend on planted trees.

There are three important conclusions to be drawn from the agriculture/forestry parallels. The first is that productivity gains are an essential strategy to conserve and otherwise make land available for parks, refuges, biological reserves and other relatively natural conditions. I am often disappointed by the failure of those who lean strongly toward the naturalistic side of forestry to recognize their stake in intensive agriculture and forestry.

The second conclusion is that a strong research and outreach program is the most cost effective approach available to increase the supply of commodities and services (including ameleration of environmental consequences) from agricultural and forest lands.

And, the third conclusion is that planted forests offer scientific and technical parallels with agriculture on such questions as sustainability, pest management, water quality and yield, soil protection and enhancement, plant breeding, and gene conservation. Protection and restoration of upstream watersheds, reduction of water and wind-caused erosion, avoidance of soil salinization and desertification, and enhancement of wildlife habitat are but a few of the practices for which agriculture and planted forests are complementary and mutually supportive of each other. We have much to learn from agriculture, and much to gain by working together.

The case studies that follow are illustrative of the applications of many lessons from agricultural sciences to the practices of growing trees for human values.

Certainly planted forests have and will continue to serve purposes beyond the parallels with agriculture, including surface-mine rehabilitation, home and farmstead protection, urban green-space enrichment, and other aesthetic and wildlife aims. But we should not forget that healthy and productive agriculture and forestry together are the most powerful strategy available to conserve wildlands for such other uses as parks, refuges, wilderness, green space, and other non-commodity purposes. If basic needs for food and a variety of forest products are not satisfied on intensively managed lands, then aggressive exploitation of undeveloped land will continue.

New Forests **18:** 5–15, 1999.

Multiple rotation yields for intensively managed plantations in the Amazon basin

KENNETH L. MCNABB[1] and LINEU H. WADOUSKI[2]
[1]*School of Forestry, Auburn University, Auburn, AL, USA;* [2]*Forestry Consultant, Curitiba, Paraná, Brazil*

Key words: Eucalyptus, Jari Cellulose, Pinus caribaea, planted forests, plantation productivity, sustainability

Abstract. Jari Celulose S. A. is a Brazilian-owned industrial enterprise located along the Jari River (the first major south-flowing tributary of the Amazon). The company produces bleached kraft pulp for international and domestic sales. The raw material of this pulp production comes from the 75,000 ha of intensively managed plantations of *Pinus caribaea* (Morelet) and *Eucalyptus* spp. harvested on 10- and 6-year rotations, respectively. All currently active Jari plantations were converted from primary forest between 1968 and 1982. Silvicultural practices have been modified over successive rotations, including plantation species changes, increased intensity of site preparation, fertilization, and chemical weed control. A review of data from the early 1970s to present indicates an increase in productivity over successive rotations and several plantation locations and soil types. The reasons for this improvement are primarily related to the higher levels of silvicultural inputs and dramatic genetic gains. The long-term sustainability of this productivity is discussed in terms of current and future inputs.

Introduction

Intensively managed plantations are increasingly employed to provide the fiber requirements of a growing world population (Sedjo and Lyon 1992). By increasing the capital inputs for genetic improvement, site preparation, weed control, and fertilization, plantation managers are able to produce more woody biomass per unit of land area. A number of countries throughout the world are expanding their plantations and relying on intensively managed areas rather than extensively managed natural forests as a source of fiber (Gladstone and Ledig 1990; Evans 1992). Recently, however, the issue of long-term sustainability of these artificial systems has been a topic of concern. Can the repeated extraction of fiber products based on intensive management decrease inherent site productivity over successive rotations? Even though fertilizer elements can be added to maintain system nutrient levels, do

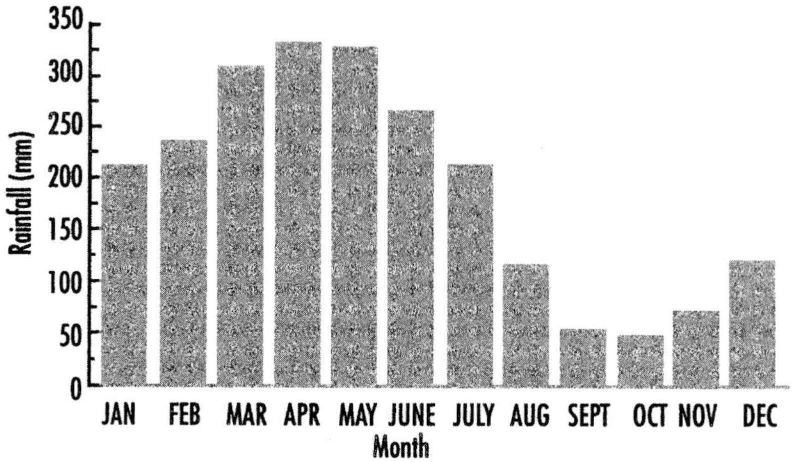

Figure 1. Average monthly rainfall at Jari Celulose S. A. (1967–1994).

cultivation, weed control, and harvesting activities somehow reduce system integrity?

Multiple rotation productivity is difficult to assess for obvious reasons. Under ideal conditions this would require identical genotypes planted on the same piece of land with the same silvicultural techniques under a relatively short rotation so that results can be obtained in a reasonable period of time. In the temperate zone, even if cultural practices could be duplicated from one generation to the next, the long rotation periods limit the practicality of long-term productivity studies. On the other hand, tropical plantations are frequently managed on shorter rotations, particularly when used for pulp and paper products. There are few intensively managed plantation enterprises in the tropics, however, that have been in existence long enough to provide multiple rotation data. One potential plantation operation in the tropics with suitable data in this regard is owned by Jari Celulose S. A.

Jari Celulose S. A. is a Brazilian-owned forest products corporation located on the banks of the Jari River, the first major south-flowing tributary of the Amazon. The company originated in 1967 when Daniel K. Ludwig, an American shipping industrialist, purchased the tract as a single 1.5 million ha unit from a group of Portuguese investors (Lins 1991). Ludwig was the owner until 1982 and during that period over 100,000 ha of rain forest were converted to single-species plantations. Since 1982, Jari Celulose S. A. has been owned by a consortium of Brazilian businesses that continues to manage plantations for the raw material of bleached kraft pulp, most of which is sold on the international markets.

The majority of Jari plantations were originally established during the early to mid-1970s by slash-and-burn techniques on a large scale. The original vegetation may be classified as semi-evergreen rain forest with a layered tree canopy, high floristic diversity, and few herbaceous species, typical of Amazonian forest types (Pires and Prance 1985). Rainfall at Jari averages 2,295 mm/yr (Figure 1), but is not evenly distributed. There is a distinct dry season when burning is feasible, and a wet season when planting is scheduled. The first plantations were established with *Gmelina arborea* (Roxb.), a species native to Southeast Asia and selected for its excellent wood properties and fast growth. It was soon discovered, however, that *Gmelina* did not do well on many Jari soils and was frequently attacked by a stem canker disease caused by *Ceratocystis fimbriata* (Muchovej et al. 1978). As a result, Caribbean pine (*Pinus caribaea* [Morelet] var. *hondurensis* [Senecl.]) was planted on many of the sandier, drier, and more nutrient-poor locations. Several native species and a variety of exotics were eventually researched. By the early to mid-1980s pine was the primary plantation species, *Eucalyptus* was increasing in area, and *Gmelina* was nearly abandoned. The Brazilian owners continued the shift to *Eucalyptus* and in 1992, 10,524 ha of *Eucalyptus* were planted. No *Gmelina* or pine were planted. Along with the shift to *Eucalyptus* was an increase in management intensity that included plantation establishment based on clonal selection and propagation, mechanical site preparation, chemical and mechanical weed control, fertilization, and mechanized harvesting (Mc Nabb et al. 1994).

Some plantation areas have been harvested and replanted through four rotations since the early 1970s. There have been many changes in the manner in which these plantations have been managed during the past 28 years. Not only have species changed, but also site preparation methods, cleaning intensity, fertilization, and other cultural practices may vary from rotation to rotation. This undoubtedly complicates the assessment of the long-term sustainability of plantation productivity. Nevertheless, there are few places, if any, in the humid tropics where multiple rotation productivities have been quantified over several rotations.

Methodology

Permanent inventory plots were routinely installed in all plantations at the time of their establishment. These plots have been maintained as a fixed number over successive rotations and can be used to study productivity trends over the history of a particular plantation. Three plantations were selected for study based on their location within a particular soil type. Jari soils have been mapped and 25 distinct soil mapping units have been identified and classified into one of eight general soil types (Tables 1–3). For purposes of this study, we selected three plantations located on heavy, medium, and light textured soils, which corresponded to high, medium, and low site productivities, respectively. Together the three selected soil mapping units make up 45 percent of the total planted area at Jari. The three selected plantations were originally cleared in 1972 and planted in early 1973.

Another plantation selection criterion was the availability of inventory data for at least ten plots in each plantation since its establishment. Because the three selected plantations had more than ten inventory plots, individual plot selection within a plantation was done at random. To calculate plantation productivity, these ten plots were averaged for a certain plantation age. The age used to calculate productivity was dependent on the availability of data for this age across all rotations. In a few cases the original plantation was later subdivided into different species and silvicultural treatments. In order to standardize species and silvicultural treatments within a plantation, some plots were dropped when calculating plantation productivity. In summary, twenty-four plots were used out of a possible thirty.

The total volume/ha, calculated on a 7-cm top and outside bark diameter, was used to calculate productivity. Annual growth/ha was used as a measure of productivity because it helps compensate for the effect of changing species and spacing, which can vary among plantations and rotations within a plantation. For example, height is often used to assess site quality but is unsuitable for comparing across species. *Eucalyptus* has a much different growth form than does *Gmelina*. The former tends to grow taller in the early stages of its life cycle, whereas the latter tends to grow relatively more in diameter. Total volume/ha/yr (growth rate) reflects both height and diameter growth.

It is important to note that this is not an attempt to calculate overall productivity across all plantations. Rather, we attempt to review and discuss productivity data for a select group of inventory plots taken from sites with a variety of soils with the objective of gaining some insight as to the general trends of site productivity over successive rotations.

Results

The three selected plantations vary considerably, not only in their soil characteristics and silvicultural histories, but also in the productivity of successive rotations. Plantation 42 (Figure 2, Table 1) was established on the most productive site, as indicated by the average 25 m^3/ha/yr growth rate during the first rotation at age 3 years. This was almost double the productivity of Plantation 43 (Figure 3), and more than three times that of Plantation 45 (Figure 4) during the first rotation. None of these plantations were fertilized

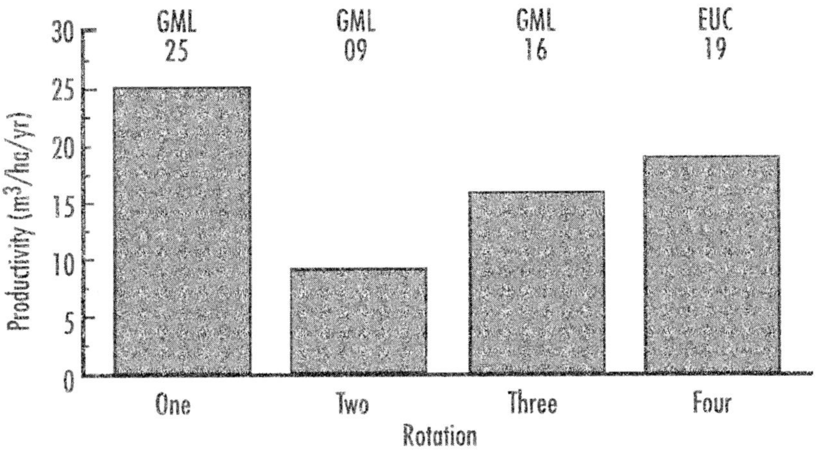

Figure 2. Productivity of plantation 42 at age 3 years (GML = *Gmelina*; EUC = *Eucalyptus*).

Table 1. Soils and silvicultural history for plantation 42 at Jari Celulose.

Terrain: Flat to gently rolling

Soil group: Well-drained, dark red to yellow red allic latosol (Corrêa et al. 1989)

Soil chemistry:

Soil horizon	Depth (cm)	Clay (%)	Carbon (%)	Aluminum (% Sat.)	Phosphorus (ppm)	Base (% Sat.)	pH
A	0–18	67	3.5	30	3	16	4.7
B	18–110	76	1.4	88	<1	2	4.6
C	110–>140	71	0.9	88	<1	3	4.5

Silvicultural history:

Rotation 1: Slash and burn native forest 1972, plant to *Gmelina arborea* 1973, several manual weedings, leaf-cutting ant control, harvest in 1980–81 at age 7 years.

Rotation 2: Coppice regeneration of *Gmelina arborea*, manual weeding, ant control, harvest 1985 at age 5 years.

Rotation 3: KG blade shear, burn, windrow, disk, plant to *Gmelina arborea*, fertilize (166 kg/ha 12-36-05), ant control, harvest 1990–1991 at age 6 years.

Rotation 4: Drag chain, burn, pile disk, subsoil, plant to *Eucalyptus urophylla* × *E. grandis* hybrid in 1991, fertilize (166 kg/ha 6-30-6), chemical and manual weed control, ant control.

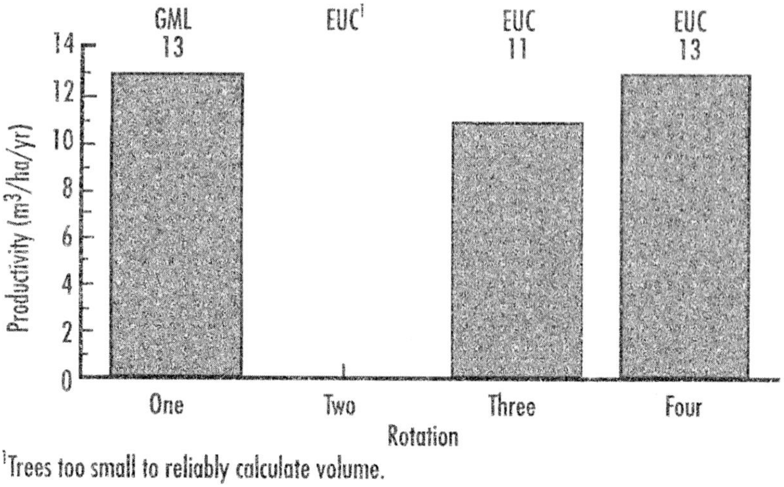

Figure 3. Productivity of plantation 43 at age 3 years (GML = *Gmelina*; EUC = *Eucalyptus*).

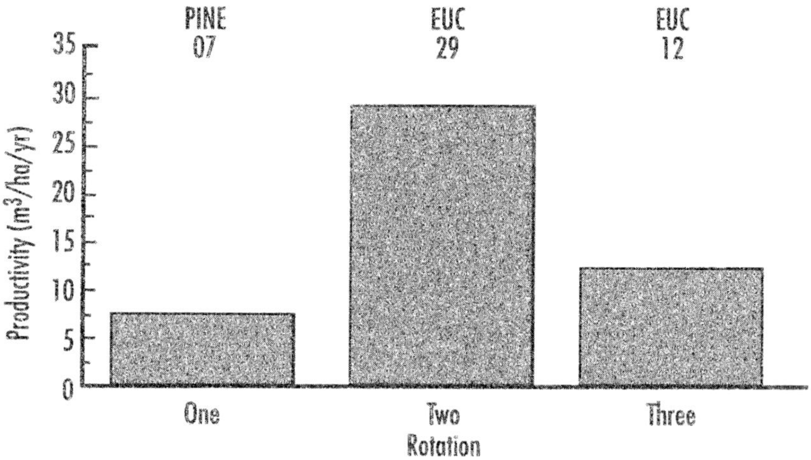

Figure 4. Productivity of plantation 45 at age 4 years (EUC = *Eucalyptus*).

during the first rotation, and productivity was largely a combination of site index and vegetative competition.

First rotation productivity was based on well-known slash-and-burn methods. This traditionally successful agricultural technique releases nutrients previously contained in native forest biomass and makes them readily available to a select crop. If the burn is not excessively intense, the soil organic matter and original soil structure are also maintained. The quality of the burn was an important factor affecting growth during the first rotation. The amount and type of vegetative competition was usually related to the quality of the

Table 2. Soils and silvicultural history for plantation 43 at Jari Celulose.

Terrain: Flat to gently rolling

Soil group: Well-drained, dark yellow to red yellow allic latosol (Corrêa et al. 1989)

Soil chemistry:

Soil horizon	Depth (cm)	Clay (%)	Carbon (%)	Aluminum (% Sat.)	Phosphorus (ppm)	Base (% Sat.)	pH
A	0–31	29	0.6	52	1	12	5.2
B	31->140	36	0.2	81	1	5	4.5

Silvicultural history:

Rotation 1: Slash and burn native forest 1972, plant to *Gmelina arborea* 1973, several manual weedings, leaf-cutting ant control, harvest in 1980 at age 7 years.

Rotation 2: Chainsaw fell, burn, plant to *Eucalyptus deglupta* in 1981, manual weeding, ant control, harvest in 1984 at age 5 years.

Rotation 3: KG blade shear, burn, windrow, disk, plant to *Eucalyptus deglupta* in 1985, fertilize (166 kg/ha 12-36-05), ant control, manual weed control, harvest in 1990–1991 at age 6 years.

Rotation 4: Drag chain, burn, pile, disk, subsoil, plant to *Eucalyptus urophylla* in 1991, fertilize (166 kg/ha 6-30-6), chemical and manual weed control, ant control.

burn. Less manual cleaning was needed after a good burn, and weeds were less of a problem. Plantation records indicate that slash-and-burn methods resulted in productive plantations. Plantation 42 productivity at 25 m^3/ha/yr is quite good for an unimproved *Gmelina* seed source in the 1970s. The same growth was not achieved in plantations 43 and 45 because of lower site quality.

There was a substantial decline in productivity during the second rotation for Plantation 42, from 25 to 9 m^3/ha/yr. This rotation was coppiced and, not only was the productivity of a coppice system less than expected, but the plantation was also severely affected by *Ceratocystis fimbriata*. The infection rate of this stem canker disease may have been accelerated by a coppice system where stumps provided an infection point. Productivity increased during the third rotation to 16 m^3/ha/yr, when the area was replanted with seedlings and fertilized. Stem canker disease was much less a factor than in the previous rotation. In the current or fourth rotation, productivity continued to improve, and increased to 19 m^3/ha/yr. This rotation had a change of species from

Table 3. Soils and silvicultural history for plantation 45 at Jari Celulose.

Terrain: Flat ot gently rolling

Soil group: Excessively well-drained, dark gray to dark yellow allic latosol (Corrêa et al. 1989)

Soil chemistry:

Soil horizon	Depth (cm)	Clay (%)	Carbon (%)	Aluminum (% Sat.)	Phosphorus (ppm)	Base (% Sat.)	pH
A	0–45	13	0.8	86	1	3	4.7
B	45–>170	20	0.2	100	<1	<1	4.4

Silvicultural history:

Rotation 1: Slash and burn native forest 1972, plant to *Pinus caribaea* var. *hondurensis* in 1973, several manual weedings, leaf-cutting ant control, harvest in 1980–1981 at age 7 years.

Rotation 2: Chainsaw fell, burn, plant to *Eucalyptus deglupta* in 1982, manual weeding, leaf-cutting ant control, harvest in 1987 at age 5 years.

Rotation 3: Chainsaw fell, burn, plant to *Eucalyptus urophylla* in 1988, fertilize with triple superphosphate, leaf-cutting ant control, harvest in 1993 at age 6 years.

Rotation 4: Drag chain, burn, windrow, bed, plant to *Eucalyptus urophylla* × *E. grandis* hybrid in 1994, fertilize with 150 kg/ha of 15-00-15 plus boron and zinc, manual weeding, ant control.

Gmelina to *Eucalyptus grandis* (Hill × Maiden) × *E. urophylla* (S. T. Blake) hybrid and was also fertilized. It did not, however, achieve the productivity of the first rotation, although it is evident that productivity has gradually risen through the last three rotations.

Plantation 43 (Figure 3, Table 2) is considerably different from Plantation 42 in that it has lower site quality. First-rotation *Gmelina* averaged 13 m^3/ha/yr at age 3 years. *Eucalyptus deglupta* (Blume) was planted for the second rotation, but did so poorly that 3-year-old trees were too small to reliably calculate volumes. Second rotation declines may be partly related to decreased expenditure for silvicultural operations in the early 1980s. Ludwig sold the company in 1982, and, for the years surrounding this date, investments for cleaning and ant control were kept to a minimum. *Eucalyptus urophylla* was planted for the third rotation and productivity increased to 11 m^3/ha/yr followed by 13 m^3/ha/yr in the fourth rotation, also planted with *E. urophylla*. Both the third and fourth rotations were fertilized. Notably, the fourth rotation has achieved the same productivity as the first.

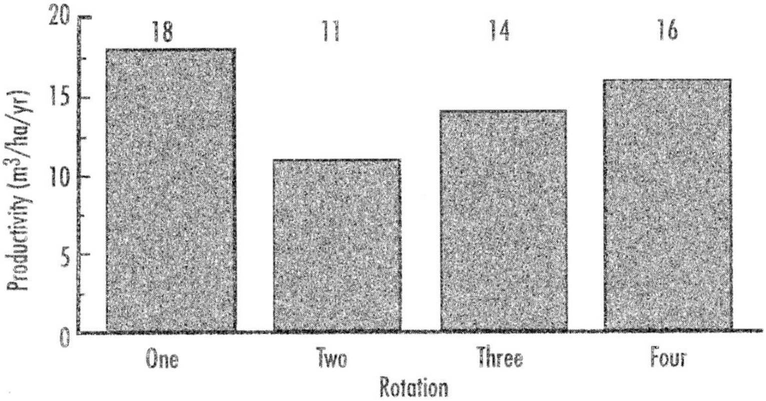

Figure 5. Average productivity of plantations 42 and 43 at age 3 years, and plantation 45 at age 4 years.

The lowest site productivities are to be found in Plantation 45 (Figure 4, Table 3). Characterized by an excessively well-drained sandy soil with low moisture retention, moisture deficits affecting plant growth could be severe from August to November. The first rotation for this site was planted to *Pinus caribaea* var. *hondurensis*. Pine productivities were low (7 m^3/ha/yr), and the second rotation was planted to *Eucalyptus deglupta*, which improved productivity dramatically to 29 m^3/ha/yr, a 4-fold increase. However, this was not maintained in the third rotation. *Eucalyptus urophylla* was planted instead of *E. deglupta* and productivity fell to 12 m^3/ha/yr even though the area was fertilized with phosphate.

The drop in productivity from the second to the third rotation in Plantation 45 most likely resulted from changing to an unsuitable genotype. *Eucalyptus urophylla* had done comparably well on other sites and, obviously, pine productivity was not very good in the first rotation. Even though *E. urophylla* did not do as well as *E. deglupta* on this site, the 12 m^3/ha/yr growth rate is comparable to *E. urophylla* in Plantation 43. The fourth rotation, planted in 1994 with the *E. urophylla* × *E. grandis* hybrid, was also fertilized, including the micronutrients boron and zinc.

Overall plantation productivity (Figure 5) averaged across sites and species, shows a general decline from the first to the second rotation, followed by increases in the third and fourth rotations. Second rotation productivity dropped from 18 to 11 m^3/ha/yr, a decrease of 40 percent. However, the third rotation increased to 14 m^3/ha/yr, although this is still 22 percent below the first rotation. Moreover, in those areas where the fourth rotation is now measurable, productivity continues to increase. The average of plantations 42 and 43 at age 3 years during the fourth rotation is 16 m^3/ha/yr. This is below overall company productivity of about 20 m^3/ha/yr (Mc Nabb 1994), which is the average across all sites and all plantations at the time of harvest. It should be noted that productivity estimates may be higher if older trees are used for calculation. The mean annual growth rate for 6-year-old *Eucalyptus*, for example, would be higher than that of 3-year-old trees.

The increases in productivity during the latter rotations are the result of three factors: (1) improved fertility management, (2) more intense site preparation and weed control, and (3) improved genotype × site matching. There is substantial evidence that biomass removal in harvesting and leaching contribute to nutrient loss (Russell 1987). Fertilization is undoubtedly an important part of the increased plantation productivity from the second to the fourth rotations. Current silvicultural prescriptions routinely include nitrogen, phosphorus, or potassium granular fertilizers. Micronutrients are added when the need becomes apparent. Like any intensively managed industrial enterprise, fertility management requires knowledge of soils and crops so that their interaction in relation to nutrient addition and availability is known. The company is currently in the process of preparing a detailed soil map of all plantations so that individual clones can be better matched to soils.

The increased intensity of site preparation and weed control has also positively affected plantation productivity. Mechanical site preparation was tried in the late 1960s, but was abandoned because of perceived problems with soil compaction. It is now known that machines can be used at Jari. As is true for their use in site preparation anywhere, however, soil disturbance must be kept to a minimum. The use of pre-emergent herbicides was made feasible with the mechanization of site preparation, and this is followed up with manual cleaning when necessary and rotary hoeing when feasible. Vegetative competition has been a serious hindrance to productivity through all rotations.

Finally, plantation productivity has been enhanced through an aggressive program of tree improvement. The *Eucalyptus* clonal selection and propagation techniques used elsewhere in Brazil have been adopted (Mc Nabb 1994). The productivity increases seen in the fourth rotation are primarily the result of improved genetics. There is no reason to believe that productivity cannot continue to increase as more becomes known regarding clone selection, propagation, and site genotype interaction. In fact, pilot experimental plots in plantations 42, 43, and 45 have attained growth rates of 42, 30, and 25 m^3/ha/yr, respectively, well above the first rotation averages.

Conclusions

Plantation productivity at Jari Celulose has increased during successive rotations, although it has not yet matched the productivity of the original plantations established with slash-and-burn techniques. The selection and manipulation of silvicultural techniques significantly influenced plantation productivity during the second through the fourth rotations. Maintaining and improving plantation productivity through these rotations have been a result of increased management intensity based on experience and research. The knowledge accumulated over nearly three decades of plantation management is the key factor in maintaining productivity. It is expected that continued investment in the production of new information through a focused research program will increase productivity beyond that of the first rotation. The combination of institutional experience and organized research has proven effective in sustaining productivity. Knowledge is the basis for sustainability.

References

Corrêa, J.J., Jocomine, P.D.T. and dos Santos, R.D. 1989. Solos do Jari: Levantamento de Reconhecimento de Média Intensidade dos Solos sob Floresta Cultivada na Área da Companhia Florestal Monte Dourado no Jari. Intelect Informatica Ltda, Rio de Janeiro.

Evans, J. 1992. Plantation Forestry in the Tropics. Oxford Science Publications, New York.

Gladstone, W.T. and Ledig, F.T. 1990. Reducing pressure on natural forests through high-yield forestry. For. Ecol. Manage. 35: 69–78.

Lins, C. 1991. Jari 70 anos de história. Dataforma, Rio de Janeiro.

Mc Nabb, K. 1994. Silvicultural techniques for short rotation Eucalyptus plantations in Brazil, pp. 89–96. In: Stokes, B.J. and Mc Donald, T.P. (Eds.) Mechanization in Short Rotation, Intensive Culture Forestry. Proceedings of the International Energy Association/Biomass Agreement Task IX, Comp. Southern Forest Experiment Station, USDA Forest Service, Auburn, AL.

Mc Nabb, K., Borges, J. and Welker, J. 1994. Jari at 25, an investment in the Amazon. J. For. 92(2): 21–26.

Muchovej, J.J., Albuquerque, F.C. and Ribeiro, G.T. 1978. *Gmelina arborea* – a new host of *Ceratocystis fimbriata*. Plant Dis. Rep. 62: 717–719.

Pires J.M. and Prance, G.T. 1985. The vegetation types of the Brazilian Amazon, pp. 109–145. In: Prance, G.T. and Lovejoy, T.E. (Eds.) Key environments: Amazonia. Pergamon Press, New York.

Russell, C.E. 1987. Plantation forestry: The Jari Project, Para, Brazil, pp. 76–89. In: Gordan, C. (Ed.) Amazonian Rain Forest: Ecosystem Disturbance and Recovery. Springer-Verlag, New York.

Sedjo, R.A. and Lyon, K.S. 1992. The Long-term Adequacy of World Timber Supply. Resources for the Future, Washington, DC.

New Forests **18:** 17–32, 1999.

Sustainable management of native and exotic plantations in Australia

JOHN TURNER[1], STANLEY P. GESSEL[2†] and MARCIA J. LAMBERT[1]
[1]*FORSCI Pty. Ltd., Beecroft, New South Wales, Australia;* [2]*College of Forest Resources, University of Washington, Seattle, WA, USA* (*[†]passed away*)

Key words: establishment, *Eucalyptus*, fertilization, monitoring, *Pinus radiata*

Abstract. The emphasis of plantation management changes as the resource and the market develop. This is especially the case when a plantation program is developing a new timber resource, is not the case with many of the *Pinus radiata* (D. Don) plantings in the southern hemisphere. Australia establishes and manages plantations of both exotic conifer and native hardwood (*Eucalyptus* spp.) plantations, and these vary in their stage of development. The tenure and objectives in establishment have varied, but some key aspects of the resources may be analyzed. Optimization of production per unit area was not a prime objective during the developmental stages of many exotic pine plantations. Currently, with increased commercial emphasis, this has changed to a greater focus on increased value through Site Specific Management and tree improvement through which gains of at least 20 percent are expected during the first stages. With a key objective of sustainability, questions regarding impacts of soils, water, and biological changes need consideration and are being addressed.

The eucalypt plantation resource in Australia is smaller in extent than is the pine resource, but of increasing importance, especially as the plantations are perceived to be more environmentally and ecologically acceptable than exotic conifers. In the past, questions of productivity, especially in relation to impacts of natural pests and diseases, have been raised. Sustainability of all plantations is a critical aspect, however, for specific issues there are different emphases with different species. For example, the relatively, high removal of calcium in smooth barked *Eucalyptus* plantations is seen as important in long term forest management.

Introduction

Countries in the southern hemisphere have initiated significant establishment programs of fast-growing, forest plantations for timber production. In many cases, the plantations have been established with the primary purpose of producing forest products for the future, whereas for others there also was the objective of alleviating pressure on native forests for timber production to allow them to be managed for other values. There are now in excess of 15 million ha planted, about 9 million ha of which are coniferous with the

Table 1. Areas of coniferous and hardwood plantations in 1993 in the southern hemisphere[a].

Region Country	Area (1,000 ha)		
	Coniferous	Hardwood	Total
Total Australasia	2,362	163	2,525
Australia	960	123	1,083
New Zealand	1,378	22	1,400
Total South America	5,419	4,509	9,928
Brazil	3,383	3,617	7,000
Chile	1,385	215	1,600
Total Africa	1,585	1,150	2,735
South Africa	842	658	1,500
Total	9,366	5,822	15,188

[a]Only countries with the larger areas of plantings have been listed individually.

remainder hardwood (Table 1). The coniferous plantings are dominated by *Pinus radiata* (D. Don) with smaller areas of southern and Caribbean pines. Hardwood plantings have primarily been of the genus *Eucalyptus*, native to Australia, with the bulk consisting of *Eucalyptus grandis* Hill ex Maiden, *E. globulus* Labill., and *E. nitens* (Deane & Maiden) Maiden. Except for the establishment of plantations of *Eucalyptus* within Australia, plantations in the southern hemisphere are composed of exotics.

The selection of species for planting has generally been made on the basis of rapid growth and the suitability of the timber for an industrial market. A relatively minor component of the overall area of southern hemisphere plantations is in Australia (Table 1), but the area is expanding as a result of the demand for timber, especially softwoods, and pressure to reduce utilization of native forests.

The aim of this paper is to present an analysis of some planting programs in Australia, to consider the impacts of research, and to indicate how the objectives of management and community acceptance change over time. The differing perspectives of exotic pine and native eucalypt plantations will be compared, together with the problems involved in maintaining a sustainable resource, by using plantations within New South Wales as a case study indicator. A central issue is whether or not plantations are sustainable, and this raises the question as to what attributes are to be sustained. Although this question may be addressed in terms of the original objective for establishing the plantation, there are further evolving expectations. This paper considers that the more intensive the intended management (e.g., in terms of inputs or site specificity) for whatever purpose, the greater the requirement for technical information and monitoring.

Historical perspective

Pine plantings commenced in Australia in 1876 (Carron 1985) to overcome shortages of softwood timber, there being only limited natural softwood resources within Australia (e.g., *Araucaria*). The first plantings in New South Wales were in 1914 on very poor coastal sands at Tuncurry (Black 1948). There was a perception that many of the *Pinus* species could be planted on depauperate soils and that rainfall, and subsequently mycorrhizae, were the primary limiting factors to growth. Although the efforts of researchers such as Stoate (1950) and Ludbrook (1942) showed the critical nature of soils and nutrition, especially in relation to the heavily weathered and often low phosphorus soils in Australia, their findings took considerable time to be incorporated into forest management. The results of these perceptions were some significant failures in plantation establishment and, even when it was recognized that nutrition was critical, there was a lengthy period when the concept of the "small quantity, quick fix" fertilizer application abounded. This concept was derived primarily from the impact on growth of small quantities of zinc applied to *P. radiata* on limited areas within South Australia (Stoate 1950).

In the 1960s, intensive research commenced on nutrients, especially phosphate deficiency, in Australian soils. The work indicated the strong influence of soil parent material on phosphorus availability, this allowing initial soil classification on the basis of nutritional status (for example, Humphreys and Gentle 1967; Humphreys et al. 1975), the magnitude of the response obtained on poorer sites, and the length of the response on some sites extending into subsequent rotations (Gentle et al. 1965, 1986). Adsorption of applied phosphate was subsequently determined and the soils on which responses to applied phosphate could be expected were defined (Ellis et al. 1975; Flinn et al. 1979; Turner 1982; Turner and Lambert 1986). The magnitude of these responses could be substantial, but only on very specific sites. However, prediction of response from site factors and foliage analysis was relatively easy (Lambert and Turner 1988). This ameliorative type of fertilizer application, although important initially, became relatively minor compared to site establishment and midrotation applications aimed at optimizing financial gain (Turner and Knott 1991). Subsequently, the critical relationships among soil parent material, location with respect to nutrient inputs (for example, maritime or industrial), and deficiencies of boron and sulfur became apparent (Humphreys et al. 1975; Lambert 1986; Lambert and Ryan 1990).

Until the post-World War II period, plantations tended to be established on poor soils or locations undesirable for agricultural pursuits, and often were established to alleviate unemployment or to support such small-scale industries as case manufacture. That is, the decisions for establishing plantations were not made primarily on an economic basis. Subsequently, an analysis was made of the cost to the nation of importing timber and the policy was proposed and accepted that Australia become self-sufficient in timber requirements. This led to the Commonwealth Softwood Agreement (Anonymous 1967). Loan funds were provided by the Commonwealth to the States for plantation establishment and land was acquired and planted; usually native forest was converted to pine plantations. However, there was minimal appreciation of site conditions, genetic properties, or establishment techniques, and there were few environmental constraints. Establishment of large areas became the main measure of success without real assessment of cost effectiveness or return. Much of the establishment of plantations was on sites being converted from native forest or woodland, and serious environmental opposition began to appear against this in the 1970s, particularly on state-owned land (Routley and Routley 1975). On economic grounds, government agencies responsible for planting resisted the pressure to stop planting on native forest sites, because the land being cleared was essentially free; the alternative was the purchase of cleared freehold land.

During this period in South Australia, further cheap suitable land became less available and the State was moving into a second rotation establishment program. Second rotation productivity declines, supporting the arguments of Routley and Routley (1975) that exotic plantations were not sustainable, were reported (Keeves 1966), and this was viewed with great concern within the pine planting industry. The soils in question were coarse-textured infertile sands where loss of organic matter, through increased decomposition or burning, had serious consequences, and they were shown to require specific management procedures to maintain nutrients and organic matter (Flinn et al. 1980; Farrell et al. 1981). However, this was not as critical for much of the planting area in Australia where soils are higher in clays and derived in situ from parent materials. A large proportion of early plantings was on sandy soils, but this was changing over time with a greater emphasis on relatively more fertile soils (Turner and Lambert 1991).

The attempt to optimize the major components, namely site, genetics, and management inputs, involved in optimum plantation output to produce a specific product has been termed "Site Specific Management." During the 1980s, the combination of environmental pressures, commercialization of Forest Services, reduced land availability, and the development of more sophisticated markets led to greater concentration on Site Specific Management to optimize productivity and financial returns from sites. There have been significant pressures to reduce or stop planting exotic species, particularly *P. radiata*. Forest managers now need, not only to reinforce that wood is a renewable crop, but also to provide evidence that this is so, e.g., through the use of appropriate indicators. The use of Site Specific Management also needs to be further evaluated in conjunction with potential environmental effects.

Eucalyptus plantations are a much smaller component of the Australian plantation resource than are pine plantations, and their development is more sporadic. These plantations have usually been used to fulfill an objective within a market, such as a pulpwood deficit. Pine plantings involve an introduced species for which there are few detrimental biological agents (predominantly they are subject to *Sirex* and *Dothistroma*), and the site interactions directly affect growth. Eucalypts are native species with many endemic pests and diseases, thus resulting in both direct and indirect impacts on productivity and value. Extension of a species away from its natural site conditions can induce stress directly affecting growth or indirectly affecting growth through biological agents. Long-term plantations have been perceived to have a much higher risk than pulpwood rotations. Despite this, eucalypt plantations are perceived by some groups, ecologists and others, as being more environmentally appropriate for Australia. The bulk of recent plantings have focused on short rotations for pulp production. Programs are now extending the plantings, but there is a need to match sites to species.

Present Australian plantation resource

Pine plantation areas were extended primarily after World War II and today there are in excess of 1 million ha planted. Very recent planting rates have exceeded 30,000 ha/yr (ABARE 1994). The areas (Table 2) indicate a concentration on *P. radiata* and reductions in some other coniferous species. The proportion of eucalypts has changed from about 4 percent of the total planted area to 11 percent, and this is expected to continue to increase.

Nationally, plantation site characteristics and species composition are changing over time. *Pinus radiata* is about 66 percent of the plantation resource, but is decreasing slightly. Of the other pine species, *P. caribaea* More. is increasing in relative importance (Table 2), whereas other coniferous species are decreasing (that is, on conversion to second rotation, these species are being replaced with others). The greatest relative increase is in

Table 2. Plantation areas (1986–1993) in Australia according to species (*Source:* ABARE 1994).

Species	Area in 1986		Area in 1993		Net annual
	(ha)	(% of total)	(ha)	(% of total)	change (ha/yr)
			Conifers		
Pinus radiata	599,476	68.8	715,353	66.0	16,554
Pinus elliottii	96,392	11.0	76,283	7.0	–2,873
Pinus pinaster	32,611	3.7	31,731	2.9	–126
Pinus caribaea	40,727	4.8	57,201	5.3	2,353
Araucaria spp.	45,704	5.2	46,774	4.3	–153
Other	17,262	2.0	27,550	2.5	1,470
Total conifers	832,172	95.5	954,892	88.2	17,225
			Broadleaved species		
Eucalyptus spp.	36,265	3.8	123,489	11.4	12,461
Populus spp.	2,262	0.3	1,277	0.1	–140
Other	2,013	0.2	3,053	0.3	149
Total broadleaves	40,540	4.3	127,819	11.8	12,470
Total	872,712	100	1,082,711	100	29,695

the eucalypt plantation resource which from 1986 to 1993 expanded from 3.8 percent of total plantation area to 11.4 percent. In 1970, approximately 33 percent of the plantation resource was established on transported sands (Table 3), such as those on which second rotation decline had been reported (Keeves 1966). However, by 1990 this had declined to 26 percent as increased plantings occurred on soil developed in situ, with higher clay contents and higher buffering capacities.

Site-specific management in pine plantations

To date, the Site Specific Management approach has been specifically applied to *Pinus* plantations, but it can be used with reference to any plantation and any desired mix of products (i.e., timber, water, soil amelioration, and aesthetics). This recognizes that sites differ in responsiveness, and that there are considerable difficulties in extrapolating results from one site to another. The objective is to optimize the output of a specified product in order to maximize

Table 3. Proportion of *Pinus radiata* plantation areas in Australia in 1970 and 1990 on soils derived from different parent materials.

Code[a] (PRC)	Rock type[b]	1970 (% of area)	1990 (% of area)	Change between 1970 and 1990 (%)
02	Quartose sandstone	1.6	1.9	2.0
022	Transported sands	32.7	26.5	21.8
03	Ferriginous sandstone	1.4	2.1	50.0
04	Limestone/dolomite	0	0	0
05	Shale/siltstone	34.9	35.5	35.9
06	Phyllites	0.2	0.2	0.2
07	Granite	2.3	2.5	9.0
08	Rhyolite	2.4	3.0	3.5
09	Granodiorite/diorite	11.2	14.7	17.2
10	Trachyte	0	0	0
11	Basalt	13.2	13.7	16.8

Note: Total area = 423,935 ha and 993,965 ha in 1970 and 1990, respectively.
[a]Soil parent material codings (Parent Rock Codes as described by Turner et al. 1990; Turvey et al. 1990).
[b]Typical example of rock types under the Parent Rock Code (Turner et al. 1990).

return, value, or other measure of output (for example, water quality). The product in most plantations at present is usually clear sawn wood, but it may be any defined product.

Site Specific Management requires the classification and ready identification of sites. This has been developed to the greatest level of sophistication for *P. radiata*. The system, when appropriately applied, recognizes climate as the major overriding factor, followed by soil and previous land use. Climatically, *P. radiata* is limited to temperate areas with rainfall in excess of 800 mm. (Southern or Caribbean pines are grown in more subtropical and tropical areas.) Soil classification has involved the development of a specific Soil Technical Classification System for *P. radiata*, and the primary factor is soil parent material, which has been classified into twelve major groupings subdivided into either in situ or transported (Turner et al. 1990; Turvey et al. 1990). The technical classification system has been found to categorize sites into productivity units and, more critically, into potential to respond to management treatments. It also provides a basis for identifying sites with potential nutrient problems (Knott and Ryan 1990; Turvey and Smethurst 1994a,b). Sites have been shown to vary significantly in productivity (Knott and Ryan 1990). For example, at age 20 years, productivity on soils derived from rhyolitic parent material (PRC 8) was on average less than 40 m^2/ha basal

Table 4. Effect of previous land use on stem volume and log quality of a 20-year-old stand of *Pinus radiata* (*Source:* Birk 1991).

Stand parameter	Previous land use	
	Native forest	Improved pasture
Volume (m^3/ha)	346	431
High Quality Logs (stems/ha)	350	42
Nonutilizable Index (%)	12	44
Mineralized Nitrogen (kg/ha/yr)	38	86

area, whereas it was more than 55 m^2/ha on soils derived from basaltic and granodiorite parent materials (PRC 9 and PRC 11). Alternative management treatments include application of fertilizers at the time of establishment and after thinning (Turner et al. 1992, 1996), but other issues such as treatment for disease may be included.

Previous land use provides additional refinement to the classification system, because it can have an effect on productivity and timber utilization. For example, where the previous land use has been improved pasture (that is, the addition of phosphatic fertilizers plus the addition of leguminous crops to cleared land), increased levels of mineral nitrogen are found, which lead to extensive stem deformity and branching (Birk 1991; Downes and Turvey 1992). The effect can increase total productivity, but deformity greatly reduces utilization and value (Table 4). The classification of site allows for the identification of risk involved in plantation management, including both risks to productivity and utilization and those which can be related to environmental aspects such as erosion (Lacey et al. 1994). Such management can increase returns per unit area, but, more critically, can increase the whole project value by identifying sites to be rejected from plantation establishment.

However, there are interactions between previous land use and soil parent material that result in direct effects on absolute productivity (Table 5). This indicates the necessity to consider multiple factors and to relate information on the potential for sites to respond to change. On the lightly buffered sands (Table 5), there has been little change from the effect of pasture improvement, especially when compared to the major differences in productivity resulting from previous land use on soils derived from basalt and sedimentary parent materials.

In addition to identification of site effects, there have been very substantial and ongoing gains as a result of genetic improvements. These improvements have had the greatest impact on plantations in the last decade, and the effect

Table 5. Comparison of productivity of *Pinus radiata* grown on sites with differing prior land use (*Source:* Skinner and Attiwill 1981).

Soil parent material	Stand age (yr)	Volume MAI (m^3/ha/yr)[a] with previous land use		Difference (%)
		Native forest	Pasture	
Basalt	15	15.3	25.8	68.6
Basalt	18	9.8	22.1	125.5
Sediments	18	10.7	17.8	66.3
Deep sand	24	20.3	21.1	3.9

[a]Volume MAI is not directly comparable between sites because of differences in stand age.

Table 6. Increases in growth in *Pinus radiata* plantations from a combination of breeding and silvicultural developments applied to large-scale plantations (*Source:* Nambiar 1995).

Main soil type	Rainfall (mm/yr)	Average MAI (m^3/ha/yr)		Range in realized gain (%)
		1950–1971	1975–1982	
Gradational clay loam	1,150	16.2	23.8	6–58
Duplex soil	750	9.1	14.2	35–59
Uniform sand	650	8.0	15.1	33–107

is increasing. The work for *P. radiata* has been carried out primarily through cooperatives, both within Australia and in cooperation with New Zealand. The value of these development programs has included the ability to focus on a limited number of desirable attributes (growth, straightness, branching), and there have been a variety of demonstrations of this in routine plantations (Table 6). The issue for managing sustainably is that of identifying realized gains from such factors as tree improvement compared to actual changes in site and environmental effects.

Environmental effects of intensive management

Conversion of native forests to plantations will lead to soil changes, such as reduction in organic matter and nitrogen (Turner and Kelly 1985). The effects of increased intensity of management of conifer plantations can result in impacts on site and other environmental attributes. Although nutrient removals may be considered to have long-term impacts, major concerns relate to the potential for increased soil erosion and compaction (Lacey 1993).

Such impacts can be identified and managed, but the actual management requirements will be site specific. More significant impacts are related to the use of water and the ability to manage long-term water requirements in terms of water quality and quantity. Conversion of previous grassland to plantation results in a greater water demand by the crop and it has been estimated that, where a high proportion of a catchment area has been converted to forest plantation, water yield may be reduced by up to 40 percent (Cornish 1989). Although this reduction may not be critical in high rainfall areas, it could be significant in more marginal areas (Turner and Lambert 1987). Conversion of native forest to faster growing forest can be expected to result in increased water use, but the extent of this will be less than where there is a change in land use. Based on a water use efficiency factor of 0.002 g organic matter/g water used, each cubic meter of wood per hectare in a radiata plantation will require approximately 40 mm of water. Without estimating all factors (for example, interception), an increase of 10 percent in the growth of a plantation with an existing mean volume annual increment of 20 m^3/ha (at 30 years) will increase annual water demand by 80 mm, and reduce runoff accordingly. Effects on water quality as a result of increased sedimentation will not be associated with higher productivity per se, but greater management activity on the area could have such an effect.

Eucalyptus plantings

Whereas there are only a limited number of species of conifers in the Australian planting program, there are a relatively large number of eucalypt species, including *E. camaldulensis* Dehnh., *E. cloeziana* F. Muell., *E. globulus*, *E. grandis*, *E. laevopinea* R. Baker, *E. maculata* Hook., *E. nitens*, *E. pilularis* Smith, *E. regnans* F. Muell., and *E. saligna* Smith. This diversifies efforts in terms of research and management practices, and reduces gains in comparison to those achieved from *P. radiata* or *P. caribaea*.

There has been a range of trials demonstrating the site specificity of some species and the general adaptability across sites of others. As an example, on a nutritionally poor site in New South Wales, a eucalypt species trial provided results for a selection of species (Table 7). *Eucalyptus dunnii* Maiden can grow on relatively better topographical positions, whereas *E. fastigata* Deane & Maiden grows moderately well in a range of situations. *Eucalyptus pilularis* is broadly successful and *E. maculata* and *E. saligna* are site selective. Apart from raising the issue of the requirements for comprehensive trial designs, this shows the range in performance of different species.

Table 7. Estimated basal area of selected *Eucalyptus* species at age 18 years on four topographical locations on a nutritionally poor site in New South Wales (*Source:* Johnson and Stanton 1994).

Species[a]	Basal area (m^2/ha)			
	Gully	Upper slope (west aspect)	Upper slope (east aspect)	Ridge
Eucalyptus citriodora	1.2	0	13.4	1.7
Eucalyptus globulus	3.7	8.4	30.0	5.6
Eucalyptus saligna	5.7	0	30.9	0
Eucalyptus maculata	8.3	15.8	37.3	9.8
Eucalyptus fastigata	21.9	32.0	31.2	26.2
Eucalyptus pilularis	37.7	33.9	44.7	26.4
Eucalyptus dunnii	41.7	28.6	48.7	5.3

[a]Ordered according to increasing productivity on the "gully" sites.

Eucalypts have been found to be very responsive to cultural treatments, such as site preparation, weed control, fertilizer application, and even insecticide treatment (Cromer and Williams 1982; Birk and Turner 1992; Stone 1993). Total biomass of a *E. grandis* plantation at 9 years without treatment was 123 t/ha. With intensive fertilizer treatment this was raised to 175 t/ha, and with intensive fertilizer treatment plus weed control and insecticide this was increased to 274 t/ha (Birk and Turner 1992). The effects of interactions among species, fertilizer use, and insect control were shown for *E. grandis* and *E. dunnii* (Stone 1993). With no insecticide treatment, *E. dunnii* produced statistically greater growth at 26 months of age (7.0 cm diameter) compared to the diameter growth of 6.5 cm in the treatment in which insecticide was applied. The response to fertilizer was the same, whether or not insecticide was applied (9.8 cm and 9.5 cm, respectively). *Eucalyptus grandis* gave a greater response when insects were controlled in an insecticide treatment (5.4 cm and 6.8 cm without and with insecticide, respectively), and particularly when fertilizer was applied (5.6 cm with no control of insects, and 10.0 when insecticide was applied).

A combination of particular eucalypt species which have high nutrient concentrations and rapid growth can lead to high nutrient accumulations in the tree component. There is potential for high nutrient removals with harvesting of rapidly grown eucalypt plantations, especially for the nutrient calcium. This is primarily a result of high calcium concentrations in bark, especially for the smooth barked species (Lambert 1981; Lambert and Turner 1991). More than 60 percent of the ecosystem calcium was found in the aboveground components for 27-year-old *E. grandis* (Turner and Lambert 1983; Lambert and Turner 1991). It has been suggested that *P. radiata* substantially drains nutrients from sites (Crane and Raison 1980); however, the related impacts associated with *E. grandis* plantations indicate a greater potential for calcium loss in eucalypt plantations. The long-term impacts of such changes on sustainability have not been measured to date.

Sustainability of plantations

The primary objective with current plantations has been to maximize productivity, yield, and value. The basis for the objective has been lack of resource for certain products and for investment. Objectives are now changing to include site amelioration (especially high water tables), mixed products, enhancement of farming land, and water quality/quantity issues. Whatever the objective, including reducing pressure on utilization of native forests, plantations require higher inputs.

Recognition of greater potential productivity, range of products, and value in plantations is balanced by a greater potential for site deterioration and loss in value as a result of increasingly more intensive management practices. For example, as discussed earlier for eucalypt plantations, there is a high risk of nutrient depletion, especially for the cations, and so higher levels of technical knowledge and precision are required in management.

The questions raised earlier were whether or not plantations are sustainable and which attributes need to be sustained. Where the objective is to maintain production of timber at a predetermined level and quality, it should be achievable, but there are costs in the process of doing this and these costs must be recognized. Increasing productivity, by whatever means (e.g., genetics, higher operational inputs), will necessitate a cost, for example, reduced water production. However, unlike management of natural forests, there is a greater chance of estimating both costs and benefits of plantations. In the forest management triangle, the outputs are determined by the management of the site, genetics, and management inputs. Thus, Product = (Genetics)(Site)(Management).

If, over successive rotations, the genetic stock is actually improved (for example, a volume improvement of 30 percent) but site factors are reduced through compaction, there could be a net volume decrease (for example, by 20 percent) with constant management inputs. The net increase in productivity is measured as 4 percent, but this is misleading because the site is in fact deteriorating. In intensively managed plantations this interaction needs to be separated out and understood. It is critical to obtain the potential in production.

Monitoring requirements

The more intensively plantations are managed, the greater the potential for increasing yield and value, but there is also increased risk for environmental or productivity-generated problems. Two types of monitoring, sustainability indicators and performance, are required.

Monitoring of sustainability indicators

Sustainability indicators provide information on the actual characteristics of forests and how they are changing under routine management. The objective is to utilize parameters which are indicators of processes within the system. The factors which need to be addressed within intensively managed plantations differ from those associated with more extensive natural forests. They include the following:

- Actual productivity measures (for example, biomass and value).
- Assessments of yield (harvested products).
- Impacts of pests and diseases.
- Environmental changes (for example, in water quality).
- Changes in soil properties (including changes over rotations).

Performance monitoring

Where there is investment, there needs to be an assurance that gains are being achieved. To do this, the individual processes (site impacts, genetics, and management) need to be addressed. These are the following:

- Changes resulting from genetic changes.
- Effects of compaction and results of amelioration (site preparation).
- Effects of fertilizer/herbicide usage.
- Effects of pest/disease control measures.
- Changes in wood quality.

Many other factors will overlap independently with sustainability, but the difference will relate to investment.

Whereas research experiments can provide information on potential effects, there is often a substantial fall down in measured effects when implemented operationally, and hence effectiveness may not be understood without appropriately designed systems (that is, performance monitoring allows for estimates of fall down to be made). The research carried out on site, on species selection and improvement, and on management inputs to fast-grown plantations shows that great gains can be obtained in productivity, and that the potential impacts can be measured. However, there have to be systems in place which will ascertain whether or not the gains are being realized, and which will enable impacts to be offset. It is this area of monitoring and interpretation at the management level that requires further research.

References

ABARE (Australian Bureau of Agricultural and Resource Economics). 1994. Quarterly Forest Products Statistics. Commonwealth of Australia, Canberra.

Anonymous. 1967. Softwood Forestry Agreements Bill. Commonwealth of Australia, Canberra.

Birk, E.M. 1991. Stem and branch form of 20-year-old radiata pine in relation to previous land use. Aust. For. 54: 30–39.

Birk, E.M. and Turner, J. 1992. Response of flooded gum (*E. grandis*) to intensive cultural treatments: Biomass and nutrient content of eucalypt plantations and native forest. For. Ecol. Manage. 47: 1–28.

Black, R.M. 1948. The history of the *Pinus* plantations in New South Wales. Forestry Commission of New South Wales Forest Recorder 1: 30–37.

Carron, L.T. 1985. A History of Forestry in Australia. Australian National University Press, Canberra.

Cornish, P.M. 1989. The effects of radiata pine plantation establishment and management on water yield and water quality – a review. Forestry Commission of New South Wales Technical Paper No. 49. Forestry Commission of New South Wales, Sydney.

Crane, W.J.B. and Raison, R.J. 1980. Removal of phosphorus in logs when harvesting *Eucalyptus delegatensis* and *Pinus radiata* forests on short and long-term rotations. Aust. For. 43: 253–260.

Cromer, R.N. and Williams, E.R. 1982. Biomass and nutrient accumulation in a planted *Eucalyptus globulus* (Labill.) fertilizer trial. Aust. J. Bot. 30: 265–278.

Downes, G.M. and Turvey, N.D. 1992. Stem deformation in Pinus radiata. University of Melbourne, School of Forestry Research Bulletin 8. University of Melbourne, Melbourne, Australia.

Ellis, R.C., Neilsen, W.A., McDavitt, J.G. and Thirkell, D.S. 1975. Early growth of radiata pine in relation to a land classification system on Forestier Peninsula in Tasmania. Aust. For. 37: 245–257.

Farrell, P.W., Flinn, D.W., Squire. R.O. and Craig, F.G. 1981. On the maintenance of productivity of radiata pine monocultures on sandy soils in south-east Australia, pp. 117–128. In: Proceedings of the Seventeenth International Union of Forest Research Organisations Congress. International Union of Forest Research Organisations, Kyoto, Japan.

Flinn, D.W., Moller, I.M. and Hopmans, P. 1979. Sustained growth responses to superphosphate applied to established stands of *Pinus radiata*. N. Z. J. For. Sci. 9: 201–211.

Flinn, D.W., Squire, R.O. and Farrell, P.W. 1980. The role of organic matter in the maintenance of site productivity on sandy soils. N. Z. J. For. 25: 229–236.

Gentle, S.W., Humphreys, F.R. and Lambert, M.J. 1965. An examination of a *Pinus radiata* phosphate fertilizer trial fifteen years after treatment. For. Sci. 11: 315–324.

Gentle, S.W., Humphreys, F.R. and Lambert, M.J. 1986. Continuing response of *Pinus radiata* to phosphatic fertilizers over two rotations. For. Sci. 32(3): 822–829.

Humphreys, F.R. and Gentle, S.W. 1967. Experience with phosphatic fertilizers in man-made forests of *Pinus radiata* in New South Wales. In: Symposium on Man-made Forests. Canberra, Australia.

Humphreys, F.R., Lambert, M.J. and Kelly, J. 1975. The occurrence of sulphur deficiency in forests, pp. 154–162. In: McLachlan, K.D. (Ed.) Sulphur in Australasian agriculture. Sydney University Press, Sydney.

Johnson, I.G. and Stanton, R.R. 1993. Thirty years of eucalypt species and provenance trials in New South Wales – survival and growth in trials established from 1961 to 1990. Forestry Commission of New South Wales Research Paper No. 20. Forestry Commission of New South Wales, Sydney.

Keeves, A. 1966. Some evidence of loss of productivity with successive rotations of *Pinus radiata* in the south-east of South Australia. Aust. For. 30: 51–63.

Knott, J. and Ryan, P. J. 1990. Development and practical application of a soils database for Pinus plantations of the Bathurst Region. Forestry Commission of New South Wales Research Paper No. 11. Forestry Commission of New South Wales, Sydney.

Lacey, S.T. 1993. Soil deformation and erosion in forestry, a review. Forestry Commission of New South Wales Technical Report. Forestry Commission of New South Wales, Sydney.

Lacey, S.T., Ryan, P.J., Huang, J. and Weiss, D.J. 1994. Soil physical property change from forest harvesting in New South Wales. State Forests of New South Wales Research Paper No. 25. State Forests of New South Wales, Sydney.

Lambert, M.J. 1981. Inorganic constituents in wood and bark of New South Wales forest tree species. Forestry Commission of New South Wales Research Note No. 45. Forestry Commission of New South Wales, Sydney.

Lambert, M.J. 1986. Sulphur and nitrogen nutrition and their interactive effects on *Dothistroma* infection in *Pinus radiata*. Can. J. For. Res. 16: 1055–1062.

Lambert, M.J. and Ryan, P.J. 1990. Boron nutrition of *Pinus radiata* in relation to soil development and management. For. Ecol. Manage. 30: 45–53.

Lambert, M.J. and Turner, J. 1988. Interpretation of nutrient concentrations in *Pinus radiata* fertilizer at Belanglo State Forest. Plant Soil 108: 237–244.

Lambert, M.J. and Turner, J. 1991. Nutrient cycling processes in *Eucalyptus grandis* forests on the New South Wales north coast, pp. 198–199. In: Ryan, P.J. (Ed.) Productivity in Perspective. Forestry Commission of New South Wales, Sydney.

Ludbrook, W.V. 1942. Fertilizer trials in southern New South Wales pine plantations. J. CSIRO Aust. 15: 307–314.

Nambiar, E.K.S. 1995. Relationships between water, nutrients and productivity in Australian forests: Application to wood production and quality. Plant Soil 168–169: 427–435.

Routley, R. and Routley, J. 1975. The Fight for the Forests. Research School of Social Sciences, Aust. Nat. University, Canberra.

Skinner, M.F. and Attiwill, P.M. 1981. The productivity of pine plantations in relation to previous land use. I. Growth responses in agricultural and forest soils. Plant Soil 60: 161–176.

Stoate, T.N. 1950. Nutrition of the Pine. Forest and Timber Bureau Bulletin 30. Forest and Timber Bureau, Canberra.

Stone, C. 1993. Fertilizer and insecticide effects on tree growth and psyllid infestation of young *Eucalyptus grandis* and *Eucalyptus dunnii* plantations in northern New South Wales. Aust. For. 56: 257–263.

Turner, J. 1982. Long term superphosphate trial in regeneration of *Pinus radiata* at Belanglo State Forest, N.S.W. Aust. For. Res. 12: 1–19.

Turner, J. and Kelly, J. 1985. Effect of radiata pine on soil chemical characteristics. For. Ecol. Manage. 11: 257–270.

Turner, J. and Knott, J. 1991. Prescribing fertilizers in New South Wales plantation forestry. In: Forest Fertilization: Sustaining and Improving Nutrition and Growth of Western Forests. College of Forest Resources, University of Washington, Seattle.

Turner, J. and Lambert, M.J. 1983. Nutrient cycling within a 27-year-old *Eucalyptus grandis* plantation in New South Wales. For. Ecol. Manage. 6: 156–168.

Turner, J. and Lambert, M.J. 1986. Fate of applied nutrients in a long-term superphosphate trial in *Pinus radiata.* Plant Soil 93: 373–382.

Turner, J. and Lambert, M.J. 1987. Nutritional management of *Pinus radiata* at Gurnang State Forest, New South Wales. Fert. Res. 13: 127–137.

Turner, J. and Lambert, M.J. 1987. Forest water usage and interactions with nutrition of *Pinus radiata*. Oecol. Plant. 8: 27–44.

Turner, J. and Lambert, M.J. 1991. Changed perspectives on soil processes associated with the changed land base of plantation forestry in Australia, pp. 15–28. In: Ryan, P.J. (Ed.) Productivity in Perspective. Forestry Commission of New South Wales, Sydney.

Turner, J., Thompson, C.H., Turvey, N.D., Hopmans, P. and Ryan, P.J. 1990. A soil technical classification for *Pinus radiata* (D. Don) plantations. I. Development. Aust. J. Soil Res. 28: 797–811.

Turner, J., Lambert, M.J., Bowman, V. and Knott, J. 1992. Two post thinning fertilizer trials in *Pinus radiata* in New South Wales, Australia. Fert. Res. 32: 259–267.

Turner, J., Knott, J. and Lambert, M.J. 1996. Fertilization of *Pinus* radiata plantations after thinning. I. Productivity gains. Aust. For. 59: 7–21.

Turvey, N.D. and Smethurst, P.J. 1994a. Soil types as classes for managing the nutrient status of planted *Pinus radiata* in Victoria, Australia. Aust. For. 57: 148–156.

Turvey, N.D. and Smethurst, P.J. 1994b. Nutrient concentrations in foliage, litter and soil in relation to wood production of 7- to 15-year-old *Pinus radiata* in Victoria, Australia. Aust. For. 57: 157–164.

Turvey, N.D., Booth, T.H. and Ryan, P.J. 1990. A soil technical classification system for *Pinus radiata* (D. Don) plantations. II. A basis for estimation of crop yield. Aust. J. Soil Res. 28: 813–824.

New Forests **18:** 33–44, 1999.

Radiata pine plantations in Chile

JORGE TORO[1] and STANLEY P. GESSEL[2,†]
[1] *Bioforest S.A., Concepcion, Chile;* [2] *College of Forest Resources, University of Washington, Seattle, WA, USA (†passed away)*

Key words: abiotic factors, biotic factors, erosion control, social impacts, sustainability

Abstract. The landscape of the southcentral part of Chile is dominated by fast-growing plantations of radiata pine (*Pinus radiata* D. Don), and in lesser extent bluegum eucalyptus (*Eucalyptus globulus* Labill.), Italian poplar, and willows. Radiata pine was first introduced in 1885 as an ornamental species. During 1940–1960 it was used to control erosion in the Coastal Range. Finally, since the 1960s it has been used for commercial plantations. The present extent of the plantation resource is 1,694,104 ha, of which 80.9 percent is radiata pine. During 1991, 1992, and 1993, this species was planted at rates of 75,416 ha, 81,868 ha, and 71,411 ha, respectively. Even though the soils are eroded to different degrees, exhibit a variety of nutritional disorders, and contain some physical restrictions, the productivity of radiata pine ranges from 18 m^3/ha/yr to 35 m^3/ha/yr. Additional stress comes from some potentially serious pests; for example, the European pine shoot moth, *Rhyacionia buoliana*, has presented a real threat to radiata pine. The main products obtained from radiata pine are exported to Japan, Korea, Middle and Far East countries, and Europe. Sawlogs, sawntimber, pulpwood, clearwood, and wood chips are some of the products comprising sales in 1994 of U.S.$1,564 billion.

The purpose of this paper is to show how radiata pine silviculture has changed since the last part of the 1980s in Chile. Also we want to show how forestry activities have produced a profound and positive change in the social and economic environment that surrounds the areas where the industry's main production areas exist.

Introduction

The landscape of the southcentral part of Chile is dominated by plantation of fast-growing forest species, where radiata pine (*Pinus radiata* D. Don) plantations cover more than 80.9 percent of the total. According to Instituto Forestal (INFOR 1994), the total area planted is 1,694,104 ha. This species was introduced in 1885 for ornamental purposes. In 1910, F. Albert, a German forest scientist hired by the Chilean government to solve the erosion problems that threatened the Coastal Range, had the opinion that this species could be planted to stop the erosive processes occurring in many regions. In the early 1940s the first commercial plantations were planted at a rate of 3,000 ha annually.

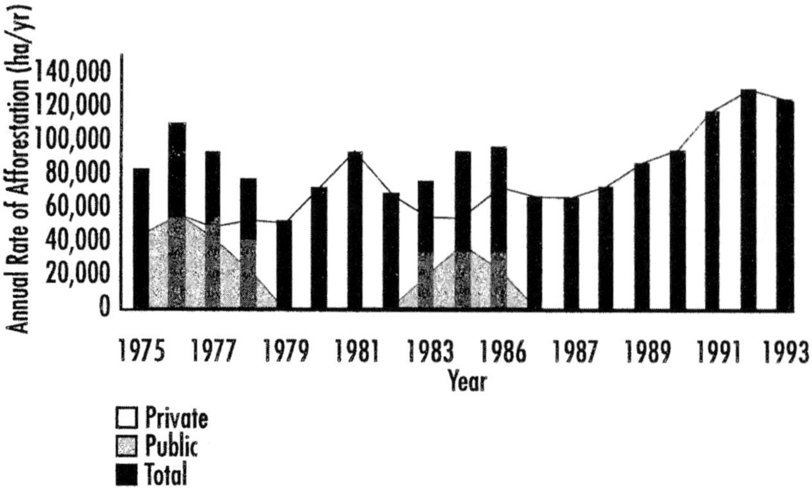

Figure 1. Annual rate of afforestation in Chile by the public and private sectors between 1975 and 1993.

It is important to emphasize the role played by the 1974 law (Decree No. 701 of Forest Promotion) through which exist subsidies of 75 percent related to forestation expenses. These are determined by the Corporación Nacional Forestal. This law will operate until the year 2010, and has permitted private forest companies to receive financial support for the establishment of new plantations and for attendant silvicultural practices, including pruning.

Figure 1 shows the annual rate of afforestation between 1974 and 1994 by public and private companies. Since 1986 only private companies have continued planting at an average annual rate of 70,000 ha. Radiata pine constitutes the base of the Chilean forest economy. The exports of this species between 1990 and 1994 have increased from U.S.$855.3 to U.S.$1,564.3 million, thus contributing significantly to the national economy (INFOR 1994). The purpose of this paper is to emphasize the role that radiata pine plantations have played in Chile, from an ecological, economic, and social point of view, especially in the areas where plantations have been established.

Some characteristics of Chile and the regions where radiata pine plantations grow

Chile is a country located in the southwest extreme of South America. Its population is about 14 million people. Its continental territory is 756,847 km^2, half of which are desert and mountains. The country is divided into thirteen

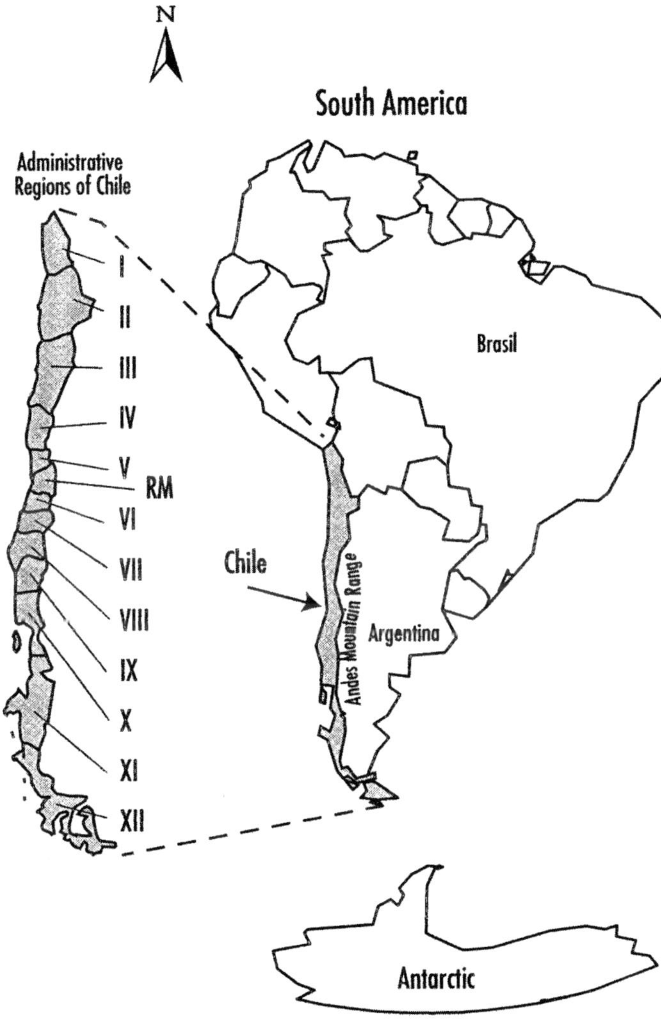

Figure 2. Administrative regions of Chile.

administrative regions distributed from 19°S to 56°S (Figure 2). The borders of the country are well defined with the driest desert in the world located in the north, the Andes Mountain Range bordering the eastern frontier, and the Antarctic and the Pacific oceans constituting the southern and the western limits, respectively. This geographical "island" condition has given the Chilean people special characteristics that make them look anxiously to the exterior. To this we should add the fact that, since the nineteenth century, powerful European colonies have contributed different cultural elements that have printed an important seal on the country. This makes Chileans different

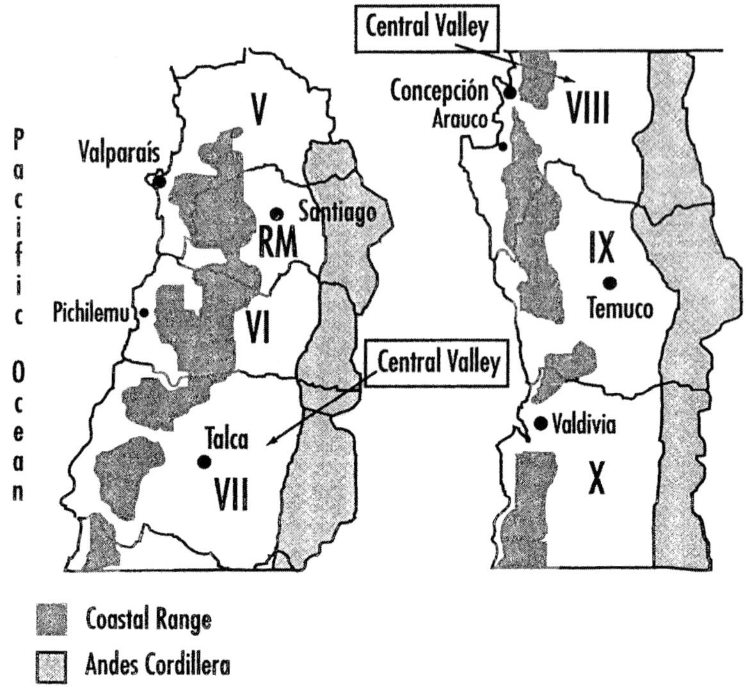

Figure 3. Main physiographic landscapes of administrative regions of southcentral Chile.

from other South Americans, even though they share a similar language and general historical background.

Characteristics of southcentral Chile

Three main physiographic landscapes dominate Chile. They extend from 19°S south to the region at 42°S (Figure 3; Fuenzalida 1967). The Coastal Range along the Pacific Ocean and the Andes Cordillera enclose the Central Valley from north to south and create a great variety of climates, soils, and forest sites. This contributes to great variation in productivities of forest plantations. Seventy-one percent of the radiata pine plantations are found in the Coastal Range, 19 percent in the foothills of the Andes, and 10 percent in the Central Valley.

Soils occupied by radiata pine plantations

The soils of the Coastal Range between the Regions V and X are derived from granitic and metamorphic rocks that compose the so-called "Batolic Coastal" (Fuenzalida 1967). They are the oldest rocks of the country with an age of approximately 250 million years (González-Bonorino 1970). The intense

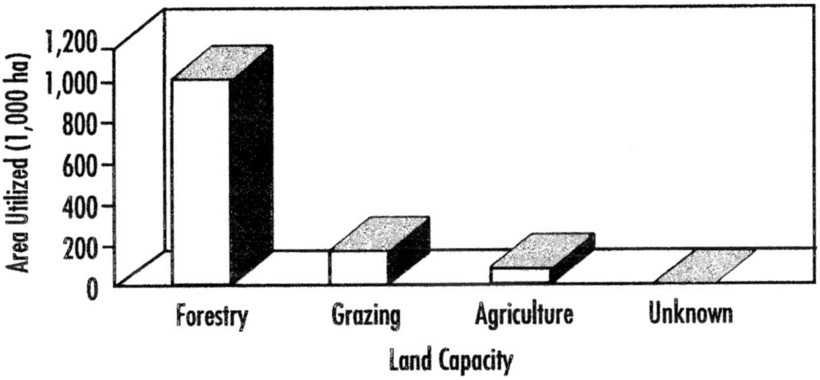

Figure 4. Previous land use of areas planted in forest plantations in regions VI–X of Chile.

agricultural and cattle-grazing activities developed during the eighteenth and nineteenth centuries on forest land provoked severe erosion in an area of over 4 million ha (Elizalde 1970). On these soils, abandoned by agriculture and cattle raising by the mid-twentieth century, 91 percent of the radiata pine plantations were established to stop the erosive processes and create the bases for an outstanding forest industry (Contesse 1990; Figure 4).

In the foothills of the Andes, soils are derived from recent volcanic deposits of andesitic-basaltic origin. Since the nineteenth century and after the intense exploitation of the forest, native forests were replaced by low-quality shrubs. *Araucaria araucana* (Mol.) K. Koch (Araucaria), *Fitzroya cupressoides* (Mol.) Johnst. (Alerce), and several species of *Nothofagus*, mainly *N. dombeyi* (Mirb.) Oerst (Coigüe), *N. obliqua* (Mirb.) Oerst var. *obliqua* (Roble), *N. alpina* (P. and E.) Oerst (Raulí), *N. glauca* (Phil.) Krasser (Hualo), were the most important commercial tree species of those forests. Between 1960 and 1990 low-quality shrubs were replaced by plantations of radiata pine on 100,000 ha (INFOR 1994).

In the Central Valley, there is a great variety of geologic materials from the Andes Mountain Range; these materials refilled the big depression created after the lift of the Andes and the Coastal Range during the Mesozoic (Brüggen 1950). A large part of the radiata pine plantations in these areas is located near the city of Los Angeles, where sandy soils exist. These sands are of andesitic-basaltic origin from the Antuco volcanic complex. About 120,000 ha of radiata pine are growing in this valley.

Radiata pine established mainly on eroded lands

There are about 12 million ha of eroded soils between Region V and Region X. Those soils exhibit different degrees of erosion, ranging from severe to moderate (Elizalde 1970). At the present time, 1.3 million ha are covered with radiata pine, this leaving 10 million ha of land without adequate forest cover to control and protect the soils (Unda and Ravera 1994). Erosion has eliminated, in many areas, all of the A horizon and part of the subsoil. This situation occurred along the Coastal Range. However, the weathering processes that affected the granitic and metamorphic rocks exposed on the surface for several million years has reached depths of from 20 to 40 m. This has permitted the root systems of radiata pine to explore fresh material and has allowed a relatively good development of the plantations. On the contrary, in the Central Valley near the foothills of the Andes, massive volcanic rocks found 40–60 cm below the soil surface restrict tree root growth.

Elements influencing the expansion of radiata pine plantations in Chile

Several factors must be considered to explain the expansion of radiata pine in Chile. In the first place, it is the only forest species of economic importance capable of growing in eroded soils, and displays good growth rates and high financial returns in a rotation of less than 30 years. Secondly, the species is easy to establish in the field, the production of plants is inexpensive, and there are few problems related to forest management. On the other hand, the native forest species of economic importance are difficult to establish on eroded soils, even though they are planted on the same sites. The rotation age for these species is longer than that for radiata pine. For instance, *Nothofagus* species rotations range between 80 and 120 years, *Araucaria araucana* has a rotation of over 500 years, and *Fitzroya cupressoides* has a rotation time of over 1,000 years.

Productivity of radiata pine plantations

The annual rate of afforestation from 1974 to 1994 was on the average 65,000 ha. During the last 5 years, there has been a great increment of *Eucalyptus* plantations, which compete with radiata pine afforestation programs. However, the annual average for radiata has been maintained at 70,000 ha. The distribution by age class, shown in Figure 5, reveals that over 50 percent of plantations are younger than 11 years, and a high percentage are in the 1–5 year age class. The latter group receives the new silvicultural regimes that include weed control, fertilization at time of establishment, pruning, and thinning.

Because of the great variety of soils and the climatic and physiographic patterns existing along the Coastal Range, Andes Mountain Range, and Central Valley, it is not surprising to find great variability in productivity

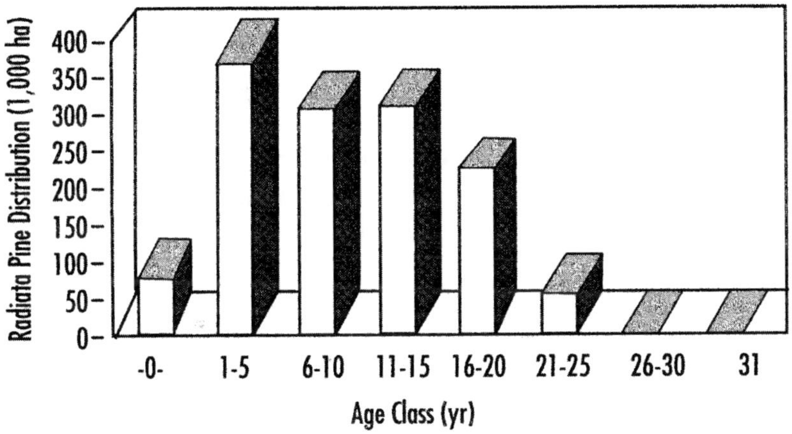

Figure 5. Radiata pine fistribution in Chile by age class.

Table 1. Productivity (mean annual increment) and physiographic features of radiata pine plantations in Chile between 32°S and 42°S (*Source:* INFOR 1987).

Feature	MAI (m^3/ha/yr)	Parent material
Coastal Range	15–36	Granitic-metamorphic
Central Valley	18–28	Sandy
Foothills of the Andes	20–34	Volcanic ash, recent and older

among radiata pine plantations (Table 1), controlled by both abiotic and biotic factors.

Abiotic factors

The severe erosion of the soils occupied by radiata pine has caused nutritional deficiencies which affect the early growth of plantations. Boron deficiency is a spreading problem within regions V–X. Nitrogen also appears in low availability in the soils of regions V–VIII (Toro 1987). Phosphorus limits growth of plantations in the foothills of the Andes, and potassium affects growth in older volcanic soils (Gerding 1993). The private companies, during establishment, apply 2 or 3 g/plant of elemental boron in order to prevent deficiencies. Trials set up in different soils currently show excellent growth responses when nitrogen, phosphorus, potassium, and boron are applied together at establishment and at first thinning.

Soil compaction is an important problem found especially in soils of the Coastal Range between regions V and VIII. This problem exists because the lower part of the subsoil is exposed as a result of erosion. Also, soil compaction is produced by harvest equipment. Gayoso and Iroumé (1984) show that radiata pine plantations growing on highly compacted soils exhibit 50 percent less volume than do plantations growing on undisturbed soils. Toro and Garrido (1991) report that, among soil bulk densities of from 1.18 to 1.46 g/cm^3, the survival percentage is about the same after the first period of growth. However, total biomass production is reduced from 50 g/plant to 13 g/plant dry weight when seedlings are grown in soils of 1.18 g/cm^3 and 1.46 g/cm^3 bulk density, respectively.

Biotic elements

Plantations are currently attacked by *Dothistroma septospora* and *Cyclaneusma* sp., fungi that affect parts of the crown of radiata pine, and therefore reduce tree productivity. These fungal problems are found between regions VIII and X. Since 1986 the European pine shoot moth, *Rhyacionia buoliana*, has severely affected plantations located in Region X. It has extended to the north, and threatens plantations in Region VI. Insecticide applications, biological control, and corrective pruning to eliminate affected branches have helped to reduce the damage produced by this economic pest (Smith and Ahumada, personal communication in 1995).

New silviculture emerges

The silviculture of radiata pine in Chile has adopted many elements, concepts, and philosophies from New Zealand. Scientists,[1] such as Will in 1981, Sutton in 1982, Knowles in 1983, and Mead in 1990, strongly influenced the Chilean silviculture of radiata pine and the forest management of this species. Additionally, scientists[2] from the United States (Gessel 1982; Zobel 1986; Kellison 1990), South Africa (Schönau 1988), and Australia (Lambert 1992; Turner 1992; Nambiar 1995) have also contributed their points of view, experiences, and ideas. The work done by Steinbrenner (1979), Turvey (1980), Gessel (1982), Turner and Holmes (1985), Grey and Jacobs (1987), and Grey (1989) has been extremely important for the classification of soils and sites related to silviculture and management options. At present, several forest companies are working on these issues.

The results from forest nutritional cooperatives in the United States (e.g., Gessel and Atkinson 1984; Allen 1987) have served to increase understanding of the possibilities of the nutritional management of fast-growing species such as radiata pine. Forest companies are making a big effort to develop these topics, considering nurseries, plantation establishment, and tending.

Table 2. Evolution of Chilean forest exports of radiata pine to different regions between 1985 and 1994 (*Source:* Chile Forestal 1995).

Region	Export (1,000 m^3)									
	1985	1986	1987	1988	1989	1990	1991	1992	1993	1994
Asia	116.6	141.9	165.0	242.9	276.5	351.8	423.6	480.9	595.6	770.6
Europe	63.8	88.4	172.7	226.2	244.4	276.0	225.4	329.9	255.6	310.9
South America	107.7	120.8	178.5	185.3	167.2	129.8	155.5	181.0	196.4	260.2
North America	31.7	30.4	44.2	59.6	61.9	58.1	79.5	102.7	155.2	199.2
Other	14.8	14.4	16.9	16.1	33.6	39.6	29.1	31.3	4.3	23.4
Total	334.6	403.1	577.3	730.1	783.6	855.3	913.1	1,125.8	1,207.1	1,564.3

Together with these silvicultural advances during the last 5 years, the genetic area is developing rapidly, with influence from the United States, Australia, and New Zealand. All of these issues benefit the development of Chilean silviculture for radiata pine.

Economic and social influence of radiata pine plantations

Forest activities did not play an important role in the Chilean economy until Decree No. 701 was established in 1974. This law stimulated afforestation programs, and made possible the current afforestation rates. Radiata pine became the main species for commercial purposes because it exhibited high growth rates, short rotations (20–26 years), and a great capability to adapt to different sites. At the present time, the forest products derived from radiata pine plantations are exported to more than 50 countries in Asia, Europe, and America. The evolution of exports to different regions from 1985 to 1994 is shown in Table 2.

Forestry activities in Chile since 1975 have transformed the life of small towns located along the coast from Region VI to Region X. Employment rates have increased, especially in the silvicultural and industrial sectors. According to the Banco Central de Chile (1993), the active working population of the country is about 4.8 million people. The forest sector occupies 2.1 percent of this population. This figure is higher than the 1.8 percent of the manpower utilized by mining industries, which is the main source of Chilean wealth. The agriculture and fisheries sectors together employ about 17.0 percent of the population.

Forest activities have contributed to the improved quality of life in rural towns located near the industry's sawmills, pulpmills, and plantations. Before

1980, people were moving to the big cities to look for employment because agricultural activities offered few opportunities and became unattractive. It is possible that the forest industry has stopped the emigration from the small towns, and has offered new life perspectives. Populations in these rural areas have increased because professionals from urban centers have moved to these smaller towns. This movement has started a cultural transformation, raising the quality of life and promoting urban development, better infrastructure, increased business activities, and improvements in education.

Conclusions

The development of radiata pine plantations in Chile has produced several conditions that could explain the success of this species in the country. First, it must be recognized that forest lands existing between regions V and VIII of the Coastal Range were destroyed by erosion resulting from the excessive agricultural use of these lands during more than 150 years. Today, a new landscape is found in these areas, where approximately 1.4 million ha are covered by radiata pine plantations. These plantations help to protect the soils against erosive processes and to reestablish destroyed ecosystems. Second, great economic expectations have been created, during more than 20 years, as a product of a large afforestation program with radiata pine. Chilean silviculturists, especially those working in the private sector, adopted experience from foresters in those countries working with radiata, or with other species that exhibited fast growth. Such concepts as sustainability of forests, and maintaining or increasing forest productivity in the long term constitute the basis of many research programs established since 1990. Third, the Chilean economy has been very well supported by the forest sector, which in 1994 exported forest products which returned U.S.$1,550 million.

Finally, radiata pine plantations have stimulated strong social development in rural areas where plantations have been installed, as well as where sawmills and pulpmills have been built. These activities have helped to reduce the migration of people toward urban zones, and have increased the number of professionals moving to small towns to work in forestry and related activities. Planted forests of radiata pine provide ecological, economic, and social values in Chile.

Notes

1. Names and years are author annotations about visits to Chile.
2. Names and years are author annotations about visits to Chile.

References

Allen, H.L. 1987. Fertilizers: Adding nutrients for enhanced forest productivity. J. For. 85: 37–46.

Banco Central de Chile. 1993. Boletín Mensual N°789. Banco Central de Chile, Santiago, Chile.

Brüggen, J. 1950. Fundamentos de la geología de Chile. Instituo Geográfico Militar, Santiago, Chile.

Chile Forestal. 1995. Exportaciones Forestales Chilenas. N°226, Año XX. Ministerio de Agricultura, Santiago, Chile.

Contesse, D. 1990. El desarrollo forestal chileno, una realidad sustentable. Lamas y Cía. Ltda, Concepción, Chile.

Elizalde, R. 1970. La sobrevivencia de Chile (segunda edición). Ministerio de Agricultura, Servicio Agrícola y Ganadero, Santiago, Chile.

Fuenzalida, F. 1967. Orografía, Capítulo II, pp. 7–33. In: Corporación de Fomento a la Producción (Ed) Geografía Económica de Chile, Texto refundido. Corporación de Fomento a la Producción, Santiago, Chile.

Gayoso, J. and Iroumé, A. 1984. Soil disturbance from logging in southern Chile, pp. 203–209. In: Proceedings of the Symposium on Effects of Forest Land Use on Erosion and Slope Stability. Environment and Policy Institute, East-West Center, University of Hawaii, Honolulu.

Gessel, S.P. 1982. Site evaluation in forest production and management, pp. 326–355. In: Schlatter, J.E. (Ed.) Evaluación de la Productividad de los Sitios Forestales. Actas Reunión de Trabajo, Universidad Austral de Chile.

Gessel, S.P. and Atkinson, W.A. 1984. Use of fertilizers in sustained productivity and Douglas-fir forests, pp. 67–87. In: Stone, E.L. (Ed.) Forest Soils and Treatment Impacts, Proceedings of the Sixth North American Forest Soils Conference. Department of Forestry, Wildlife, and Fisheries, The University of Tennessee, Knoxville, TN.

Gerding, V. 1993. Estabilidad de las plantaciones de *Pinus radiata* D. Don. Extracción y pérdida de elementos nutritivos asociados al manejo. Boletín Sociedad Chilena de la Ciencia del Suelo 10: 55–63.

González-Bonorino, F. 1970. Series metamórficas del basamento cristalino de la Cordillera de la Costa, Chile Central. Publicación N°37. Departamento de Geología, Universidad de Chile, Santiago, Chile.

Grey, D.C. 1989. A site-growth study of *Pinus radiata* in the southern cape. S. Afr. For. J. 150: 32–39.

Grey, D.C. and Jacobs, E.D. 1987. The impact of harvesting on forest site quality. S. Afr. For. J. 140: 60–66.

INFOR (Instituto Forestal). 1987. Especies forestales exóticas de interés económico para Chile. Corporación de Fomento a la Producción, Santiago, Chile.

INFOR (Instituto Forestal). 1994. Estadísticas Forestales. 1993. Boletín Estadístico N°32. Corporación Chilena de la Madera, Santiago, Chile.

Steinbrenner, E.C. 1979. Forest soil productivity relationships (chapter XII), pp. 199–229. In: Heilman, P., Anderson, H. and Barmgartner, D. (Eds.) Forest Soils of the Douglas-fir Region. Extension Service, Washington State University, Pullman, WA.

Toro, J. 1987. La fertilización forestal en Chile. Avances logrados hasta 1987. Chile For. N°143, Septiembre 1987: 27–30.

Toro, J. and Garrido, J. 1991. La mecanización de faenas forestales y su relación con la degradación de suelos. Actas III Taller de Producción Forestal. Fundación Chile, Grupo de Producción Forestal, Concepción, Chile.

Turner, J. and Holmes, G.I. 1985. Site classification of *Pinus radiata* plantations in the Lithgow District, New South Wales, Australia. For. Ecol. Manage. 12: 53–63.

Turvey, N.D. 1980. A forest soil survey. II. The application of soil survey information to forestry operations. Aust. For. 43: 172–187.

Unda, A. and Ravera, F. 1994. Análisis histórico de sitios de establecimiento de las plantaciones forestales en Chile, pp. 13–22. In: Espinoza, M., Knockaert, H. and Millán, J. (Eds.) Seminario: Erosión, Cultivos y Recuperación de Suelos. Corporación Chilena de la Madera, Santiago, Chile.

New Forests **18:** 45–57, 1999.

Small-scale planted forests in Zululand, South Africa: an opportunity for appropriate development

GUY A. CELLIER
Forest Solutions, Hawaii, USA

Key words: contract forestry, *Eucalyptus*, extension, sustainability, woodlot model

Abstract. Forestry in South Africa is sharply divided between commercially driven private enterprise which controls 1.3 million ha of plantation forests, and government and nongovernment organizations which promote a variety of social forestry programs.

One area of crossover is that of small-scale (1–2 ha) commercial woodlots being promoted by private timber companies as an additional source of fiber for their pulpmills, but grown and managed by individual farmers. Inputs, such as plants and fertilizers, as well as expertise in the form of extension foresters, are provided by the company, whereas the farmers supply the land and the labor. The company guarantees to buy the timber at the time of harvest. Under such arrangements, the risk is spread between the farmers and the company; the farmers assume most of the risks of production, and the company takes on the marketing risks.

The possibilities for appropriate and sustainable development do exist under such a program, but issues involving both ecological and social sustainability must be addressed. This paper suggests that four key characteristics are needed to achieve appropriate development in Zululand. These characteristics are participation, flexibility, empowerment, and commitment. When compared to the current arrangement, the company would have to make a fundamental paradigm shift to achieve these goals. However, the long-term support of the farmers, and ultimately more fiber in the mills, would be the reward.

Introduction

Forestry in South Africa is sharply divided. Private industry owns 1.3 million ha of commercial forest land, which is intensively managed for a variety of industrial purposes. The species grown are primarily *Eucalyptus*, *Pinus*, and *Acacia* spp., all of which are exotic to the country. On the other hand, there are numerous nongovernmental organizations involved in various forms of social forestry with the general goal of providing a wide variety of tree products to rural users, generally for individual family and community consumption, rather than for sale. The government forestry organization is split along these same lines, with their own pine and eucalypt plantations commercially managed for profit, and an extension service that aims to provide plants and information for social goals.

The timber industry experienced exceptional growth during the 1980s. The bulk of this growth was in the pulp and paper sector, and was achieved mainly through additional plantation establishment, primarily of *Eucalyptus* species, in the Natal province.[1] Pulpwood production experienced an annual growth rate of 6.6 percent for the period from 1979/1980 to 1991/1992, and during this time 82,710 additional ha of *Eucalyptus* species were planted in Natal province (Forest Owners Association 1993).

Mondi Forests (Mondi) responded to the general optimism of the late 1980s by considering an expansion of its Richards Bay pulpmill in 1989. The proposed expansion was to take the form of an additional pulp line which would require 1.5 million t fiber/yr. Mondi management considered land in neighboring KwaZulu to be ideal for the establishment of additional plantations, but, because this land was not available for purchase, the company implemented its Khulanathi ("grow with us") program.

Under this program individual farmers in KwaZulu are encouraged to grow *Eucalyptus* timber for sale to Mondi. All of the material inputs (such as plants and fertilizer) are supplied by Mondi, whereas the farmers are responsible for planting and tending the trees. The costs of the inputs are considered to be a loan to the farmer and are deducted at the time of harvest; he or she is then paid the balance of the market price for the timber. Interest is also charged on the money loaned.[2] As part of the program, Mondi has established woodlot grower committees in each area of operation as a means of communication between the company and the farmers.

The Zululand region

The Zululand coastal region (20 km wide, on average) lies along the east coast of South Africa, between the Tugela River in the south and Lake St. Lucia in the north. In terms of forestry production, this area is generally considered ideal, particularly for *Eucalyptus* species, which under "average" conditions achieve mean annual increments greater than 25 m^3/ha/yr when the mean annual precipitation is 1,000 mm or more (Herbert 1993a). The climate is humid-mesothermal, with a mean annual precipitation of 1,200 mm and a mean annual temperature of 22 °C. The deep sandy soils are aeolian in origin, whitish to light gray, generally fine-grained, and consist of almost pure silica (Noble et al. 1991). Since 1986 Mondi has also been working on a tree-breeding and clonal program for both pines and eucalypts. The eucalypt program has yielded a number of *E. grandis* Hill and hybrid clones that appear to offer exceptional production potential (in excess of 50 m^3/ha/yr) for the Zululand region[3] (Denison and Kietzka 1993a,b).

Social forestry

The Mondi program has been described as social forestry, commercial farm forestry, community forestry, agroforestry, and a form of contract farming. It is necessary to define these terms and to determine what exactly the program is, as there is currently considerable confusion. In South Africa (as elsewhere) claims are made by companies in order to gain "politically correct" points with agencies promoting rural development, for whom most of the terms have very positive connotations.

It is useful to note from the start that there is no universally accepted definition for social forestry. Gregersen, Draper, and Elz (1989, p. 3) use the term "social forestry" interchangeably with "farm forestry", "community forestry", and "forestry for local community development", and loosely define social forestry as "a broad range of tree- and forest-related activities that rural landowners and community groups undertake to provide products for their own use and to generate local income." They stress that the focus must be on people and community involvement, and believe that this is where social forestry differs from conventional production forestry.

Magno (1986) also emphasizes community involvement, i.e., that social forestry should focus on people and aim to provide goods and services for the benefit of rural communities. Cernea (1991) believes that there is now a broad consensus about the content of social forestry programs, which he states should do three things – namely, motivate people to plant trees; promote the kind of trees that will provide fuelwood, timber, grasses, and income; and increase benefits to the poorer groups.

Community forestry, on the other hand, is defined by Foley and Barnard (1984, p. 107) as "programs ... based upon growing trees on public or community land as opposed to private farms." They point out that community forestry is intended to benefit the community as a whole, a sentiment shared by Blair and Olpadwala (1988).

Blair and Olpadwala (1988, p. 11) believe that the two main components of social forestry are community and farm forestry. They believe that farm forestry "consists of landowners cultivating trees on their own land." This distinction is also made by Foley and Barnard (1984, p. 75), who define farm forestry as "programs which promote commercial tree growing by farmers on their own land." The woodlots that are the focus of this study are commercial in nature, grown by individual farmers on their own landholdings for profit. Certainly, by the above definitions, they fall into the category of farm forestry, but only marginally into the concept of social forestry because they provide only one product.

In *The greening of Africa*, Harrison (1987, p. 204) states, "Agroforestry is arguably the single most important discipline for the future of sustainable development in Africa." Von Maydall (1979, p. 18) defines the same term as follows: "Agroforestry implies sustained, combined management of the same land for silvicultural, agricultural and/or animal crops leading to an overall increase of production in comparison to single-crop management." Agroforestry is generally taken to be concerned with one or more forms of agriculture in conjunction with trees, and with the idea that any such system should be sustainable, and ideally should produce more than would be possible from a monoculture of one of the component species.

Contract farming

Contract farming is a production and marketing arrangement in which the company "replaces or supplements company production of agricultural commodities with purchases from local farmers through contracts" (Glover and Kusterer 1990, pp. 1–2). The control that the company exercises can vary considerably. For example, the company may merely pay a market price for the crop, and exercise little or no control over the production process. This is often the case for a nonperishable commodity and where the market price does not fluctuate very much. Alternatively, prices may be fixed at planting time, and the company may exercise almost constant supervision over the production process, often providing all the inputs and planting and harvesting equipment (or else doing the work itself) (Glover and Kusterer 1990).

Advantages and disadvantages for the company

There are several advantages for a company entering into a contract farming arrangement with small farmers. In the case of commercial timber in KwaZulu, the prime factor is good timberland which does not have to be (or could not be) purchased close to the processing plant. Besides land, labor is a major consideration for any contract farming arrangement. The company avoids conflict issues which may arise with a unionized labor force, and has access to labor (family and other) which may not normally be available to it.

Another factor is a steady supply of raw material to a processing plant; this can be better accomplished through contracts than by buying on the open market. Although Glover (1987) suggests that this is particularly important for food processing companies, it is no less a consideration for a pulpmill with large fixed costs which must keep a constant throughput at predictable prices to run economically. Besides quantity, this type of arrangement also allows the company a degree of quality control over the production process.[4]

Through contract farming the company is able to transfer most of the production risk to the farmers. Through not owning the land, the company may also avoid the risk of expropriation, while indirectly benefitting from financial aid provided to growers by the government or an external donor agency (Glover and Kusterer 1990). In Mondi's situation, there is no real aid going to the farmers from an external source, but calls for land reform in South Africa do present a very real danger of expropriation.

From a public relations point of view, the company can look good by promoting a progressive image and involving local producers. The company's own wages and benefits can also be made to appear superior in comparison to those offered by local farmers (Glover and Kusterer 1990). On the other hand, the security of the company's supply is one of the major drawbacks of contract farming from the perspective of the company. After investing management and capital into the raw material, the company runs the risk that the grower will default. Glover and Kusterer (1990) state that this is most common when the market price rises above the contract price.[5] They also state that it is difficult for the company to control the quantity and quality of the product, particularly if growers do not follow technical advice (Glover and Kusterer 1990).

Advantages and disadvantages for the farmer

Proponents of contract farming refer to the arrangement between the company and the small-scale farmer as "a dynamic partnership for development" (Williams 1984). The most readily identified advantage for the farmer is the opportunity for increasing net cash income on the farm. This is done through access to technologies, credit, and markets which would not normally be available to farmers.[6] Because the company supplying the inputs has a direct interest in seeing that the product is delivered, it will see that the right inputs are supplied at the right time, including the necessary extension. The company's access to large (often international) markets means that prices can be fixed and the risk from highly volatile perishable markets can be reduced (Glover and Kusterer 1990). There is also the potential for the development of small-scale contractors and other secondary employment and development opportunities, related either directly to the production process or indirectly through the injection of cash into the local economy.

However, one of the most common criticisms of contract farming is that the company maintains a monopsony[7] on the market and a monopoly on the inputs of production (Clapp 1988; Levin 1988; Glover and Kusterer 1990).

This is partly a question of control, and critics observe that the companies tend to control the most profitable sector, the processing and marketing of the commodities, and restrict the farmers to the less profitable production process (Levin 1988). It is also argued that various forms of dependency are created in contract farming arrangements (Clapp 1988). Companies often provide sophisticated technical inputs (as Mondi does with its clonal hybrids), and Glover (1987) points out that excluding small farmers from such technologies prevents them from capitalizing on their country's or region's comparative advantage. Credit is another form of dependence that can arise, and Clapp (1988) cites examples from South America where farmers accumulate debt rather than capital after entering into contracts.

Levin (1988) notes that, although the company is free from labor management, the farmers are under pressure to increase the length of the work day, or to use family labor to meet requirements. Another major criticism is that resources are diverted from staple food to cash crop production, which does little or nothing in terms of food security (Glover 1987; Glover and Kusterer 1990).[8]

Access to equipment (for harvesting, for example) may be limited for small growers, particularly if they rely on company equipment, because the company will first harvest its own crop (Glover 1987). The company may also manipulate inspection standards to control deliveries, if, for example, it is oversupplied (Glover and Kusterer 1990). Growers may receive poor technical assistance. They cite examples of contracts which require growers to follow the company's advice, but absolve the company from poor crop results. This is compounded by the fact that the farmer must assume the risks of weather, insects, and disease, because the company only pays for the produce it receives (Clapp 1988).[9]

As Glover and Kusterer state (1990, p. 130): "Contract farming is fundamentally a way of allocating risk between the company and its growers. The company tends to assume marketing risks while the growers bear most of the production risks." In forestry, there is the very real risk of fire, which can destroy a long-term (6–8 years for eucalypts grown for pulp) crop overnight. They point out that, although a contract may specify exactly how the risk is distributed between the company and the grower, the actual implementation of the contract may differ considerably. They believe that the distribution of risk will change over time, and essentially depends on three factors: bargaining power, availability of alternatives, and access to information.

The biggest risk the company takes is that of growers defaulting, either by not tending and harvesting the crop, or by selling to another buyer.[10] Through entering into contracts with many growers, the company is able to spread this risk for itself. It similarly spreads the risk of fire; because the woodlots are generally far apart, and even though a grower may lose 100 percent of his or her crop, the loss of timber to Mondi is small. The Khulanathi program is clearly a form of contract farming, incorporating distinct advantages and disadvantages for both the farmers and the company.

Contributions to appropriate development

Appropriate development should be both ecologically and socially sustainable, with ecological sustainability being perhaps the more easily understandable (if not necessarily achievable) goal. Considerable work has been done in South Africa on the ecological aspects of eucalypt plantations, particularly with regard to water use, as eucalypts are well known as a "water-hungry" tree species. From this work, some general recommendations can be made to promote ecological sustainability. A patchwork pattern of woodlots which are kept at the correct distances from water sources will allow for recharge and should not impact negatively on the long-term sustainability of water supplies. Individual plots should be protected against erosion through correct planting patterns and the use of cover crops for mulching purposes. At harvest, as much biomass as possible in the form of bark, leaves, and branches should remain on the site. The temptation to harvest whole tree lengths and debark at a central depot or the mill should be strongly resisted. Site rotation (planting trees on another site to allow recovery of the first) should be considered after the second or third coppice crop in light of any new evidence.

Ecological sustainability is just one aspect of appropriate development. As Lele (1991) argues, local social conditions can greatly influence ecological sustainability, but social sustainability is important in itself. Holmberg et al. (1991) argue that development does not happen without participation in the design and execution of the process by the people meant to benefit. This concept is now widespread in the development literature, and Lele (1991) suggests that participation has taken over from equity as a primary objective. There is sufficient evidence from the present study to suggest that the woodlots have not been promoted in a manner that enhances participation, and thus fundamental changes to the overall approach by the company are needed to achieve this objective.

Essentially, the company needs to realize that rural farmers have limited options available to them; many of these farmers accept that commercial woodlots are their best alternative at present. However, precisely because so little else is happening on their land, the woodlots are important to the farmers and they want to be involved in the whole process. One hectare to a company that controls over 500,000 ha of plantations is somewhat insignificant; this is not so to a farmer who has given over one-third or one-half of his or her entire landholding to trees.

Equity can be encouraged through commercial woodlots. Mondi has sufficient evidence that some large woodlots owned by prominent individuals are being neglected, whereas small woodlots owned by poor farmers are being carefully tended. The extension foresters need to avoid the temptation of pandering to the more powerful elite within the community. There is little, however, that woodlots can do for the landless poor apart from a limited amount of work associated with contracting. It is also possible that, by increasing the value of previously underutilized land, unequal allocations will be made by the local *Inkosi* (local authority) to individuals in his favor.

An appropriate woodlot model

Based on the evidence presented in this study, an appropriate woodlot model should have certain characteristics. The four that are considered absolutely crucial are participation, flexibility, empowerment, and commitment (shown schematically in Figure 1). First, participation is vital. There is sufficient evidence in the literature, as well as compelling evidence from this study, to indicate that growers really want to be involved in the whole process, from conceptualization through all the stages of implementation, and that this involvement is vital to the long-term success of the program. A participatory process will of necessity proceed slowly. For the company involved, this will require an understanding that initially, at least, fewer trees will be planted as mutual trust and understanding are established. The company will have to ensure that, from the start, potential growers are informed and know what they are getting into, in terms of financial returns as well as possible risks. The growers, too, will have to accept that a slower planting rate will mean that fewer plots are planted initially, and an equitable arrangement will have to be found within the community for deciding whose plots are planted first.

Participation will be encouraged by promoting strong and representative growers' committees. This will mean support from the company involved for organizational development, and a willingness to address the issues raised by these committees. This will also mean that the people appointed by the company to work as extension foresters are actually suitable for the job at hand. Their social skills will have to be as good as their technical skills, and they will have to be convinced, right from the start, of the need for participation.

1. PARTICIPATION
Commercial woodlots should involve a process with ownership by both the company and the farmers. Strong, representative growers' committees are vital

2. FLEXIBILITY
The process must be flexible, adapting to the particular circumstances of the growers in specific areas. Alternative systems involving different species should be encouraged; they will bring long-term sustainability.

3. EMPOWERMENT
Empowerment must be encouraged through a process of ongoing, on-farm research and training that is participatory and seeks appropriate alternatives.

4. COMMITMENT
The farmers are making a major, long-term commitment. The company must do the same. This commitment has to be to the whole process of rural development, and not just to commercial timber.

Figure 1.

Eberhard (1992), however, offers a word of caution against increasing the burden of the rural poor through too much participation. People with limited resources and numerous responsibilities may not be able to participate at every stage of the process. Sensitivity on the part of the company, as well as the removal of obstacles or the commitment of additional resources, will go a long way toward relieving this situation. For example, ensuring that meetings are centrally located and using the forester's vehicle to transport growers from farther afield would cut down on the time spent in meetings. The optimum level of participation is that which is desired by the people of a specific area.

Second, an appropriate woodlot model will be less prescriptive, thus allowing for greater flexibility depending on the circumstances of individuals or communities. This is obviously linked closely to the concept of participation, but there are differences. Discussions with growers have shown that some want to participate fully in the process, whereas others have other commitments, and want as little to do with the trees as possible. The company should accommodate these differences, i.e., encourage those farmers who want to do it all themselves and arrange contractors for those who prefer a "hands-off" approach. Flexibility is the key.

The company involved in supporting commercial woodlots should allow for alternative forestry systems which build linkages between the commercial orientation of the eucalypts and other priorities that the farmers may have. This may appear to be paradoxical, but, by being part of a more appropriate (and probably sustainable) system, the company would be building a more solid foundation for ongoing cooperation. Support by the company for fruit trees is a good example. Farmers in a particular area may want to set up an agricultural cooperative for better access to inputs or for marketing purposes. The company can help to facilitate this process, and offer collateral for a cooperative tractor, for example, when it can be seen that there is sufficient forestry work in an area to justify the expense of a machine that may well spend much of its time on nonforestry crops.

Third, empowerment must be encouraged through a process of appropriate training and ongoing research. Commercial woodlots will not result in appropriate rural development unless the growers "own" the whole process, and growers have indicated that they want and need technical training before they will feel adequate to the tasks at hand. With forestry training, many growers will plant and manage their own trees, and with harvesting training, some growers will harvest the trees themselves. Training is also a priority for the harvesting contractors, most of whom face technical problems and would like management, harvesting, or mechanical training.

The training and research that is implemented should be "on-farm" and again should have a participatory emphasis. Growers must be able to set the goals and develop the whole training and research effort in conjunction with the company. Strong growers' committees could play an important part in coordinating and deciding on priorities for an overall training and research strategy, as well as negotiating prices with the company or contractors and securing funding.

Fourth, an appropriate woodlot model would have the long-term commitment of the company involved to the development of the farmers. Some of the necessary elements, i.e., participation, appropriate on-farm research and training, and greater flexibility, have been discussed. However, this commitment would also manifest itself in specific issues raised by the farmers. For example, will the company guarantee a minimum price that will be paid for timber? Similarly, fire insurance is an issue of concern to the growers that should be addressed. Even though there is no insurance company willing to underwrite woodlot insurance at the present time, the timber company could again show good faith by matching an amount raised through timber levies, for example, to cover a portion of the loss to a woodlot through fire or some other disaster. This would show a commitment to the long-term well-being of individual farmers, and an understanding of the risk assumed by the farmers themselves.

Rural development utilizing trees is a slow and difficult process. Before a company embarks on a woodlot program, it needs to take the time to ensure that it has the support of the local farmers, and in turn is willing to commit itself to the process for the long term.

Conclusion

Commercial woodlots that are both appropriate and sustainable in the long term are entirely feasible in rural KwaZulu at the current time. Companies must realize, however, that the process is slow and will require creativity and a long-term commitment on their part. There is no "quick-fix" or predetermined procedure, and, although support is quickly fostered among rural people with few options, this support and trust can easily be lost if the growers feel that they are being neglected by the company.

Participation is a key element in the process, and private companies need to understand that absolute control (including ownership) is not essential to secure timber. Eucalypts grown on woodlots often receive more care and attention than those grown on company plantations, and growers also indicate that they are loyal to the company that offered the initial assistance, and would sell the trees to that company even if offered a better price by another purchaser.

A fundamental paradigm shift will be required for timber companies to allow commercial *Eucalyptus* woodlot growers to achieve sustainable and appropriate rural development. However, the companies have little to lose and everything to gain in their own long-term interests by allowing the growers to be flexible.

Notes

1. Prior to the general elections of 1993, Natal was one of the five provinces in South Africa, and KwaZulu was one of the ten so-called "homelands", choosing to remain "nonindependent" rather than "independent" (such as Transkei, Ciskei, Bophuthatswana, and Venda). All of these homelands, as put succinctly in The Urban Foundation's Urban Debate 2010 series, were "primarily a political artifact of the policies of grand apartheid, conceived in the 1950s and implemented with the fullest force during the 1960s and '70s" (Urban Foundation 1990, p. 16). The country map of South Africa has been redrawn and the whole northeastern seaboard is now officially KwaZulu/Natal, although the general region has traditionally been, and still remains, known as Zululand.
2. The rate in 1993 was 10 percent simple interest, whereas the prime bank rate was around 15 percent.
3. Herbert (1993b) cautions against assuming optimistic averages in Zululand; severe problems can be experienced during both wet and dry cycles, and there are questions as to whether or not specific sites can sustain these yields.
4. Mondi, for example, specifies minimum and maximum dimensions for the timber it purchases, and also insists that the timber be relatively dry, although this is hard to control.
5. Mondi may well face this danger. Although there is no agreed-on contract price with the growers and Mondi has agreed to pay a "market price", other timber purchasers who have not invested in the timber could afford to pay more than Mondi for the timber (assuming that Mondi plans to recover some of its investment).
6. Mondi supplies clonal hybrid eucalypts, which are not commercially available, to its Khulanathi growers.
7. A monopsony is a single purchaser of some type of input. The market power that the monopsonist has means that it can reduce the price paid without losing all of the input. A monopoly, on the other hand, is generally defined as the "sole producer of some product that has no close substitutes" (Browning and Browning 1983, p. 303).
8. In the case of timber production, the labor requirements are small, but the issues of available time and food production were raised during the field work of this study.
9. Although trees are not generally as susceptible as more perishable food crops, the severe drought conditions in 1992/1993 in Zululand have resulted in eucalypts losing substantial growth, and even dying. The trees are also susceptible to a number of pests and diseases.
10. Mondi has an informal agreement with Sappi (a competitive forestry company) not to buy timber from each others' small growers (Kewley 1993).

References

Blair, H.W. and Olpadwala, P.D. 1988. Forestry in Development Planning: Lessons from the Rural Experience. Westview Press, Boulder, CO.

Browning, E.K. and Browning, J.M. 1983. Microeconomic Theory and Applications. Little, Brown and Company, Boston.

Cernea, M.M. (Ed). 1991. Putting People First: Sociological Variables in Rural Development (second edition). Oxford University Press for The World Bank, New York.

Clapp, R.A.J. 1988. Representing reciprocity, reproducing domination: Ideology and the labour process in Latin American contract farming. J. Peasant Stud. 16(1): 5–39.

Denison, N.P. and Kietzka, J.E. 1993a. The development and utilisation of vegetative propagation in Mondi for commercial afforestation programmes. S. Afr. For. J. 166: 53–60.

Denison, N.P. and Kietzka, J.E. 1993b. The use and importance of hybrid intensive forestry in South Africa. S. Afr. For. J. 165: 55–60.

Eberhard, A. 1992. Community empowerment or exhaustion? New Ground 9: 29.

Foley, G. and Barnard, G. 1984. Farm and Community Forestry. Earthscan Energy Information Programme. International Institute for Environment and Development, London.

Forest Owners Association. 1993. Forestry and Forest Products Industry Facts 1979/80 to 1991/92. Forest Owners Association, Sandton, South Africa.

Glover, D.J. 1987. Increasing the benefits to smallholders from contract farming: Problems for farmers' organizations and policymakers. World Dev. 15(4): 441–448.

Glover, D. and Kusterer, K. 1990. Small Farmers, Big Business: Contract Farming and Rural Development. MacMillan, London.

Gregersen, H., Draper, S. and Elz, D. 1989. People and Trees – the Role of Social Forestry in Sustainable Development. The World Bank, Washington, DC.

Harrison, P. 1987. The Greening of Africa: Breaking Through in the Battle for Land and Food. An International Institute for Environment and Development – Earthscan Study. Paladin Grafton Books, London.

Herbert, M.A. 1993a. An initial assessment of the sandy soils on the Zululand coastal plain. Institute for Commercial Forestry Research Newsletter 5/93.

Herbert, M.A. 1993b. Principal Research Officer, Institute for Commercial Forestry Research, P.O. Box 375, Pietermaritzburg, 3200, South Africa. Personal communication.

Holmberg, J., Bass, S. and Timberlake, L. 1991. Defending the Future: A Guide to Sustainable Development. Earthscan, London.

Kewley, H.C. 1993. Regional Manager, Zululand, Mondi Forests, P.O. Box 35, Kwambonambi, 3915, South Africa. Personal communication.

Lele, S.M. 1991. Sustainable development: A critical review. World Dev. 19(6): 607–621.

Levin, R. 1988. Contract farming in Swaziland: Peasant differentiation and the constraints of land tenure. Afr. Stud. 47(2): 101–120.

Magno, V.C. 1986. Community forestry handbook. Rome: Food and Agriculture Organization of the United Nations. Cited in E. Sato. 1991. Social forestry: An evaluation of social forestry projects. M.S. thesis. University of Washington, Seattle.

Noble, A.D., Ramsden, R. and McInnes, C. 1991. The influence of two organic fertiliser sources on the rehabilitation of an under-performing stand of E. grandis, pp. 705–713. In: Schonau, A.P.G. (Ed.) International Union of Forestry Research Organisations Symposium on Intensive Forestry: The Role of Eucalypts. Southern African Institute of Forestry, Pretoria.

Urban Foundation. 1990. Rural development: Towards a new framework. In: Policies for a New Urban Future: Urban Debate 2010 (volume 4). Urban Foundation, Johannesburg.

Von Maydall, H.J. 1979. Agroforestry – a combination of agricultural, sylvicultural and pastoral land-use. Plant Res. Dev. 9: 17–23.

Williams, S. 1984. Agribusiness and the small-scale farmer: A dynamic partnership for development, pp. 1–11. In: Williams, S. and Karen, R. (Eds.) Agribusiness and the Small-scale Farmer: A Dynamic Partnership for Development. Westview Press, Boulder, CO.

New Forests **18:** 59–73, 1999.

Sustainability issues in Switzerland's forests

WOLFGANG ORTLOFF
Swiss Federal Institute for Forest, Snow and Landscape Research, Birmensdorf, Switzerland

Key words: age structure, consumption, forest functions, forest history, regeneration

Abstract. Basically there are three methods of regenerating a forest: sowing, planting, and natural regeneration. In the eighteenth and nineteenth centuries, large forested areas of Switzerland were devastated by exploitation and grazing within the forest borders. A significant percentage of these areas was regenerated by planting seedlings. Thus, the Swiss landscape is the result of both land use and cultural history. Today practically no old virgin forests remain.

Forests in Switzerland are required to meet diverse expectations. They function as a means of protection against avalanches, rock falls, and landslides; a source of wood; a natural habitat for animals and plants, and areas for recreation and tourism. Forest management aims to ensure all of these functions on a long-term, sustainable basis. Several problems exist with regard to sustainability in Swiss forests. Data from the Swiss National Forest Inventory reveal that the age structure is off balance, i.e., that there are too many old and not enough young forests. In addition, only two-thirds of the potential Swiss timber yield is harvested, whereas wood is imported from foreign regions where no control over the type of forest management is possible. Without a reduction in the demand for and consumption of wood, this practice, over the long term, is not a sustainable management option. The matter also raises ethical questions, e.g., is it correct to practice nature conservation in your own country while at the same time using wood produced under unsustainable conditions elsewhere?

Introduction

This paper gives a short overview of issues in Switzerland's forests which in one way or another relate to sustainability. In order to prevent the term sustainability from becoming an empty formula, a word without a concrete meaning, I have associated it with some forestry issues. It is not my objective, however, to ultimately define sustainability. Switzerland, in this case, merely serves as an example; most of the points mentioned can also be found in or applied to many other countries.

Planted

I refer to "planted" as one possible form of forest regeneration. To my understanding – and I am sure there are many formal definitions to distinguish between them – a planted forest and a plantation are two very different things. The major difference, in my opinion, is that the main purpose of a plantation is to produce biomass, whereas a planted forest can fulfill a number of different functions, which I will demonstrate with examples from Switzerland.

Basically there are three methods of regenerating a forest: natural regeneration, sowing, and planting. Because natural regeneration has both ecological as well as economic advantages, it is the management measure of choice under certain circumstances. These circumstances include seed trees that are autochthonous or at least highly adapted to the site, and seed trees that are of the desired species for the future stand at the respective site. If these prerequisites are not met or if areas without forest are to be afforested, active management measures are needed for regeneration. These measures include either sowing or planting. Sowing on a larger scale took place as early as the thirteenth century in the forests around the city of Nuremberg (Germany) in order to reafforest the vastly exploited area. Planting has some advantages over sowing. For example, germination and seedling growth take place under controlled conditions in nurseries, weed and pest (insects, fungi) controls are easier and more cost efficient in a nursery than in the field, and seedlings, once planted, have a competitive advantage against weeds and grasses. Regeneration can also take place in a combined pattern, with high-density plantations created on the best sites and the poorer soils colonized by natural regeneration (Combe 1994). An often neglected, but very important, prerequisite is that the planted trees match the site – not only in terms of the right species, but also in terms of genetic constitution.

Sustainable

One of the most frequently cited definitions of sustainable development is that given in the report of the World Commission on Environment and Development (WCED 1987), known as the Brundtland Report: "Sustainable development is development which meets the needs of the present without compromising the ability of future generations to meet their own needs."

A point often neglected is how very important it is to distinguish between sustainable development and sustainability. This is clearly expressed in a definition by Dovers and Handmer (1992): "Sustainability is the ability of a human, natural or mixed system to withstand or adapt to endogenous or exogenous change indefinitely. Sustainable development is therefore a pathway of deliberate change and improvement which maintains or enhances this attribute of the system, while answering the needs of the present population .... An important aspect here is the explicit separation of sustainability as a difficult and distant goal, and sustainable development as a variable process of moving toward that goal. They are not the same thing." Therefore, a sustainable society would be a society which is in such a process of moving toward sustainability.

Several initiatives evolved in the aftermath of the United Nations Conference on Environment and Development (UNCED or "Earth Summit") in Rio de Janeiro in 1992; they aimed at reaching an agreement on how sustainable management or sustainability should be defined. One of these initiatives was the Helsinki Ministerial Conference on the Protection of Forests in Europe in 1993, which provided the definition: "Sustainable management means the stewardship and use of forests in a way, and a rate, that maintains their biodiversity, productivity, regeneration capacity, vitality and their potential to fulfill, now and in the future, relevant ecological, economic and social functions, at local, national, and global levels, and that does not cause damage to other ecosystems."

This emphasizes that the objective of the UNCED conference, and of other international initiatives, is ecological sustainability. Over the past years there have been a number of initiatives aimed at defining criteria for sustainable forest management, and several *measurable* indicators for each of these criteria (Forest Stewardship Council 1994; Hahn-Schilling et al. 1994; IWGF 1994). The most comprehensive so far is the recently published report of The Montreal Process (1995).

Forest history

Over the centuries Switzerland has witnessed several phases of extensive forest exploitation. The reasons, among others, were usage of wood for construction material, for fuel, and for railroad ties; extensive grazing within forest borders; nutrient export through litter raking; and, in the eighteenth and nineteenth centuries, urban growth and the boom of such industries as iron, brick, glass, and charcoal works. These exploitations resulted in increasing resource damage through erosion, landslides, avalanches, and flooding, events often referred to as "natural hazards." The Swiss realized that domestic supplies of timber could be improved only if size, condition, and yield capacity of the forests were known, and if the forests received proper treatment. Early forest regulations recognized the threats posed to people by overexploitation of natural resources. The Zurich Forest Decree of 1702 mentioned the risk caused to future generations by excess cutting, and this forced many citizens to leave the country because of timber shortage. Other regulations and decrees recommended that the forests be treated more appropriately, i.e., more in line with forest science. Another publication called for proper planting, treatment, and utilization of forests "for their own and their successors' welfare" (Schuler 1984).

At the beginning of the nineteenth century, academically trained foresters (mainly in German forestry schools) initially influenced Switzerland's forestry. "They were convinced that forests and their management should not be looked at separately from anything else happening in the country. They were further sure that plans for forest tending and utilization were necessary if forests were to fulfill their functions and purpose to the profit of the whole country over a long period of time and therefore on a sustained basis" (Schuler 1984). Knowing the German and French schools of thought, Swiss foresters critically adapted these theories to the special conditions of Switzerland. The desire to systematically study and, thus, better understand the processes within forest ecosystems led to the establishment of a forestry school at the Federal Polytechnikum in Zurich in 1855.

As mentioned, the number of catastrophic events, e.g., floodings, avalanches, and rock falls, peaked in the mid-nineteenth century. Around that time a report was compiled to document the often poor state and exploitative utilization of the Swiss forests, especially those with a protective function, and the relation between forest exploitation and these catastrophic events (Landolt 1862). As a result, the first Swiss Forest Law was established in 1876, and set the foundation for a nationwide forest policy. It called for the conservation of existing forest area and the reafforestation of areas without sufficient forest cover. It must be stated, however, that these goals would not have been reached if the railway system had not been established at that time. This made it possible to import large amounts of coal from Germany and, thus, to reduce the demand for wood. Because there were no systematic forest inventories in those days, no exact data exist regarding the proportion of reafforestation carried out by planting trees. The exploited status of vast areas suggests that regenerating the forests by planting must have taken up a large proportion.

Thanks to the "pioneer" foresters of the nineteenth century, a formal forestry organization has developed, thus allowing Swiss forests to be managed on a sustainable basis and causing a boom in reafforestation. Of course a comprehensive ecological sustainability has not yet been achieved because it is not comprehensively defined. In comparing the time span of development of this "controlled forestry" to that of a tree, one recognizes that this development has to be looked at as a historical process which is still evolving. Although many ideas proposed by foresters in the nineteenth century were not realized at the time, and might not even be today, one has to acknowledge their broad perspective on long-term forest management. They pertained, not only to timber production, but also to the important role that the many functions of forests play for a society. In conclusion, it can be expected that the change in paradigms from sustainable multipurpose forestry as practiced today to true ecological sustainability will promote an explosion in knowledge similar to that promoted by the principle of sustained-yield forestry two centuries ago (Glück 1994).

Sustainability and sustained yield

When studying the term "sustainability" in connection with forestry, one always encounters the concept of "sustained yield." Yield is an obvious criterion for ecosystem productivity and has several indicators, e.g., standing volume and annual increment, which are relatively easy to measure. The idea of sustainability was developed very early. In Germany, the first written source of the term "Nachhaltigkeit", generally used as the synonym for "sustainability", dates back to 1713. In a book from 1806 by Zschokke on the Swiss mountain forests, the concept of sustained yield was considered to be the "first main principle of forestry", i.e., the forest itself was the capital which should not be reduced (Zschokke 1806; after Schuler 1984). Yet a fundamental problem existed: the volume of sustained yield was unknown. Therefore, fellings were often excessive or insufficient, and, thus, one way or the other, ruined the forests. Another early forester with an academic education was Kasthofer, who also attributed the devastation of the mountain forests to the fact that no forestry rules were applied. He understood sustained-yield management to be "felling not more per annum, but also not less, than nature produced annually ..." (Kasthofer 1818; after Schuler 1984).

Kasthofer was the first forester in Switzerland to use the term "Nachhalt", an abridged form of the German term "Nachhaltigkeit." In this context, he spoke explicitly of the different functions of the various forests and emphasized the importance of forests in keeping the country heritable and fertile. Consequently, he saw selection forestry as a proper silvicultural method to be practiced in mountain forests because the most important forest function there is that of protection. Only in forests treated in such a manner is a permanent forest cover and, therefore, the maximum protective function guaranteed. In his opinion, any attempts to regulate on the basis of formula and normal conditions were, especially in the mountain forests, "useless, costly, and sometimes even ridiculous." This makes clear that the term "Nachhaltigkeit" not only pertains to timber production but also includes other values and relevant forest functions which are to be provided on a long-term, sustained basis.

Several centuries ago, in the early stages of forestry science, models were developed to help determine the use of forests on a sustained yield basis. The first models predicted that the same area be cut yearly in a certain sequence, given by the predominant wind direction (Schuler 1984); later models determined that a certain timber volume be cut each year, because growth, and therefore yield, might be different on different plots. Both models can be seen as attempts to secure a steady supply of wood. For a respective forest area, the total standing stock divided by the rotation period (in years) results in a volume to be cut each year. If the total forest area being managed is big enough, clearcutting may be used to provide the sustained yield of timber. However, this does not take into account the effects of this type of forest management on the ecosystem. On the other hand, the management system in which sustained yield is realized on the smallest spatial extent is "Plenterung", which can be translated as selection cutting. Thus, when talking about sustainability, it is necessary to give the spatial extent referred to, as will be shown in a later section.

The Swiss forest today

Current data

The data on Swiss forests were provided by the National Forest Inventory section at the Swiss Federal Institute for Forest, Snow and Landscape Research. The forest cover in Switzerland is 29 percent of the total surface area and 37 percent of the total stockable area, i.e., excluding glaciers, high mountain areas, and so forth. Spruce is the dominant species (Table 1), although it is claimed that its proportion would be considerably smaller if left to nature. The proportions of the different methods used for regenerating the forests are different in the five main regions (Figure 1). For example, planting is used more often in the Central Plateau, whereas natural regeneration predominates in the mountainous regions. The age structure is off balance (Table 2), with too many old and not enough young forests.

Old growth/virgin forests

There are areas, especially in the mountains, where the last direct disturbance by humans took place several decades ago. These forests could, therefore, be called "old growth" if one allows the definition of old growth to include human disturbances. "Virgin" or "primeval" forests, i.e., forests without any human influence, do not exist in Switzerland because of the very intensive land use of the last few centuries. Depending on how strict the definition

Table 1. Tree species in Switzerland's forests (*Source:* EAFV 1988).

Common name	Species	Standing volume (%)
	Conifers	
Spruce	*Picea abies* (L.) Karst.	49
Silver Fir	*Abies alba* Mill.	15
Larch	*Larix decidua* Mill.	5
Scots Pine	*Pinus sylvestris* L.	3
Cembran and Mountain Pine	*Pinus cembra* L., *P. mugo* Turra	1
Other conifers		0
	Broadleaved trees	
Beech	*Fagus sylvatica* L.	16
Sycamore Maple	*Acer pseudoplatanus* L.	2
Ash	*Fraxinus excelsior* L.	3
Sessile and Pedunculate Oak	*Quercus petraea* Liebl., *Q. robur* L.	2
Chestnut	*Castanea sativa* L.	1
Other broadleaved trees		3

of "human influence" is, one could make the assumption that there are no virgin forests left on the earth because human influence exists even in the most remote areas, e.g., through anthropogenically elevated atmospheric concentrations of CO_2.

Forest functions

Forests in Switzerland have to fulfill a variety of different functions, and a general overview of the environmental functions of forests is given in Table 3. Because of the rough topography in Switzerland, especially in the mountainous regions of the Alps, forests are relied on as a means of protection against avalanches, rock falls, and erosion, and they also prevent flooding in the lower parts of the country. Forests are a source of wood and many other, non-timber resources (e.g., mushrooms, berries, and game). Furthermore, the forests themselves, as well as in combination with such other elements of the landscape as meadows, agricultural lands, and rivers and creeks, offer possibilities for recreation and tourism which are major components of the Swiss economy. Studies in landscape aesthetics reveal interesting insights into the expectations of people (locals, as well as tourists) concerning preferred landscapes.

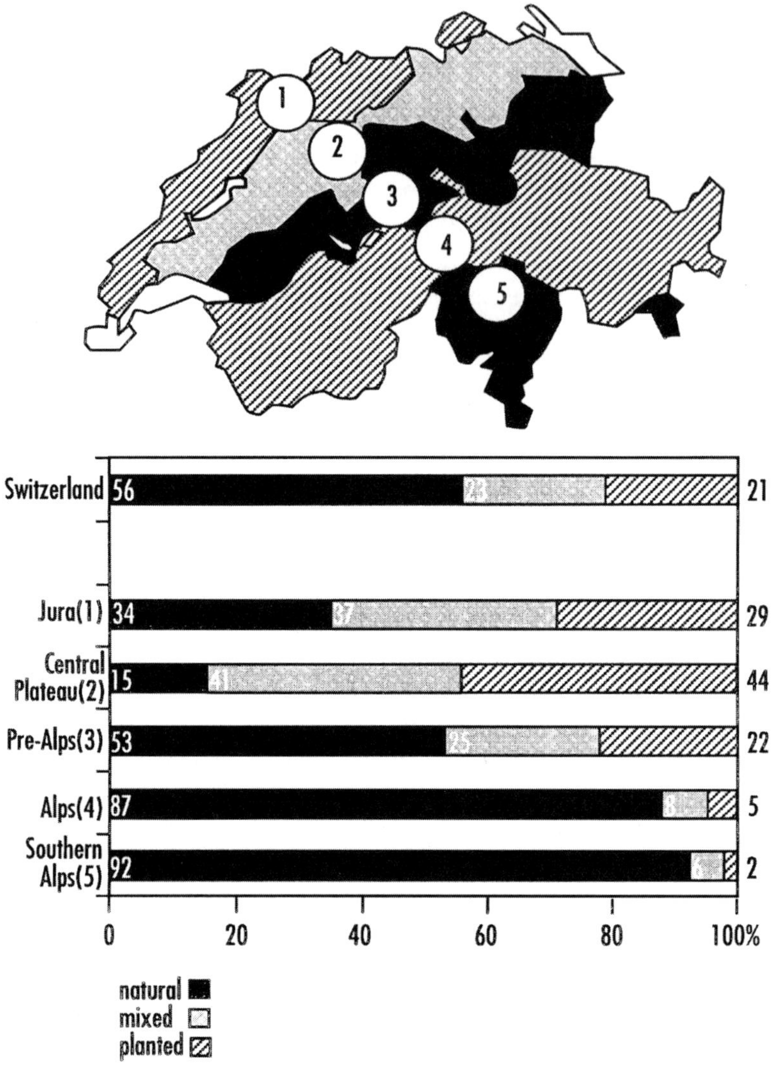

Figure 1. Methods of regeneration in the five main regions of Switzerland. (*Source*: EAFV 1988).

Another increasingly important aspect of forests is their ability to provide natural habitat for animals and plants. Biodiversity, similar to sustainability, is a term often used in the recent scientific literature and in political statements (e.g., the CIA World Factbook 1994 on the Internet informs us that a current environmental issue in Switzerland is "loss of biodiversity"). Without any parameters, this type of information is useless. The usage of the term "biodiversity" is often inadequate, i.e., it is often used only in the sense of

Table 2. Actual and modeled (economic ideal) age structure of Switzerland's forests (*Source:* EAFV 1988).

Age class	Forest area (%)		Difference
	Actual	Model	
Uneven-aged	25	25	
0–30	12	15	–3
31–60	10	15	–5
61–90	18	15	+3
91–120	17	15	+2
121–150	9	10	–1
151–180	5	5	0
over 180	4	0	+4

Table 3. Environmental functions of forests (*Source:* Pearce and Moran 1994).

Sources of materials and services	Sink for wastes	General and life support
Timber	Absorption of waste	Genetic pool
Fuelwood	Recycling nutrients	Climate regulation
Other business products	Watershed protection	Carbon fixing
Non-wood products	Protecting soil quality and erosion resistance	Habitat for people, flora, and fauna
Agricultural production		Aesthetic, cultural and spiritual source
Recreation and tourism		Scientific data

species diversity. But biodiversity also includes the diversity of landscapes (i.e., the pattern or mosaic of different ecosystem and land-use types), as well as the genetic diversity within species. This point was clearly emphasized in the Helsinki process and also in the Rio conference. In Switzerland the obligation of maintaining genetic diversity is fulfilled by several genetic inventory projects which provide valuable information about where and how to install genetic reserves. The principle underlying these activities is that of conservation of genetic reserves in situ.

Another function that forests fulfill worldwide is that of carbon sequestration and storage. The national balance for CO_2 and other greenhouse gases reveals that forestry in Switzerland is the only branch acting as a sink for these

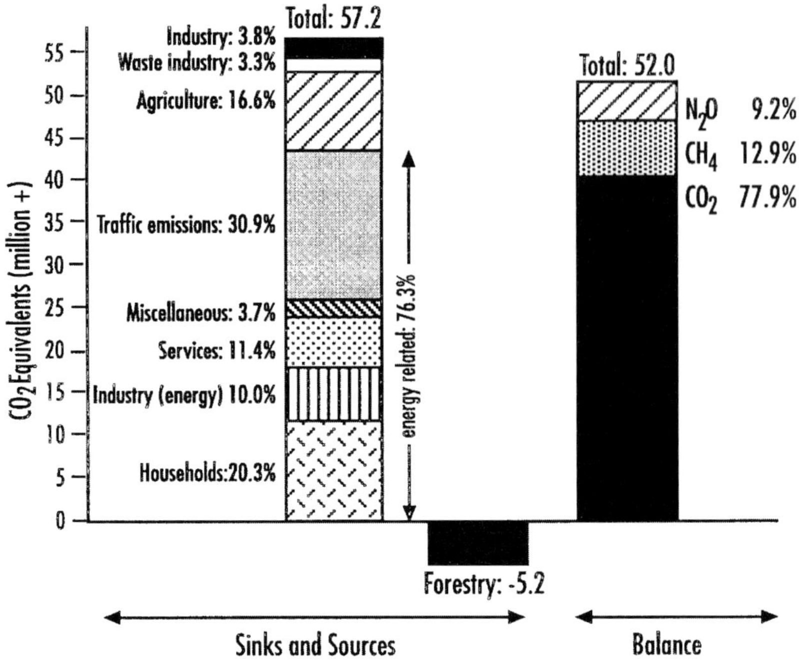

Figure 2. Greenhouse-gas balance emissions of Switzerland for 1990 in CO_2-equivalents (*Source*: after Federal Office of Environment, Forests and Landscape 1995).

gases (Figure 2). In this context, it is also worth mentioning that using wood for energy production and for heating of buildings has advantages over fossil fuels because wood is CO_2-neutral. The use of forests for these purposes is widespread in Switzerland. Nevertheless, various assessments show that 40–70 percent of the potential yield of wood for energy is not used (Brändli 1994). The federal program "Energy 2000" aims at doubling the country's consumption of wood for energy by the year 2000. Using wood for construction purposes is also a good option because the CO_2 fixed in the wood is stored in the buildings for decades.

Issues for consideration

Insufficient fellings

The annual yield in Swiss forests is about 7–8 million m^3, but only 4.5 million m^3 are harvested. Nevertheless, total wood consumption is about 7 million m^3, i.e., about one-third of the wood consumed in Switzerland is imported from foreign countries.

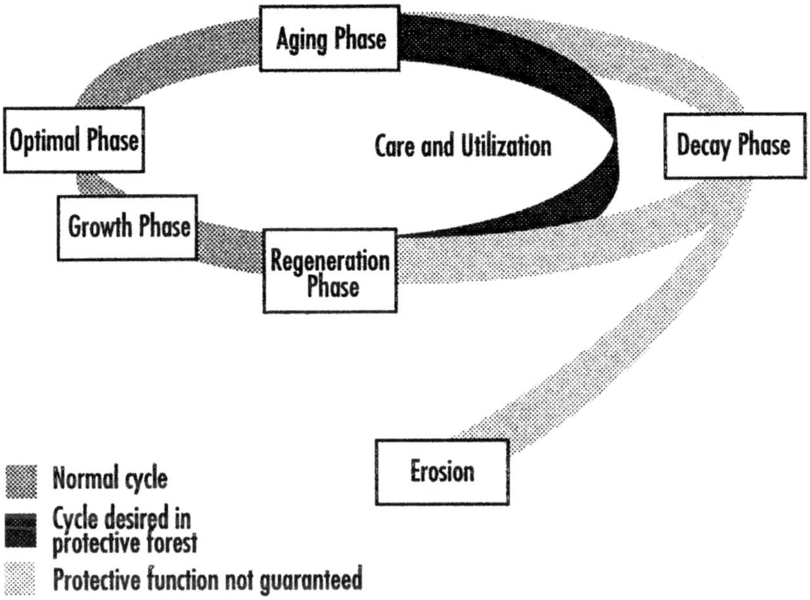

Figure 3. The natural cycles in forests and the influence of silvicultural practices in order to Ensure the Protective Function of Mountain Forests (*After*: Combe 1994).

There are several reasons for the insufficient fellings in Switzerland, and among the most important are the high costs for harvesting (Brändli 1994) and the setting aside of forested areas for conservation. The primary reason for high harvesting costs are high wages. Often, the price for timber is too low to even cover the harvesting costs. This leads to a reduction in such management actions as thinning and regeneration, which might, in the longer term, have negative implications for the stability (e.g., storm events) and the protective function of some forests, as well as the future wood supply. Very little information exists on whether or not the protective function can be maintained in mountain forests which are allowed to enter the natural decay phase (Figure 3).

When discussing sustainability, the spatial context has to be made clear. At first glance, it might seem sustainable to conserve forests in a country and not use them anymore. Without a reduction in the demand for and consumption of wood, this practice of chronic insufficient fellings while importing wood from countries where no control of the method of forest management is possible is, over the long term, not a sustainable management option. The matter also raises ethical questions, e.g., is it correct to practice nature conservation in your own country and at the same time use wood produced under unsustainable conditions?

Unbalanced age structure

Data from the Swiss National Forest Inventory (EAFV 1988) reveal that the age structure is off balance, i.e., there are too many old and not enough young forests (Table 2). Forest management can deal with such a situation by means of careful planning. In the long term, however, this situation poses a serious threat to sustainability, e.g., through lack or scarcity of habitat for species depending on early-stage succession forests. In addition, a steady supply of wood for different kinds of uses, such as energy, construction material, and pulp and paper, is not guaranteed.

Tree species not matching the site

There are debates in progress on what the proportions of different tree species would be if forest development had not been anthropogenically influenced after the last ice age. Today almost 50 percent of the standing volume in Swiss forests is made up of spruce (*Picea abies* [L.] Karst.). Spruce is not an exotic tree species to Switzerland, but some claim that many stands of spruce are located on unsuitable sites. It is not known if this will, over the long term, have detrimental effects, e.g., on soil and groundwater.

Besides soil, geologic substrate, and water availability, the term "site" in forestry also includes the influences of weather and climate. These latter influences underlie a constant change, i.e., it is impossible to define normality for these factors. Nevertheless, the ongoing discussion and the majority of the scientists in the field of global change predict a general warming during the coming decades as a result of the anthropogenically altered composition of the atmosphere. The regional effects of this predicted global warming are not yet clearly understood.

What does this expected global warming mean for forest management and silviculture? Should only warm-tolerant species which are more adapted to water stress be planted? According to a number of Central European forestry scientists (see Sperber 1994), the main goal of forest management should be to maintain structurally, biologically, and genetically diverse forests, because they are the most apt to have the capacity to deal with change, be it biotic or abiotic, climatic, or anthropogenic. There is no justification for the use of genetically altered plant material. Nature develops under constantly changing conditions, thus genetically rich species which match the respective site are the best guarantee of tolerance against any kind of stress.

Damage by deer and other ungulates

Damage by deer and other ungulates is a widespread problem in Central Europe, and represents "an unholy alliance between hunters and misguided animal lovers" (Giordano et al. 1993). Hunters want more game to shoot, and animal lovers with their "bambi-mentality" refuse to accept that natural predators, such as wolves and bears, have become extinct over large areas of Europe. Depending on the point of view, the latter can be seen as positive or negative, but it is the reality today and needs to be dealt with – no matter what its cause.

Excessive deer populations to a large degree hinder regeneration, be it natural or planted. The problem exists in many European countries and is not a new one. Aldo Leopold visited Germany in 1935 and was shocked by the damaging effect deer had on the forests there (Leopold 1936a,b). Because regeneration by planting involves a considerable investment, the forest owner must fence the regenerated plots for several years at considerable expense in order to keep the deer out. This issue is a serious one which, in my opinion, will, in the long term, have effects more detrimental than such other factors as air pollution and climatic stress.

Conclusions

- Natural regeneration is the method of choice wherever certain prerequisites are met: the seed trees are autochthonous or at least highly adapted to the site, and the seed trees are of the desired species for the future stand at the respective site. Because of the long history of exploitation and the following boom of reafforestation in the nineteenth century, a significant proportion of the forests stocking Switzerland today were planted. One has to always bear in mind that today's forests and the landscape in general are shaped by a long history of anthropogenic land use. These processes are still going on, i.e., forests are not static, although in many discussions on sustainability this seems to be implied. Forests are, on the contrary, always changing, and any discussion on sustainability has to realize this aspect of constant change (Dovers and Handmer 1992).
- Forests in Switzerland are required to meet many diverse expectations. They function as a means of protection against avalanches, rock falls, and landslides; a source of wood; a natural habitat for animals and plants; and an area for recreation and tourism. Sustainable forest management in the broadest sense is apt to ensure that all of these and maybe additional functions are met on a long-term sustainable basis.

- Chronic insufficient harvest levels threaten the principle of sustainability if at the same time wood is being imported from foreign countries where no control over the harvesting methods is possible. Rich countries can afford nature conservation, and buy parts of the timber resources they need from other countries, which might be financially dependent on the exploitation of their natural resources. By importing wood, environmental problems are exported. Sustainability not only implies *intergenerational equity*, as it is called, e.g., in the Brundtland Report, but also *intragenerational equity*, i.e., "justice within the current world population" (Dovers and Handmer 1993).
- One of the main tasks of forest policy is the promotion of sustainable forest management in accordance with UNCED principles. This should not cause major problems because these principles are very much in line with those already accepted and established in Swiss forestry (Combe 1994).
- Ecological sustainability has yet to be well defined. We need better criteria and, even more importantly, measurable indicators in order to decide if forest ecosystems are moving toward or away from what we define as sustainability. Moreover, forestry needs to use public relation mediums more effectively (Halder 1992) in order to gain publicity for the principle of sustainability in natural resource management. Surveys in Germany have shown that most people do not even know what the term "sustainable forest management" means (Schanz 1995).
- One of the major contributions of forestry to society is that, centuries ago, forestry was the first field to attempt to implement the principle of sustainable action. The principle of acting on a sustainable basis originates from forest management and forestry sciences. This is a result of the long-term nature of any forestry management action. Whatever is done in the forest will most probably affect succeeding generations more than it does the present generation. In my opinion, this fact must always be kept in mind when discussing sustainability.

References

Brändli, U.B. 1994. Swiss foresis and their potential for energy wood on the basis of data from the National Forest Inventory, pp. 1–9. In: Richardson, J. (Ed.) Silviculture Tuned to Nature and Wood Energy Production. Natural Resources Canada, Canadian Forest Service, Ottawa.

Combe, J. 1994. Silvicultural practice in Switzerland – a nature-tuned philosophy, pp. 10–12. In: Rickardson, J. (Ed.) Silviculture Tuned to Nature and Wood Energy Production. Natural Resources Canada, Canadian Forest Service, Ottawa.

Dovers, S.R. and Handmer, J.W. 1992. Uncertainty, sustainability and change. Glob. Envirn. Chang. 2(4): 262–276.

Dovers, S.R. and Handmer, J.W. 1993. Contradictions in sustainability. Envirn. Conserv. 20(3): 217–222.

EAFV (Swiss Federal Forestry Research Organisation). 1988. Schweizerisches Landesforstinventar: Ergebnisse der Erstaufnahme 1982–1986 (Swiss National Forest Inventory: Results from the First Survey 1982–1986). Eidgenoessische Anstalt fuer das Forstliche Versuchswesen, Report No. 305. Flück-Wirth, Teufen.

Federal Office of Environment, Forests and Landscape. 1995. Klimapolitik der Schweiz. Informational brochure.

Forest Stewardship Council. 1994. Forest Stewardship Principles and Criteria for Natural Forest Management. Forest Stewardship Council, Oaxaca, Mexico.

Giordano, E., Glück, P., Hummel, F., Kurth, H. and Kuusela, K. 1993. Policy implications of the UN-ECE/FAO 1990 Forest Resource Assessment. European Forest Institute Working Paper 1. European Forest Institute, Joensuu, Finland.

Glück, P. 1994. Von der ökonomischen zur ökologischen Nachhaltigkeit (From economic to ecological sustainability). Österreichische Forstzeitung 8: 24–28.

Hahn-Schilling, B., Heuveldop, J. and Palmer, J. 1994. A Comparative Study of Evaluation Systems for Sustainable Forest Management (including principles, criteria, and indicators). Bundesforschungsanstalt für Forst- und Holzwirtschaft, Hamburg.

Halder, F. (Ed.). 1992. Public relations fuer den Wald (public relations for the forest): ein Workshop mit Diskussionsbeitraegen aus Wissenschaft und Praxis. Fischer, Munich.

IWGF (Inter-Governmental Working Group on Forests). 1994. Final Report. Second Meeting. Ottawa/Hull, Canada, October 10–14, 1994. Ottawa/Hull.

Landolt, E. 1862. Bericht an den hohen schweizerischen Bundesrat über die Untersuchung der schweizerischen Hochgebirgswaldungen (Report to the honored Swiss Federal Council on the study of the Swiss mountain forests undertaken in the years 1858, 1859 and 1860). Weingart, Bern.

Leopold, A. 1936a. Deer and Dauerwald in Germany I. History. J. For. 36(4): 366–375.

Leopold, A. 1936b. Deer and Dauerwald in Germany II. History. J. For. 36(5): 460–466.

Montreal Process. 1995. Criteria and Indicators for the Conservation and Sustainable Management of Temperate and Boreal Forests. Canadian Forest Service, Hull, Quebec.

Pearce D. and Moran, D. 1994. The Economic Value of Biodiversity. IUCN, London.

Schanz, H. 1995. Zur Bedeutung von "Nachhaltigkeit" in der Forstwirtschaft (The importance of "sustainability" in forestry). Der Wald Berlin 45: 114–117.

Schuler, A. 1984. Sustained-yield forestry and forest functions, as seen by Swiss foresters in the nineteenth century, pp. 192–201. In: Steen, H.K. (Ed.) History of Sustained-yield Forestry: A Symposium. Forest History Society, Santa Cruz, California.

Sperber, G. 1994. Das Klima ändert sich – was können wir Förster tun? (The climate is changing – what can we foresters do?), pp. 115. In: Verband Weihenstephaner Forstingenieure (Ed.) Waldökosysteme im globalen Klimawandel (Forest ecosystems and global climate change). Economica Verlag, Bonn.

WCED (World Commission on Environment and Development. 1987. Our Common Future. Oxford University Press, Oxford.

New Forests **18:** 75–88, 1999.

Ecosystem management and plantation forestry: new directions in British Columbia *

CLARK S. BINKLEY
Hancock Timber Resource Group, Boston, MA, USA

Key words: extensive management, forest practices, intensive management, land-use, multiple-use

Abstract. Historically British Columbia's (B.C.) forests were managed under the implicit assumption that virtually the whole forested land base would, one day, be available for timber production. The B.C. Forest Service and licensees incorporate non-timber values into timber production plans through a process of "integrated resource management" which attempts to consider wildlife, riparian habitat, recreation, water flows, grazing, and other forest uses in each decision about each hectare where logging is to occur. Under this extensive form of management, silvicultural investments are low. This policy has clearly failed either to satisfy legitimate demands from the environmental community or to produce the predictably high levels of timber harvest needed to sustain the forest products industry and industry-dependent communities. The core problem is that, despite a vast forest estate in British Columbia, land has become scarce. It is, therefore, logical to substitute capital, labor, and knowledge for land in forest production processes. Implementing this general economic prescription requires a change in forest management approach to zone the landscape and manage each zone intensively for a specific purpose. For the bulk of commercial timber production, planted forests represent the best technological option. New directions in B.C.'s forest policy – land-use zoning, a new forest practice code, and the dedication of capital for silvicultural investments – generally move toward this objective, but implementation remains uncertain. Major impediments include dysfunctional forest tenure arrangements and a comparatively poor information base.

Introduction

No reasonable person doubts that the forests of a region should accommodate a variety of uses and values. Ecosystems where human influence is absent or nearly so provide critical spiritual, ecological, cultural, and economic values. At the same time, wood is one of the world's most ubiquitous and important

* This paper was derived from an article previously published in Forestry Chronicle 73(5): 553–559. © Canada: Canadian Institute of Forestry.

industrial raw materials and sources of energy. The critical issue facing forest managers – and indeed society more broadly – is how best to accommodate these diverse values across the world's vast forested landscapes.

This paper considers the challenges facing British Columbia (B.C.) in responding to this conundrum. Although circumstances are probably similar in much of Canada, specific ecological, institutional, and economic circumstances always are consequential for forest policy, so generalizations may be misleading. We begin by considering the economic and ecological challenges. Past forest management approaches in British Columbia have set the stage for massive economic dislocations with no certainty for sustaining the province's enormous store of biological diversity. As a prelude to crafting new policies, it is useful to understand exactly why the "integrated resource management" approach failed. Most simply, as a result of increased wealth and population, forest land has become scarce. Our traditional management approaches have inhibited the logical economic responses to scarcity: substitution of technology, capital, and labor for land. An alternative forest management approach of intensive management for primary uses across the landscape would enhance the environmental and industrial uses of the forest. The province has taken the first, tentative steps toward this new approach, but much remains to be done and the potential for recidivism is great.

Forest stewardship challenges in British Columbia

British Columbia faces the predictable problems associated with the transition between reliance on natural old-growth forests and second-growth forests created through human stewardship. In brief, these are as follows (Binkley 1997):

1. *Increased wood costs*

Delivered wood costs have risen as harvesting pushes into increasingly remote sites, and as timber rents increase – either as a matter of government policy or through normal market forces (Lyon 1981). Increasing wood costs reduce operating profits for firms selling into large, international markets where the costs or production decisions of a specific region have little or no effect on the prices of final products. Just as increased wood costs in the United States opened the door for expanded production in the B.C. Interior, increasing wood costs in Canada support expanded timber supplies in such other, competing locations as New Zealand, Chile, Brazil, and perhaps eventually Siberia.

2. *Substitution of capital for timber and capital for labor*

As a result of massive investments in state-of-the-art technology, the best B.C. sawmills are extremely efficient. Labor productivity – measured as output per person hour – has increased, thus making high wages sustainable. But the mathematical inverse of labor productivity is employment per unit of output, and this social indicator has fallen by half in the last 30 years in British Columbia (Nixon 1991). More significantly, in the absence of rapid technical innovation, the "iron law of diminishing returns" means that, as more and more capital is used in the sector, the return on additional capital investment will decline. Lower returns obviously mean less investment, and lower levels of capital investment will initiate a self-perpetuating downward economic spiral in the forest sector.

3. *Upward pressure on product prices*

Canada is large enough in many of the markets it serves so that higher domestic costs will tend to push product prices up. Although higher product prices will help offset upward pressures of costs, they will also encourage substitution away from conventional wood-based products. Econometric studies suggest that each 1 percent increase in the price of softwood lumber produces a 0.30 percent increase in the use of structural steel, a 0.15 percent increase in the use of cement, and about a 0.65 percent increase in the use of bricks (Binkley 1994). Increases in the use of these substitute products will reduce the demand for wood products and will limit the potential increases in their prices.

4. *Increased demand for environmental services of the forest*

The demand for natural environments increases as per capita income increases. Good empirical evidence supports this claim for some features of the environment – air quality, water quality, and outdoor recreation – and it is probably more broadly applicable as well. At the same time that demand for these services is increasing, their supply has been diminished by logging, especially in places where environmental regulations have historically been lax.

5. *Institutional failure for non-market forest values*

Most environmental services operate outside formal markets, either because they are true public goods (e.g., aesthetically pleasing landscapes or carbon sinks), or because society has chosen not to allocate them through markets (e.g., clean water flowing from a forested watershed and outdoor recreation).

It is a simple truism that such goods are systematically undervalued in forest consumption and production decisions. Governments will act – through either regulations or other means – to ensure that private decisions adequately reflect these increasing environmental values. In the context of forest management, this means allocating more land to the production of environmental services, and regulating industrial forest management activities to reduce their collateral environmental damage.

6. *Inadequate, poorly focused research and development*

An abundance of high-quality, old-growth timber combined with the typical Canadian forest tenure arrangements provide little economic incentive for investment in research and development. Forestry-related research and development, sponsored as part of general public-sector research activities, tends not to be tightly linked to the solution of specific forest-sector problems. Adoption of research results in management practice is slow. As a consequence, Canada generally lags our major competitors in such key forest management technologies as tree improvement and our capacity to match forest management regimes with industrial strategy (Binkley 1993, 1995).

These factors collectively imply massive economic dislocations in British Columbia. Recent comprehensive reviews of timber supply across the province suggest that the current policy regime will result in about a 23.5 percent long-term reduction in total harvest levels provincewide, with a significantly greater impact on the coast than in the interior (Binkley 1997). All else equal, harvest-level reductions of this magnitude would mean a loss of over 90,000 jobs, nearly a $5 billion reduction in GDP in the province, and an increase in the provincial budget deficit of about $2 billion (Binkley et al. 1994). The current round of timber supply assessments does not incorporate full implementation of planned land withdrawal for parks, or new environmental regulations on timber harvests. As a consequence, fully responding to apparent societal demands for environmental protection will produce even greater dislocations.

The failure of "integrated resource management" and the danger of proposed low-intensity replacements

Historically B.C.'s forests have been managed extensively under the implicit assumption that virtually the whole forested land base would, one day, be available for timber production. The B.C. Forest Service and licensees incorporate non-timber values in the timber production plans through a process of "integrated resource management" which attempts to consider wildlife, riparian habitat, recreation, water flows, grazing, and other forest uses in each decision about each hectare where logging is to occur. Investments in silviculture have been low. Although licensees are now required to regenerate all areas logged to a "free to grow" stage and massive reforestation efforts under the various federal/provincial agreements have virtually eliminated the backlog of "not satisfactorily restocked" (NSR) lands, B.C.'s use of silvicultural technology lags that in virtually every country with which it competes.

This approach to land management has clearly failed. Current controversies about forest management indicate that it does not satisfy those concerned with the non-timber values of the forests. The recent timber supply review demonstrates that this approach is a clear prescription for reduced harvest levels, and responds neither to the commercial needs of the forest sector, nor to the economic needs of communities, nor to the financial needs of the provincial government.

Collectively, the economic and ecological challenges faced in Canada's forests today imply that land has become scarce. Each hectare seems to face multiple demands – from local residents, from the province as a whole, and from the international community. Without changing forest management policies, the only way to offset harvest reductions is to push the extensive margin still further into unlogged lands. Yet those concerned with the increasingly valuable environmental services of forests claim those same lands for parks and preserves. Increased scarcity of land logically implies that other productive factors – knowledge, capital, and labor – should be substituted for land to produce the desired outcomes, whether environmental values or timber production. But our current forest management approach (and some of the putative improvements) proscribes these kinds of adjustments.

Nearly two decades ago, the U.S.D.A. Forest Service "Roadless area-intensive management trade-off study" (Fight et al. 1979, p. 150) documented the problem in the national forests of the Pacific Northwest:

> From a timber production standpoint under nondeclining yield, the opportunity apparently exists to make up, through management intensification, much of the loss for leaving the roadless areas undeveloped on these two [national] forests. This increased harvest, however, is not possible under the multiple-use guidelines for protecting non-timber resources.

Similarly, in a study of the Revelstoke (B.C.) Forest District, Sahajananthan (1994) found that, through moderately increased management intensity, about 40 percent of the land base would produce the same amount of timber as would 100 percent of the land base under the current rules for integrated resource management.

Interestingly, the foresters in New Zealand reached a similar conclusion at least a decade ago. Commenting on the problem with integrated management, career Ministry of Forests civil servant Andy Kirkland noted (Birchfield and Grant 1993, p. 34):

- Strictly commercial decisions were not possible because major environmental and/or social trade-offs had to be made.
- There was no satisfactory common yardstick for measuring commercial and noncommercial returns.
- There was no pragmatic way of recognizing whether or not maximum combined benefit had ever been achieved.
- Without meaningful measures, achieving the best balance becomes a matter of judgment.
- Because it is impossible to completely fulfill any one objective, one (or more) special interest group will always be dissatisfied.
- Unsatisfied clients will constantly criticize management so that the job becomes a thankless task.
- Because the highest attainable goal in client servicing is a "state of moderate dissatisfaction", the agency is prone to adopt a fortress mentality as it attracts criticism from one side or the other, each seeking compromise in its own direction.
- In response to attacks on their professional integrity, managers accused of usurping popular decisions encourage public participation in one form or another.
- This provides opportunities for motivated, politically aware, or articulate groups to highjack a project for their special interest.
- Consequently, decisions become political and end up back with the politicians.
- Politicians should not have to, nor do they wish to, continually adjudicate between competing claims, so they resort to alternative conciliation mechanisms.
- Whatever the mechanism, it must recognize that multiple-use management is not so much a technical or administrative process as a political one, and, therefore, will inevitably involve political intervention.
- Political intervention is necessary in allocating public goods and services. It is unnecessary and inefficient in providing and allocating commercial goods and services.

Kirkland concludes his lament by noting that this approach satisfied no one in New Zealand, neither the conservationists nor those interested in efficient timber production.

The evident failure of the "integrated resource management everywhere" approach to forest management has created a small industry manufacturing new alternatives carrying such titles as "new forestry" (Franklin 1989) or "natural forest landscape management" (Booth et al. 1993). Although the details of these alternatives may differ, they share the idea of "going lighter on the land." These approaches reduce commercial timber yields per hectare, increase logging costs, and require larger road systems (with all of the associated negative environmental effects). And there is no guarantee that they will actually sustain the natural biodiversity of unmanaged forests. After examining these new approaches to forest management, Alverson et al. (1994, p. 159) comment:

> These approaches imply that we know, or will know in the near future, how to actively manipulate forest stands for timber production while sustaining all valid aspects of biodiversity on the same acreage. Because claims of sustainable harvesting in other temperate and tropical ecosystems have been shown to be invalid or overly optimistic once they were seriously evaluated scientifically . . . , we urge readers to carefully evaluate similar claims made in this present context.

In the Canadian context, Booth et al. (1993, pp. 144–145) propose to put "more total land area in use . . . . The opportunity costs of land under forest cover in the country are very low, i.e., their values for alternative uses are rather limited . . . ." Were this truly the case, land-use conflicts would not persist. Because competition for forest land implies that the opportunity costs of land are high, management approaches should logically conserve on the use of this scarce factor of production rather than use it profligately.

A better way: Zoning and intensively managed plantations

Government ownership and control of most forest land in Canada means that the price system does not signal the increased scarcity of land, so all of the parties involved quite logically proceed under the false assumption that land is abundant. Resolving the resulting conflicts requires a mechanism to allocate land among uses. The absence of a price system and the presence of important non-market environmental values make land-use zoning a logical approach. To be effective this approach requires (1) designation of specific areas for specific uses, (2) clear management rules appropriate to the societal values to be produced on each zone, and (3) effective institutional arrangements to ensure that those responsible for managing each zone face appropriate incentives.

Designation of land-use zones

Recent research (Bowes and Krutilla 1989; Swallow et al. 1990; Vincent and Binkley 1993) confirms the old idea that the multiple benefits of forests are best provided by zoning the forest into a series of special-use areas corresponding to the range of forest values society demands – from wilderness to timber production. Such a policy might involve three general kinds of zones: intensive timber production areas, strictly protected areas, and integrated management joint-use areas to form the transition between the other two landscapes.

Making such a strategy work requires first that a core of protected areas is selected in a way that includes reasonable representation of the great variety of B.C. ecosystems in blocks large enough to sustain landscape-scale processes. British Columbia's current Protected Areas Strategy is well suited to this task. Imagining for a moment that the province devotes some 20 million ha to zones where timber production is a significant or dominant use, then a vast area remains to support other forest uses. Urban and agricultural lands comprise a fairly small portion of the province, so perhaps 65 million ha could be devoted, more or less exclusively, to sustaining the environmental values that forests provide. This area would include at least 23 million ha of productive forest, 20 million ha of savannas and other vegetated land, and 20 million ha of "rocks and ice." As a point of comparison, the area of France equals about 55 million ha. In other words, through a policy of zoning and intensive, dominant-use management, British Columbia could devote an area larger than France to sustaining non-timber values of the forest.

British Columbia has moved tentatively toward land-use zoning through a process established by the Commission on Resources and the Environment (CORE). In the three areas studied (Vancouver Island, the Cariboo-Chilcotin, and the Kooteneys), CORE defined zones ranging from protected areas to intensive timber production areas. Some of these designations were obvious and easily accepted by all parties, whereas others were contentious and required tough negotiations. The details of management in each zone remain to be developed, but, if the management rules actually do permit intensive management on some of the land, then it may be possible to offset much of the planned reduction in harvest levels.

To do so would require sustaining current harvest levels of about 75 million m^3/yr from a land base of some 10–20 million ha. Is this technically feasible? A definitive response requires detailed, site-specific information, but consider one benchmark. In the late 1970s the Weyerhaeuser Company studied the biochemical efficiency of trees in turning sunlight into wood, and modeled maximum biologically possible timber yields (Farnum et al. 1983). They applied this model to two sites in the United States – one in the

Pacific Northwest with Douglas-fir (*Pseudotsuga menziesii* [Mirb.] Franco), and a second in the Southeast with loblolly pine (*Pinus taeda* L.) – where they practice as intensive forestry as can be found anywhere in the world. For example, in the Southeast, the study plantations were site-prepared, bedded, fertilized, planted with genetically improved/mycorrhizal-inoculated seedlings, optimally spaced after planting, and repeatedly thinned and fertilized. Yet the production of these stands achieved only 40–50 percent of the theoretical yields. Natural stands in the same locations grow only about 10–25 percent of the theoretical yields.

Forests in British Columbia are managed much less intensively than are these study sites. As a consequence, there appears to be considerable latitude for increasing timber production/ha in the province. With intensive silviculture, yields of from two to five times the levels attainable in natural stands generally appear to be economically feasible.

There is some local empirical evidence that these kinds of increased yields are indeed feasible in British Columbia. Tolnai's (1991) analysis of Weyerhaeuser's TFL 35 near Kamloops indicates that, with more intensive management, harvest levels on this license could be sustainably increased by 70.1 percent. In some stands the increase is even more dramatic, rising from 2.3 m^3/ha/yr under the current management regime to 8.3 m^3/ha/yr in a more intensive management regime. And, remarkably, these management regimes involve only a better understanding of actual site productivity and prompt restocking with desirable species (in this case, lodgepole pine, *Pinus contorta* Dougl. ex Loud.) at an appropriate density. Increased yields from genetic improvement of planting stock and site emendations could be added to these gains.

Clear management rules for each zone

Management rules must be carefully crafted to support the distinct management objectives of each zone. Just as industrial intrusion on protected areas should be strictly limited, regulatory intrusion on intensive management areas should be assiduously avoided.

The joint-use areas must be managed in a way that responds to environmental needs. This will occur nominally through the B.C. Forest Practice Code, but there may be serious environmental problems associated with the approaches it specifies. The Code will reduce the average size of clearcuts, and will require that logged areas "green up" (i.e., reach a specified age or cover condition) before adjacent areas may be logged. Although this sounds innocuous – if not beneficial – the net effect of these two provisions is to scatter the harvest across the landscape. This pattern of harvests will fragment the forest with predictable consequences for biological diversity

(Harris 1984). Empirical evidence that the landscape corridors prescribed to offset the effects of fragmentation will function as intended is slim. The area of forest edge habitat will increase, and the area of forest interior habitat will decline. These landscape changes will tend to favor early successional species, whereas much of contemporary public concern focuses on late successional ones (e.g., the spotted owl).

The Forest Practice Code will also require more rapid development of the road system, and maintenance of a greater amount of it each year. The greater length of an active road network will provide more opportunity for illegal hunting. By building the road system faster, British Columbia forecloses the future option of deciding that an area should remain roadless. Given that the very worst environmental problems associated with timber production in British Columbia come from the failure of roads, it is ironic that new legislation purporting to protect environmental values actually demands more road building.

Finally, the intensive management areas must be managed very intensively for timber production to make up for the harvests lost through the creation of protected areas and low-intensity management areas. The Forest Practice Code logically should contain a very different set of standards for these intensive management areas than for the other zones. On the intensive management areas, control of the collateral environmental damage frequently associated with logging should occur primarily through land-use designations and not through management regulations. In the intensive management areas such policy-driven constraints as minimum rotation ages, green-up/adjacency, and visual quality objectives should be relaxed if major increases in productivity are to be expected.

Effective institutions

Once land-use zones have been defined and zone-specific Forest Practice Code rules have been written, it will be possible to craft a set of institutional arrangements and land tenures that more productively serve society. These arrangements logically involve significant public control in those zones where public values dominate, and significant private control in those zones where private values are most important. In the case of parks and protected areas, institutional reform will require a much strengthened parks agency to handle the capital investment and management activity needed to ensure that the protected areas and very low-intensity zones do, in fact, provide the environmental values anticipated from them. In the case of integrated resource management zones, institutional reform will require either strengthened direct public management activities, or a much more sophisticated set of license documents than are currently used. Economic instruments which bring licensee and public interests into line merit careful attention (e.g., pricing systems for recreation, water flows, or site degradation).

In the case of intensive timber management areas, ample evidence from British Columbia and elsewhere in the world suggests that strengthened property rights lead to higher levels of private investment in forest management. For example, Zhang (1994) examines expenditures on silvicultural activities on different kinds of B.C. forest tenures while holding constant a variety of factors that might affect such investments – e.g., biogeoclimatic zone, site quality, and location. After controlling for all of these factors, silvicultural investments are strongly correlated with the strength of property rights. Taking the weakest form of tenure – forest licenses – as the base, expenditures/ha on tree farm licenses are 27.4 percent greater, and expenditures/ha on private land are 81.0 percent greater.

In British Columbia stronger property rights could be achieved either through outright privatization of those lands in the intensive management zones created by CORE, or through the sale of long-term leases along the lines of New Zealand's recent policy (Birchfield and Grant 1993). Existing tenure holders might retain their current lands, or might be satisfied simply to buy logs from others. New kinds of organizations would probably enter the field. For example, pension funds now own timberland in the United States, New Zealand, and Chile worth more than $4 billion (Binkley et al. 1996). The largest of the timberland management organizations serving pension funds – the Hancock Timber Resource Group – now owns and manages about 30,000 ha in British Columbia. Stronger private property rights would not only increase the pool of capital available to invest in forests themselves, but would also provide greater flexibility in responding to the rapidly changing circumstances now found in British Columbia.

Experience from elsewhere in the world – especially New Zealand, Sweden, and the United States – will provide useful guidance on the advantages of different kinds of private ownership and of different mixes among the kinds of private ownership (small and large; institutional, industrial, and individual), but the unique circumstances in British Columbia will no doubt require unique approaches to strengthening private property rights in forest land. Forest products companies may be willing voluntarily to relinquish existing licenses in return for stronger property rights on a fraction of their current operating areas.

Conclusions

Current approaches to forest management in British Columbia will produce massive economic dislocations, and yet are unlikely to satisfy legitimate societal demands for protection of environmental values. The leading alternative approaches are based on the assumption that forest land is abundant. Signals from markets (where they exist) and the political process clearly indicate that this assumption is false. Because forest land has many competing uses, rational forest management approaches would substitute knowledge, capital, and labor for land. A forest management system involving land-use zoning and intensive management for each important forest value is one way to implement this approach.

The vanguard of the environmental movement understands the wisdom of zoning the landscape for separate, intensive uses (Alverson et al. 1994). For example, Jeff McNeely, Chief Conservation Officer of the International Union for the Conservation of Nature (IUCN), comments (McNeely 1993):

> It appears the best way to maintain biodiversity in the late twentieth century is through a combination of strictly protected areas ..., multiple-use areas managed by local people, natural forests managed primarily for logging (but with other benefits accommodated to the extent possible), and forest plantations intensively managed for the wood products needed by society.

The "Forestry Principles" adopted by the United Nations Conference on Economy and Development (UNCED) in Brazil reiterated the importance of intensively managed forest plantations. Principle 6d says in part:

> The role of planted forests ... as sustainable and environmentally sound sources of renewable energy and industrial raw materials should be recognized, enhanced and promoted. Their contribution to the maintenance of ecological processes, to offsetting pressure on primary/old-growth forests, and to providing regional employment and development ... should be recognized and enhanced.

The use of plantations as part of an overall strategy for resolving conflicts between economic and environmental objectives requires more than land-use zoning alone. Along with the explicit delineation of intensive management zones must come a set of forest practice regulations which permit or even encourage intensive management. Institutional reform is required to support management on each of the landscapes. This includes stronger governmental agencies to manage parks and preserves, innovative public/private approaches to the mixed-use zones, and stronger property rights in those areas where investments of capital and knowledge will be used to offset harvest reductions elsewhere. Applied research is needed to understand how to sustain valuable ecosystem conditions efficiently on those areas designated for those purposes, and how sustainably to increase the production of commercial timber products on intensive management zones.

Debates about forest ecosystem management frequently and logically reflect on the ethical responsibilities associated with forest management. Perhaps the greatest challenge facing modern society is not creating wealth through technological advancement, but rather using the wonderful creativity of science and technology to ensure that parts of the natural world escape industrial exploitation. This is what the forester Aldo Leopold (1970, p. 240) called a "land ethic":

> A land ethic of course cannot prevent the alteration, management and use of [forest] 'resources', but it does affirm their right to continued existence, and, at least in spots, their continued existence in a natural state.

As global population and wealth expand, the area of the earth untrammeled by humans steadily decreases. In the temperate zones, only Canada and Russia still harbor great tracts of true forest wilderness. The Russian Nobel-laureate in literature, Vladimir Soloukhin, in *A time to gather stones*, laments that what he calls a "bacchanalia of accessibility" has left the Russian landscape desolate and ugly, with ruinous effects on the relationship between humans and nature.

The forest products industry can continue to push into this untapped reserve, thus fighting both the tough economics of the extensive economic margin where the last tree cut always carries no value, as well as the tough politics of environmental concern and the bureaucratic regulations that are inevitably associated with it. The alternative is to deploy our forestry skills to minimize the land area required to produce the forest products needed by society. By doing so we will help to preserve the wild places all foresters love, and to sustain the long-term economic prosperity of those communities and regions whose economies depend on the forest products industry.

References

Alverson, W.S., Kuhlman, W. and Wallen, D.M. 1994. Wild Forests: Conservation Biology and Public Policy. Island Press, Washington, DC.

Binkley, C.S. 1993. Creating a knowledge-based forest sector. For. Chron. 69: 294–299.

Binkley, C.S. 1994. Adapting to global supply constraints or: Timber famine and six reasons it won't occur, pp. 35–44. In: The globalization of wood: Supply, processes, products and markets, Proceedings No. 7319. Forest Products Society, Madison, WI.

Binkley, C.S. 1995. Designing an effective forest sector research strategy for Canada. For. Chron. 71: 589–595.

Binkley, C.S. 1997. A cross road in the forest: The path to a sustainable forest sector in British Columbia. BC Stud. 113: 39–61.

Binkley, C.S. and Zhang, D. 1994. The impact of timber-fee increases on B.C. forest products companies. Unpublished manuscript. Faculty of Forestry, University of British Columbia, Vancouver, British Columbia.

Binkley, C.S., Percy, M., Thompson, W.A. and Vertinsky, I.B. 1994. A general equilibrium analysis of the economic impact of a reduction in harvest levels in British Columbia. For. Chron. 70: 449–454.

Binkley, C.S., Raper, C. and Washburn, C.L. 1996. Institutional investment in U.S. timberland: History, motivation and implications for forest management. J. For. 94(9): 21–28.

Birchfield, R.J. and Grant, I.F. 1993. Out of the woods: The restructuring and sale of New Zealand's state forests. GP Publications, Wellington, New Zealand.

Booth, D.L., Boulter, D.W.K., Neave, D.J., Rotherham, A.A. and Welch, D.A. 1993. Natural forest landscape management: A strategy for Canada. For. Chron. 69: 141–145.

Bowes, M.D. and Krutilla, J.V. 1989. Multiple Use Management: The Economics of Public Forestlands. Johns Hopkins University Press, Baltimore, MD.

Farnum, P., Timmis, R. and Kulp, J.L. 1983. The biotechnology of forest yield. Sci. 219: 694–702.

Fight, R.D., Johnson, K.N., Connaughton, K.P. and Sassaman, R.W. 1979. Can intensive management make up the harvest lost when roadless areas are left undeveloped. J. For. 77: 148–151.

Franklin, J.F. 1989. Towards a new forestry. Am. For. 95: 37–44.

Harris, L.D. 1984. The Fragmented Forest. University of Chicago Press, Chicago.

Leopold, A. 1970. A Sand County Almanac. Ballantine Books, New York.

Lyon, K.S. 1981. Mining of the forest and the time path of the price of timber. J. Envirn. Econ. Manage. 8: 330–344.

McNeely, J.A. 1993. Lessons from the past: Forests and biodiversity. Unpublished manuscript. IUCN, Gland, Switzerland.

Nixon, R. 1991. Comparative data charts explain forest management policies. For. Plan. Can. 7: 32–45.

Sahajananthan, S. 1994. Single and multiple use of forest lands in British Columbia: The case of the Revelstoke Forest District. Report submitted to B.C. Ministry of Forests, Revelstoke Forest District.

Swallow, S.K., Parks, P.J. and Wear, D.N. 1990. Policy relevant nonconvexities in the production of multiple forest benefits. J. Envirn. Econ. Manage. 19: 264–280.

Tolnai, S. 1991. Addition value to our heritage through silviculture. Paper presented to Western Silvicultural Contractors Association. Vancouver, British Columbia.

Vincent, J.R. and Binkley, C.S. 1993. Efficient multiple-use forestry may require land-use specialization. Land Econ. 69: 370–376.

Zhang, D. 1994. Implications of tenure for forest land value and management in British Columbia. Ph.D. dissertation. Faculty of Forestry, University of British Columbia, Vancouver, British Columbia.

New Forests **18:** 89–96, 1999.

Ideology and culture in the Oregon country: the landscapes of a planting society

WILLIAM G. ROBBINS
College of Liberal Arts, Oregon State University, Corvallis, OR, USA

Key words: Pacific Northwest, social consequences, biological consequences, settlement, human-induced change

Abstract. Ideologies, and especially the choices that are rooted in ideology, have profoundly influenced the shaping of the Oregon country as it exists in the late twentieth century. "All human groups," the historian William Cronon argues, "consciously change their environments to some extent." During the last two centuries, the successive cultural groups occupying the Pacific Northwest inhabited landscapes that have been shaped, to some degree, by previous dwellers. Each of these cohorts, operating from a deeply rooted cluster of beliefs and values, has shaped the land around them. It is impossible to understand the vast biological and other modifications that have taken place in the Oregon country without looking far beyond the region. Of particular interest is the introduction of a sometimes spectacular system change – the making of the landscape of a planting society – as human cultural practices moved from subsistence to commercial behavior, and as the Northwest became integrated into national and international market relations.

The ideologies associated with the Western industrial world, especially the choices that are rooted in its cluster of values, have profoundly influenced the shaping of the landscape of the Pacific Northwest as it exists in the late twentieth century. This region, as well as much of the temperate zone in both the northern and southern hemispheres, also replicates what the historian Alfred Crosby has called "Neo-European characteristics," areas around the globe where European plant, animal, and human species have successfully colonized. The secret to the success of those invading biological species lay in their mutual and symbiotic adaptation to similar climatic regimes around the globe. The Europeans, imbued with a particular fix on the material world, brought with them an alien and exotic collection of biological species which they introduced, sometimes forcibly, into cultural and natural worlds occupied by indigenous peoples. But what was planted in the Neo-Europes included, not only plant and animal species, but also ideas, especially those associated with the industrial capitalist ethos which, of course, originated in

Europe. Both in a political and in a biological sense, and at least for the nineteenth century, I call these "planting" societies.

The temperate forest region of the North Pacific Coast, extending from northern California into southern British Columbia, provides a superb setting for studying the social, cultural, and biological consequences of planting societies. In its "natural" condition, this region has the coherent and defining features common to a bioregion or a "greater" ecosystem. To the degree that the Douglas-fir country comprises a "natural geographic system," therefore, it provides a wonderful place to examine human relationships with the natural world. As a geographical, biological, and cultural place, the Pacific Northwest is unique because it possesses certain distinct attributes. The European intrusion into this part of North America is very recent – it extends back in time little more than a century and a half. In other words, there is an abundance of documented evidence about the cultural, biological, and topographical character of the region at the time the invading hosts appeared on the scene.

The record also reveals that the historical achievements and accomplishments of planting cultures had a darker side in disease, warfare, the dispossession of native peoples, and the destruction of all sorts of indigenous flora and fauna. Those circumstances existed in the Pacific Northwest as well. With the exception of the explosion of Mount Mazama (about 4,000 years ago), environmental conditions in the region had remained relatively stable for at least 10,000 years, acknowledging, of course, that native peoples profoundly influenced their surroundings through the use of fire. But that millennia of relative demographic, cultural, and ecological stability came to an end sometime in the 1830s, a decade that I call "the great divide," when the larger region passed through a watershed of sorts.

What is striking about that decade is the number of witnesses who had some sense of the great transformation that seemed to be underway. These witnesses related accounts of incredible human tragedy and military campaigns against native peoples, and some of the localized forays can be classed as wars of genocide. When it was over, the survivors were forcibly confined to small reserves of land in their once vast homeland. For the Willamette Valley and the lower Columbia River country, largely untouched by warfare, the principal cause of the drastic decline in the Indian population was a series of seasonally recurring malaria outbreaks between 1830 and 1833. Those European-introduced diseases killed approximately 90 percent of the native peoples of the Willamette and lower Columbia valleys. The anthropologist Robert Boyd estimates a pre-epidemic native population of about 14,000 people for the Willamette Valley (Boyd 1990). When Lieutenant Charles Wilkes visited the area in 1841, he counted approximately 600 survivors (Wilkes 1899).

In retrospect, we can say with some certainty that epidemic diseases had thinned the native population at a critical period in the post-Columbian history of the Willamette country. What followed in the wake of that catastrophic decline was a great demographic shift that ultimately contributed to the disruption of existing human and ecological communities. To explain those connections, Alfred Crosby (1986) uses the seemingly apolitical metaphor of folklore, the delightful old song about Sweet Betsy from Pike County, Missouri, who crosses the western mountains "with her lover, Ike, with two yoke of oxen, a large yellow dog, a tall shanghai rooster, and one spotted hog." The symbolism in Betsy's party, Crosby points out, is the host of colonizing species in her entourage (mixed in gender, of course!). "Betsy came not as an individual immigrant," he argues, "but as part of a grunting, lowing, neighing, crowing, chirping, snarling, buzzing, self-replicating and world-altering avalanche."

For the Pacific Northwest, it is fair to say that Sweet Betsy took along that familiar set of animals and much more in her cultural and ideological baggage. The European newcomers, who gradually extended their control over the Pacific Slope as the nineteenth century advanced, brought with them a cultural blueprint, a prescription for their recently adopted homelands that embraced the wholesale transformation of those landscapes into ecological settings that were both more orderly and populated with familiar flora and fauna. Those were the characteristics that marked Neo-European planting communities in the Northwest and elsewhere. In that sense, the resettlement of this region was carried out by groups considerably more conservative and less venturesome than one might believe. European settlers, as Richard White (1980) puts it, "husbanded the familiar"; that is, where circumstances of climate and soil permitted, they became planters of wheat and other familiar crops of European provenance that survived well in temperate climates. Planting societies to a large degree were imitators rather than innovators.

But that is not to imply that planting cultures did not adjust and, in many cases, adapt extraordinarily well to their new surroundings. It is well known, for instance, that wheat culture flourished in the Willamette Valley in the wake of the California gold rush. So did other species of European origin: hogs, cattle, horses, sheep, and domestic fowl of various kinds. We can say with some degree of certainty, then, that the new cultural groups that became dominant in the Douglas-fir bioregion promptly set in motion the transformation of their immediate surroundings and subsequently the introduction of spectacular, systemic changes everywhere. Beginning in about the mid-nineteenth century, the regional landscape increasingly began to reflect the cultural values and hard work of planting society. That tendency accelerated as human cultural practices moved from subsistence to commercial behavior, and as the Northwest became more closely integrated into national and international market relations. It is important to remember that for the last 200 years, larger global forces, especially those associated with an expanding world capitalist system, have increasingly influenced the direction and evolution of planting societies.

By 1843 on what is now the American side of the border, newly named Oregon City at Willamette Falls had 75 dwellings, as well as sawmills and gristmills – processing technologies that reflected various degrees of ecological disturbance to places far beyond the sites where logs were sawed into boards and wheat was ground into flour. In the valley's hinterland south of Oregon City, settlers were putting to the plow prairie environments made ready through the annual burning by the native Kalapuya people. At the close of the first railroad decade, say, the 1880s, the old Indian-maintained prairie landscape was characterized by cultivated fields, orchards, farmhouses and outbuildings, and carefully patterned and fenced boundaries.

That remodeling of the regional landscape – imposing preconceived notions about how agricultural lands, forests, and rivers should be managed – initiated the slow and then dramatic alteration of the physical contours of the region. European settlers destroyed the Indian Northwest, Richard White (1980) argues, and shaped it into something much different, a natural world that had little function other than to serve human needs, and eventually human profits as well. The cultural values brought to bear on the North Pacific Slope were at one with the kind of thought world that pervaded most Neo-European societies around the globe: progressive, secular, thoroughly materialistic, and steeped in acquisitive values. The European newcomers brought with them a unique cultural vision – a social imagination and core of beliefs, an "economic culture," Donald Worster (1979) calls it – that viewed the natural world as capital, that obliged humankind to use that capital for self-advancement, and a conviction that the social order should promote activities associated with those values.

By the close of the nineteenth century a vast complex of cultural forces – economic, industrial, biological, and scientific – increasingly shaped human behavior toward the forests, streams, and valley bottomlands along the North Pacific Slope. The ever-more powerful and intrusive technologies of the industrializing world were fused with the commercial practices of planting settlements to increase the productions of nature. The coming of steam power, first to the Douglas-fir, spruce, and cedar forests of the region, brought an end to the use of the slow and laborious bull teams for hauling logs to the water's edge. The steam donkey yarding machine and the railroad quickened the pace of activity in the woods – and disturbance to the ecology of the forests.

George Emerson, a Grays Harbor lumberman and the first to use the steam donkey north of the Columbia River, was enthusiastic about its performance (Ficken 1987):

> When one considers [that] they . . . require no stable and no feed, that all expense stops when the whistle blows, no oxen killed and no teams to winter, no ground too wet, no hill too steep, it is easy to see they are a revolution in logging.

With "no ground too wet, no hill too steep," the donkey engines began what Richard White (1980) calls "The Creation of a New Forest." From the perspective of a century's hindsight, we can say that the steam donkey offered mixed blessings. It greatly increased log production, and it dramatically expanded cultural disturbances in forest ecosystems.

As the most visible symbol of industrialism in the western United States, the railroad had an even greater and more dramatic influence in altering the forested landscapes of the Northwest. Railroads provided access to the region's natural abundance, and linked planting settlements with industrial centers; the revolutionary new technology also increased the dependency of those resource communities on distant sources of capital and markets. Railroads liberated the transportation of logs and other freight from the restrictions of natural geography, thus enabling operators to build lines into hitherto inaccessible terrain. In that sense, logging railroads replaced waterways as transportation arterials, brought sweeping changes to the landscape, vastly stepped up production in mill communities, and extended the scope and degree of industrial intrusion into forest ecosystems. "The railroad," historian William Cronon (1991) remarks, "left almost nothing unchanged: that was its magic."

The entrance of the machine into the forest also fostered a rationalized system of production, one in which the economics of production eventually led to a new kind of science centered on the planted forest and the further systematizing of nature. As the twentieth century advanced, loggers and foresters developed new harvesting practices to accommodate the ever-more efficient and productive equipment they used in the woods. Clearcutting the entire area within reach of the yarding machine, a method once considered avant-garde, became the order of the day. Forestry schools at the University of British Columbia, the University of Washington, and Oregon Agricultural College (now Oregon State University) developed "scientific" arguments – critics would call it pseudoscience or an incomplete science – to show how clearcutting and then replanting cut-over lands replicated natural phenomena such as forest fires. In effect, as cutting practices were adapted to technology, the machine became the chief determinant in reshaping the ecology of the forest.

The new, highly productive technology unleashed in the early twentieth century also contained the seeds of social and ecological destruction: when the timber was gone – as happened in Washington's Grays Harbor County by the mid-1930s, the consequences were treeless landscapes and equally devastated communities. In a rationalized effort to counter the effects of that production-driven social and ecological wreckage, industry progressives and a few USDA Forest Service officials began pushing a variety of reforestation and sustained-yield proposals in the 1920s and 1930s. In effect, these proposals were newly constructed cognitive maps fashioned to fit dramatically altered cultural relations with the natural world. In terms of the larger issues before this symposium, these initiatives established what supporters believed was a solid ideological argument in support of the sustainable economic health of nearby communities. Planted forests, in that sense, reflected a scientific and economically based strategy for achieving and perpetuating sustainable societies.

The theoretical components of sustained-yield forestry elaborated in the years following World War II were consistent with Neo-European, rationalized, production-driven schemes to reorder nature. For the public forest lands of the Pacific Northwest, it was argued that the application of machines, labor, chemicals, and capital would contribute to endless economic growth and maximized production. The centerpiece to that new-thought world was intensive management, an assembly of scientifically reductionist, and, by the 1960s, computer-driven models that provided a blueprint for the regimenting of nature. In his recent book, *A conspiracy of optimism*, Paul Hirt (1994) argues that "intensive management was essentially . . . a flexible collection of 'techno-fix' theories and practices that allowed foresters to continually say yes to ever-increasing demands for commodities like timber."

The point to be emphasized is that the "science" of forestry and "theories" about sustained-yield were driven for the most part by economic and political concerns, and not by dispassionate inquiries about the health of forest ecosystems or the stability of dependent communities. Naive and innocent believers, who, in my view, were too mesmerized by production-driven objectives and Cold War rhetoric, surely existed among that group. The real purpose and function of planted forests in the post-World War II years, however, had to do with considerations of an overheated market for lumber and production and profits. Theoretical calculations about sustained yield coming from the industry side, according to Paul Hirt (1994), "simply provided a useful ethical facade over short-term economic objectives." I would add that those formulations also ignored a host of unanticipated ecological and economic issues that now appear as headlines in our daily newspapers.

That brings us to the present state of affairs: dwindling supplies of public and private timber, declining anadromous fish runs, what many people believe are badly disrupted ecosystems, and deeply troubled communities that once relied on that natural wealth for their livelihood. The resource-dependent settlements of an industrialized planting culture are, in fact, confronting the scarcity of the very thing that brought them into existence in the first place. What has been happening, in part, is a steady process of deindustrialization as the world of global capitalism continually alters the marketing and financial parameters of human relationships with forests, streams, and urban settings.

Human-induced environmental changes, largely driven by the market and production-driven demands of industrial cultures, have brought about the spectacular transformation of the landscape of the North Pacific Slope in the last century. We have some understanding, I believe, about the biological impact of Sweet Betsy from Pike County, Missouri, her lover Ike, and their accompanying flora and fauna. We have been less observant – and I include Alfred Crosby in this group – about the greater environmental influences of industrial capitalism and its variety of spin-offs from the Industrial Revolution. Because capitalism has exercised such a pervasive influence on the global level, those who strive for some form of future ecological and economic stability have much to learn from anti-Marxist Peter Berger (1986), who argues in a recent book that capitalism "has radically changed every material, social, political, and cultural facet of the societies it has touched, and it continues to do so."

That influence extends, of course, to the Douglas-fir bioregion, a geographic setting whose cultural and biological character today reflects much of the precision of Neo-European science, economics, and industrial activity. But the problems associated with planted forests – with the production-driven scientific ethos that has prevailed for the last half century or more – are of a piece with other difficulties: with our rivers, with our declining salmon runs, with our urban airsheds, with our tainted underground water supplies – indeed, with our decaying industrial infrastructures as well. We tend to lose sight of the interrelatedness of these issues and the manner in which they reflect our cultural values. It would be wise for us to understand better the ideas that Sweet Betsy and her companion Ike brought with them.

References

Berger, P. 1986. The Capitalist Revolution: Fifty Propositions about Prosperity, Equality, and Liberty. Basic Books, New York.

Boyd, R. 1990. Demographic history, 1774–1874, p. 137. In: Sturtevant, W.C. (Ed) Handbook of North American Indians (volume 7). Smithsonain Institution Press, Washington, DC.

Cronon, W. 1991. Nature's Metropolis: Chicago and the Great West. W. W. Norton, New York.
Crosby, A. 1986. Ecological Imperialism: The Biological Expansion of Europe, 900–1900. Cambridge University Press, New York.
Ficken, R.E. 1987. The Forested Land: A History of Lumbering in Western Washington. University of Washington Press, Seattle.
Hirt, P. 1994. A Conspiracy of Optimism: Management of the National Forests Since World War II. University of Nebraska Press, Lincoln.
White, R. 1980. Land Use, Environment, and Social Change: The Shaping of Island County, Washington. University of Washington Press, Seattle.
Wilkes, C. 1899. The correspondence and journals of Captain Nathaniel J. Wyeth, p. 148. In: Young, F.G. (Ed) Sources in Oregon History. University Press, Eugene.
Worster, D. 1979. Dust Bowl: The Southern Plains in the 1930s. Oxford University Press, New York.

New Forests **18:** 97–109, 1999.

The integration of planted and natural forests in a regional landscape

KATHLEEN KAVANAGH[1], GEORGE STANKEY[2] and JIM BOYLE[3]
[1]*Forestry Extension Faculty, College of Forestry, Oregon State University, Astoria, OR, USA;*
[2]*Pacific Northwest Research Station, USDA Forest Service, Corvallis, OR, USA;*
[3]*Department of Forest Resources, College of Forestry, Oregon State University, Corvallis, OR, USA*

Key words: Oregon, Douglas-fir, social forces

Abstract. The 10,700 km^2 region of northwestern Oregon, USA, is dominated by mountainous forested landscapes fringed by agricultural lands and rapidly expanding urban areas. The Douglas-fir/western-hemlock trees, admixed with other species, in the mild, moist regional climate with rich soils are among the most highly productive of temperate forests. Timber harvest has been the dominant land use for much of this century. Many current forest stands are planted, and have the potentials to be managed and shaped for a variety of traditional and evolving forestry objectives. The ages, resilience and productivity of these forests and mosaics of land ownerships permit a variety of future scenarios of forested landscapes, constrained largely by capacities of social organizations to plan and execute management for desired future conditions.

Introduction

Northwest Oregon is a landscape of forests that blanket coastal mountains from the farmlands and expanding urban areas on the east to the Pacific Coast on the west. Planted forests are a major component of this forested landscape: the 147,307 ha (364,000 A) Tillamook State Forest stands as a tribute to human planting efforts to rejuvenate a landscape impacted by fires in 1933, 1939, 1945 and 1951; extensive corporate and state forest lands extend to the north, and private and Federal forests to the south; planted pines stabilize dunes that frame the Pacific Coast; hybrid cottonwoods line the banks of the Columbia River and intermingle with farmlands and Christmas tree farms on the fringes of the Willamette Valley. The complex mosaics of planted forests among natural forest stands and human-created land patterns are products of sculpting by successive generations of people interacting with bio-physical forces. The variety, nature and significance of these planted and natural forests and tree farms create a complex web of interdependencies that challenge the rhetoric of sustainability: How to maintain ecological, economic and social "balance" in the face of dynamic demographics, technologies, knowledge and ecological forces in a regional, forested landscape?

The landscape that is the subject of this paper is the Northwest corner in the State of Oregon, encompassing Clatsop, Columbia, Washington, Tillamook and Yamhill counties, an area of 1,069,677 ha (4130 mi^2; 10,696.7 km^2). We chose this area as an example because of the dominance of forests on the landscape along with a socio-cultural and economic gradient. The forests of the Coast Range, which rises from sea level to 1220 m (0–4,000 ft), dominate the land. Even on the mountain tops, there are few rocky promontories in this forested sea of green. In the absence of development, the western forested slopes meet the sea with only a narrow band of sand or steep rocky cliffs separating the waves and trees. To the east, this region is bounded by the Willamette Valley and urban Portland. The north boundary is defined by the Columbia River, along the Oregon – Washington border, and to the south the area of interest is arbitrarily halted at the Lincoln and Polk county borders.

Regional biological and physical forces

Parent materials for the soils in the Oregon Coast range are dominated by sandstones and siltstones, interspersed with numerous areas of basaltic and other volcanic materials. The soils are heavily weathered and although susceptible to landslides, many are deep, nutrient-rich and well-drained making them highly suitable for forest growth. The slopes of the coastal mountains range from gentle rolling to steep and rocky with few flat valleys or benches. The terrain is highly dissected by streams and rivers that are noted for their rapid increases in peak flow rates during winter storms, and are highly valued as habitats for numerous anadramous fish populations.

The Pacific Ocean exerts a strong influence over the climate. The majority of the area is characterized by relatively warm winters, cool summers and high rainfall. Even higher elevations are too warm to maintain a snowpack throughout the winter. Annual rainfall amounts range from 1000 to 3200 mm, with up to 5000 mm on local mountain tops. High winds dominate the physical forces exerted by the climate, particularly along the ocean slopes. This combination of a cool, wet, yet moderate, climate is the dominant reason for the extreme productivity and unique vegetation of the region, some of the most productive temperate forests of the world (Grier et al. 1989; Waring and Franklin 1979).

Coastal rainforests have dominated this landscape for over 2 million years. Maintenance of diversity and adaptation of the resident species over evolutionary time has resulted in an extremely productive ecosystem that is well adapted to the climate and natural disturbances associated with it (Waring and Franklin 1979). Currently, in northern Clatsop county, primary succession is still occurring on sand recently deposited along the mouth of the Columbia River. Initial invasion is by various beach grass species in conjunction with shore pine (*Pinus contorta var.contorta* Dougl. ex. Loud.). As the pines shade out the grass, sword fern (*Polystichum munitum*) and Sitka spruce (*Picea sitchensis* (Bong.) Carr.) emerge completing the transition toward a coastal forest ecosystem.

Years of evolutionary change in concert with the unique climatic features have resulted in tremendous accumulations of biomass (Waring and Franklin 1979). The tree-generated biomass that occurs is dominated by relatively few species of conifers. Four major conifer species; Douglas-fir (*Pseudotsuga menziesii* (Mirb.) Franco), western hemlock (*Tsuga heterophylla* (Raf.) Sarg.), western red cedar (*Thuja plicata* Donn ex D. Don), and Sitka spruce account for approximately 90% of the conifer biomass. The remainder is composed of grand fir, (*Abies grandis* (Dougl. exD. Don) Lindl.), noble fir (*Abies procera* Rehd.), Pacific silver fir (*Abies amabilis* Dougl. ex Forbes) shore pine, and Pacific yew, (*Taxus brevifolia* Nutt.). Deciduous species, about 15% of the total tree biomass, are dominated by red alder (*Alnus rubra* Bong.) and big leaf maple (*Acer macrocarpa* Pursh). The overwhelming dominance of conifer species is unusual compared to coastal forest regions of the world at similar latitudes that have either mixed or largely deciduous forest types.

The very favorable growing conditions for plants result in a high degree of inter and intra species competition for growing space and light. Therefore, the selective pressures have been for large, long-lived individuals that are able to outcompete and outlive invading vegetation. Of the conifer species that occur here, the oldest and largest representatives have historically been located in this region. Publications and vegetative surveys from as early as 1908 state that many of the largest individuals had been felled, so it is hard to predict what the largest potential size for many of these species really is (Sudworth 1908). Estimates of maximum ages for western red cedar exceed 1400 years. (Pojar and MacKinnon 1996). Sitka spruce is thought to have a maximum diameter of 525 cm and 95 meters in height. (Pojar and MacKinnon 1996) Currently the *average* projected tree height in this region at 100 years of age is 55 meters (Pojar and MacKinnon 1996). Aside from the large diameter and extreme height the density of stems per hectare or basal area is very high, often exceeding 77 sq. meters per hectare (300 sq. ft per acre) (Burns and Honkala 1990).

Fire and wind have been the most prevalent landscape-scale disturbance mechanisms influencing forest succession in this four county region, while landslides and flooding regularly disturb small to medium areas of tens to hundreds of square hectares, with occasional "catastrophic" events affecting tens of square kilometers. The slopes facing the ocean are periodically battered by high winds often exceeding 130 km/hr (80 mph). Occasionally, hurricane force winds in excess of 160 km/hr (100 mph) sweep across the headlands into the interior, causing tree blowdown, particularly along ridges, shoulders of hills and unprotected areas. The interior Coast Range and valley slopes have been subject to infrequent but catastrophic fires ignited by lightning and more recently, human-caused. In contrast, riparian areas are subject to frequent disturbance from debris flows, slope failure and high stream flow generated by the perennially high winter rainfall and unstable soils.

Pathogenic organisms did not leave their mark through the historical record so it is difficult to predict their impact in evolutionary times. In more recent history, there have been a number of pathogens identified with these forests. Gap level disturbances are associated with several root rot pathogens that are common in the region. Outbreaks of hemlock looper (*Lambdina fiscella lugubruso* (Hurst.), a defoliating insect on hemlock, have been noted several times in the past 100 years but long-term historical records are not as clear. Douglas-fir bark beetle (*Dendroctonus pseudotsugae*) and Sitka spruce tip-weevil (*Pissodes strobi* (Pk.)) have historically been present as well. Within the past ten years, Swiss needle cast (*Phaeocryptopus gaeumannii*), a needle pathogen on Douglas-fir has been severely impacting Douglas-fir on the western edge of the region. Overall the area is relatively free of major mortality-causing insect and pathogenic diseases that generally play such an important role in the elimination of early seral species in many forested ecosystems.

The origins of the forests found here at the time of first human settlements, in about 6000 BC, were products of dynamics of vegetation advances and retreat to and from refugia during Pleistocene ice ages and subsequent climate fluctuations. Followed by cycles of more minor disturbances and secondary succession. The infrequent fire, interspersed with frequent windstorm, landslides, floods and pathogenic disturbances created a mosaic of species, age classes and patch sizes across the landscape.

Human influences on forest structure-the recent past

The forests we currently view in this landscape are a result of human influence interacting with ecosystem response to disturbance. Well before the arrival of Europeans, the Native Americans of the region shaped and modified the landscape in ways to sustain and support their way of life. Fire was their principal management tool and it led to changes in the nature of the species, their predominance, and their distribution. Indeed, many observers conclude that our lack of understanding of the profound influence of Native Americans on the regional landscape continues to plague our ability to understand natural historic conditions.

With the arrival of European settlers in the 1800s, the pace of change in the bio-physical character of the region quickened. Ironically, the forests of the region were the source of both anxiety as well as wealth. For many, the vast, sprawling forests were a source of threat, both in terms of personal safety as well as a spiritual reminder of the fragile hold civilization held on the wilderness. As settlers explored accessible valleys, surrounding forests were cleared and burned to make room for crops and pastures. Tree felling, burning, stump removal, plowing and sowing increased. Fire, set to clear lowlands, often spread into undisturbed forested uplands resulting in catastrophic fires which reburned some areas more often than historic frequencies.

This "land taming" was a legacy of centuries of attitudes, buttressed by religious edicts, that pressed for the subjugation of the wilderness and its conversion to civilization. At the same time, these forests were the source of wealth and the foundation of the region's development. Taken together, and further abetted by government policies and programs designed to facilitate the economic development of the region, the vast forests were rapidly exploited by early settlers and increasingly large proportions of the region changed from public domain to private ownership.

The majority of early settlement and logging occurred close to the rivers which supplied log transport and power to run the mills. By 1850, there were active lumber producing operations located throughout the lower Columbia River, and lumber loaded aboard schooners was bound for California. As markets and technology for wood removal expanded, timber operations spread throughout the region. In 1911, 100 years after the initial settlement in Astoria, the state legislature enacted a law creating a state forestry department and appointed a state forester. The main interest at this time was fire protection as harvest practices and continual land clearing were threatening what was now deemed a valuable forest resource.

Even with this new-found interest in fire protection, human-caused, large forest stand-replacement fires burned through a large portion of this region in 1933, 1939, 1945 and 1951, leaving a large non-forested area referred to as the Tillamook Burn. The origin of the fires was ignition by logging equipment, and fires were fueled by large amounts of logging debris and dry winds. Adjacent standing forests also burned, with a cumulative total of over 140,000 ha of forest affected during those years. Reforestation on these sites was severely limited by lack of seed trees and by use of the newly established shrublands and grasslands for cattle grazing. These sites remained largely non-forested until the mid-1950's when a massive reforestation effort was undertaken using aerial seeding and hand-planting techniques. Successful reforestation efforts on the Tillamook Burn – now the Tillamook State Forest – soon led to replanting of vast areas, previously logged and/or burned as well as abandoned farmland and pasture.

The cumulative effects of land clearing, timber harvest techniques and fire combined with ecosystem responses and forest planting and management to create the forests that currently blanket this region. For example, early forest harvesting objectives were to remove only merchantable tree species in the vicinity of major waterways. Logging focused on large diameter, knot-free logs leaving the remainder as slash on the site. In the absence of fire, the forests that resulted from these slash-filled, selective "cut and run" harvest techniques favored the more shade-tolerant western hemlock, cedar, and Sitka spruce. Douglas-fir was not well regenerated even following slash fires, probably because often the fires were very intense, large and at frequent intervals, eliminating adequate seed sources.

The region's economic infrastructure reflected the changes in land uses. Capital investments in transportation and wood processing made the region a major focus of economic activity. The creation of employment led to the immigration of a work force not only from across the United States, but from overseas as well. The social-cultural, as well as economic structure and character of Northwestern Oregon was a direct product of the region's natural resource base; at the same, these human qualities began to lead to changes that would inevitably be reflected in the region's natural resources and the policies and programs governing their management.

In the 20th century, the socio-cultural and economic structure of northwest Oregon began to change. To the east, the city of Portland grew rapidly, and its economic base was highly diversified. It served as a focal point for the transportation of goods from the hinterland to the south through the Willamette Valley and by ship to California and beyond, in addition to being an important trans-shipment point for goods and products from eastern Oregon. With economic growth came an expanded demand for a host of other activities in the manufacturing, service, and industrial sectors. As Portland's economic presence grew, so too did the demand for an expanded workforce. In the middle decades of this century, Washington County, on the eastern edge of our four-county region, increasingly became a "bedroom community" to the Portland area with simultaneous losses of agricultural and forest lands. For example, the Washington County population grew 12-fold between 1930 and 1994 (and in fact, doubled between 1970 and 1994); during this same period, the population of the remaining three counties has approximately doubled. In addition, while Clatsop, Columbia, and Tillamook counties today remain close to the rural-urban distribution reported in the 1950 Census (53, 21, and 20 percent urban, respectively), Washington County's urban population shifted from 31% in 1950 to 90% in 1990.

The dramatic change in the amount and distribution of population in the four-county region was matched by an equally impressive shift in the proportion of persons employed in various sectors. In the 20-year period from 1972–1992, the number of persons employed in the lumber and wood products industries fell, on average, more than half in all four counties; in Tillamook, it dropped from 38% of the total workforce to 9%. It is also worth noting that Federal forest ownership in the four-county region is low; the shift in employment reflects the wood product industry restructuring and the adoption of new, more efficient technologies more than it does reductions in supply of timber from federal lands in this region (this is not necessarily the case for other areas of Oregon).

In the western two-thirds of this landscape, the early land clearing is no longer readily apparent, particularly where climate restricted modern agriculture. Once abandoned, the agricultural lands were rapidly colonized by forest ecosystems, and today the only indication of clearing is an occasional apple tree, daffodil flower and a noticeable lack of tree stumps. In the eastern part of the landscape, bordering the Willamette Valley, agricultural crops were much more productive and the bottom lands remain largely agricultural except where replaced by urban development. In general, the percentage of land that has remained forested is much higher in the western counties where the value of the land for timber production is higher than for other uses. In the eastern counties, on a fringe of a major city, there are many more options for financial return from the land such as housing, agriculture and commercial development.

The major impact of recent European settlement has been to restart the successional clock on more frequent intervals. Currently, there is only a small remnant of forests that existed prior to the 1800's and a majority of the forested area has been through two or three harvest/fire cycles in the past 170 years. In the coastal regions, the forests that seeded in following the initial logging and fires are primarily later successional species such as western hemlock, cedar and Sitka spruce. The current stocking in these stands is generally very high, and the stands are just entering a phase where competition and wind-caused mortality reduce stem densities thereby prompting timber harvest

Today, in accordance with biophysical and socio-cultural forces, Northwest Oregon has a variety of forest types: 1) western hemlock, Sitka spruce, red alder and Douglas-fir forests that seeded in naturally following selective timber harvests in the early-mid 1900s; 2) dense Douglas-fir forests that were planted and seeded following stand replacement fires in the early-mid 1900s; 3) young stands of Douglas-fir recently planted following harvest and now sharing their growing space with naturally regenerating spruce, hemlock and red alder; 4) forests which either invaded or were deliberately planted on land originally cleared for agriculture.

The pattern of and relative abundance of the mosaic of planted and naturally regenerated forests are dispersed throughout the landscape, and their relative abundance is directly tied to the land use practices following European settlement. Human interaction is currently responsible for the relatively young age class of both the planted and natural forests throughout this region.

Human influences on forest structure-the future

In short, what this brief overview reveals is a region in transition. It is a transition not unlike that found in other regions, not only in the Pacific Northwest, but indeed, around the world. It is a transition driven by complex, interwoven forces. Changes in public attitudes, shifts in technology, changing markets, and so forth have led to profoundly different ways in which resources are valued, perceived, and managed. The resulting changes are painful, especially for those whose livelihood is tied to a way of life and a means of sustaining themselves that finds itself in decline. Yet our capacity to respond to change reflects our only ability to meet the future

In the future, the types of forests that occur across this landscape will be shaped by the decisions and policies in accordance with fluxes in biophysical forces. The interaction of these factors has shaped the forests ecosystems we are currently managing and their imprint will remain across the landscape coupled with new and evolving forces. The rate at which change is occurring will probably have the largest impact. Using a geological reference, fluxes in climate and rates and intensity of disturbances influencing forest composition were likely very different from current rate of change, and ecosystem responses will result in additional widespread changes in the shifting mosaic of forests across the landscape.

As we move through time, the percentage of forests regenerated from seed following harvest or other disturbance will probably decline. Increased value of timber products, as well as land costs, and concerns about soil erosion will encourage landowners to invest in the rapid reforestation of their harvested lands through planting. The cycle of harvest and abandonment that produced so many of the forests across our current landscape are a product of the past "pioneering" mentality. Lessons learned from transferring seedlings out of the region to which they are locally adapted will result in more careful guidelines for matching seedling origin to site characteristics.

Afforestation will continue to take place, particularly as agriculture aggregates in the more productive locations. Agricultural production across much of this landscape is in decline because increasing land values, urbanization pressures, and cost to market are resulting in shifting economic opportunities. In the western end of this region farmers and ranchers are continuing to look at alternative crops such as hybrid poplar. In many of the lowlands marginal for dairy and pasture, production will likely shift to high yield poplar plantations to be managed much like a farm crop.

The biophysical response of the forest ecosystems to current management practices will shape many of our future forest types and hence management options. The historic and current timber harvests across this landscape have perpetuated young forests – the majority now younger than 80 yrs. Looking back several thousand years, this shift toward early successional age classes is not unusual in this landscape. However, there are ecosystem impacts already observed with keeping stands in early successional stages over several rotations. One example of this is the Sitka spruce tip weevil. The weevil infects young spruce under 18 meters tall resulting in dead terminal leaders and stunted height growth (Furniss and Carolin 1977). Studies have shown that this beetle is not highly migratory and spread of infection can be very slow, particularly if there are no susceptible spruce to move to (Wright 1960; McMullen et al. 1987). However in our current landscape with a large percentage of the spruce in the susceptible age-class the weevil has become so prevalent that the western hemlock and Douglas-fir are outcompeting the now stunted Sitka spruce trees. Although far from being eliminated across its native range, Sitka spruce will provide constrained future management options limited by poor growth and shifting markets.

As our colleague silviculturist John Tappeiner points out (personal communication), planted forests of Douglas-fir in this region are "versatile" and resilient, can be very long-lived – to more than 500 years, and can continue to produce impressive wood growth at old ages. Rotation ages for wood production of 100 years are not unreasonable, and could be extended to 150 years or more if desired for management objectives. Rapid growth can produce trees of 1.5–2 m in diameter at age 120 to 150 years. (Of course, individual trees can live to 300 to 400 years age.) During long-rotation cycles relatively frequent thinnings and stand tending for wood production would be possible and desirable. Douglas-fir planted forests can be managed in any of a variety of ways: for industrial wood production at relatively high densities for relatively short (40–80 years) rotations; at lower stand densities and for longer rotations, e.g. to 150 years, to provide large trees and old forest characteristics; in combinations of these or other scenarios. Individual stands can be grown for wood production for about 50 years, then repeatedly thinned and shifted to development of old forests. This seems to be the path that some public land management organizations are curently following.

In a landscape so obviously dominated by forested ecosystems the origins and age class distributions of the forests will have a significant effect on the total region. The mosaic of forest types across this landscape is currently a product of complex and rapidly changing bio-physical and social forces linked with policies and economics. One of the fundamental questions that has not been addressed in this descriptive narrative is whether the origins of the forests are as important as their potential contributions to desired near- and distant-future landscape conditions. If the landscape in question is seen as an integration of forests and humans, then the desired landscape condition will include features which are present to benefit just the forests, features which are beneficial to the humans, and features which are mutually beneficial. Future conditions of forests, regardless of origin, should be designed with multiple values in mind, with continually evolving forest management practices and monitoring. Forests of northwestern Oregon have the potentials and resilience to provide myriad human values and can be managed to preserve many options for the future.

The challenge of sustainability

Our purpose in this brief review of northwest Oregon has been to use it as illustrative of both the circumstances and history of many formerly timber-dependent areas as well as to help frame the kinds of challenges facing such regions in the future. To conclude, we address five key issues confronting policymakers, citizens, and members of the forestry profession. Using the framework of biophysical forces in accordance with human influence outlined in this paper to stimulate thinking, we challenge readers to review these key issues with an appreciation for the complexity and potential variety of solutions associated with them. First, growing public interest in the concept of sustainability has resulted in increasing attention given to the questions of both resource production and consumption. Such interest is welcome because such discussions are critical to identifying appropriate choices for the future. However, we need to be cautious in our optimism. Sustainability is, in many ways, an excellent example of the concept of a "guiding fiction"; i.e., images and constructs that serve to provide a sense of identity, community, and purpose at any abstract level, but, like the end of the rainbow, fade as we attempt to define and clarify them (Shumway 1991). As rhetoric, sustainability has had the desirable effect of encouraging discussions about issues of common concern; e.g., our future, the relationship with the environment. However, such discussions often fail to come to grips with the extraordinarily complex and difficult social choices that a sustainable future will necessarily demand. Questions of what should be sustained (Gale and Cordray 1991), for whom, through what means, and over what spatial and time scales (Stankey 1995) must be resolved before sustainability can become more than rhetoric.

Second, resolving a complex issue such as sustainability is further confounded by the growing separation of production and consumption typically found in modern industrialized societies. Today, more than eight out of ten Americans reside in urban areas (similar figures can be found in other industrialized societies). The resulting "disconnection" between a population that demands and consumes a variety of goods and services and the resources and resource management activities that make that demand and consumption possible raises difficult questions about the capacity of society to make informed choices, given that such choices lead to consequences that are often only poorly identified, let alone understood. Given that population projections generally foresee a continuation of the urbanization of the country, it is clear that major challenges confront the natural resource professions in helping the wider society understand the implications and consequences of choices relative to natural resource management.

This leads to the third challenge. Political scientists speak of the "quandary" facing modern societies: i.e., "how can the democratic ideal of public control be made consistent with the realities of a society dominated by technically complex policy questions?" (Pierce and Lovrich 1983, p. 1). That is, most western industrialized societies rely upon a democratic form of government; political power is vested in the citizenry. However, an increasingly complex world confronts such societies with a major challenge; how to retain democratic traditions and processes while, at the same time, making informed, intelligent decisions about technically complex questions. Do we simply turn the major prescriptive questions before society over to the "experts" (e.g., foresters, engineers, physicians)? It is tempting to simply call for more education and while this clearly is important, such a response probably oversimplifies the issue. Foresters and other natural resource specialists have been, and will continue to be, challenged by such questions, and innovative approaches to engaging the wider society in discussions about the future of our forests are sorely needed.

Fourth, there is mounting evidence of the need for new institutional structures to manage natural resources in an increasingly complex future and in response to growing demands and societal preferences. The inadequacy of current institutions has drawn increasing attention. For example, most commentators on the emerging notion of "ecosystem management" point to the inability of current institutions to contend effectively and appropriately with what would appear to be the central tenets of such a management approach (e.g., Grumbine 1994; Freemuth 1996). This would include the need to manage across long-time scales, over large, multi-jurisdictional spatial scales, and for a wide range of values. However, institutions often possess powerful inertia (Wilkinson 1992); changing the variety of formal and informal organizations, behaviors, and norms that govern forest management today will likely encounter strong resistance from many quarters.

Finally, it is clear we face major challenges to integrating planted forests, natural forests, and society. Despite the growing interest in integrated resource management, it, like sustainability, remains more in the realm of rhetoric than reality. Many reasons underlie this, not the least of which is simply that thinking and behaving in an integrated fashion (i.e., an inclusive, holistic manner) is extremely difficult, given the complexity of the systems with which we are involved and our inadequate understanding of second- and third-order effects. However, more subtle, yet formidable barriers also limit our ability to manage on an integrated basis. Functionally-based budgeting systems and disciplinary-bounded organizations, for example, operate to differentiate rather than synthesize. Yet, the kinds of issues and pressures that will surely characterize future forest management will make an integrative approach to management necessary. Resource professionals, including those in the academic institutions, will be challenged to promote innovative processes and structures to better achieve true integration.

The region of northwestern Oregon, our backyard, will be one of many testing grounds for responses to these challenges. The process is intense and contentious as we write, with uncertainty as to whether timber production should be explicitly stated as the primary goal for certain state lands which are mandated to be managed for the "benefit" of the residents of the counties. That issue may soon be "resolved" by a decision, but the debates and forest management challenges will long continue as the landscape evolves in response to biophysical, social, economic and political forces, acting in this, another El Niño year, and beyond.

References

Burns R.M. and Honkala, B.H. 1990. Silvics of North America, Vol 1. Conifers. Agriculture Handbook 654, USDA Forest Service. 877 p.

Franklin, J.F. and Dyrness, C.T. 1988. Natural Vegatation of Oregon and Washington. Oregon State University Press, Corvallis, OR (reprinting, with new bibliographic supplement of 1973 U.S. Forest Service publication). 452 p.

Freemuth, J. 1996. The emergence of ecosystem management: Reinterpreting the gospel? Soc. and Nat. Res. 9(4): 411–417.

Furniss, R.L. and Carolin, V.M. 1977. Western Forest Insects. Misc Publication No. 1339 USDA Forest Service. 654 p.

Gale, R.P. and Cordray, S.M. 1991. What should forests sustain? Eight answers. J. of For. 89(5): 31–36.

Grier, C.C., Lee, K.M., Nadkarni, N.M., Klock, G.O. and Edgerton, P.J. 1989. Productivity of Forests of the United States and Its Relation to Soil and Site Factors and Management Practices: A Review. USDA Forest Service Pacific Northwest Station PNW-GTR-222. 51 p.

Grumbine, R.E. 1994. What is ecosystem management? Con. Biol. 8(1): 27–38.

McMullen, L.H., Thomson, A.J. and Quenet, R.V. 1987. Sitka Spruce Weevil (*Pissodes strobi*) Population Dynamics and Control: A Simulation Model Based on Field Relationships. Information Report BC_X_288 Pacific Forestry Center, Canadian Foresty Service. Victoria B.C.

Pierce, J.C. and Lovrich, N.P. 1983. Trust in the technical information provided by interest groups: the views of legislators, activists, experts, and the general public. Pol. Stud. J. 11: 626–639.

Pojar, J. and MacKinnon, A. 1994. Plants of the Pacific Northwest Coast. B.C. Ministry of Forests and Lone Pine Publishing Vancouver B.C. 526 p.

Shumway, N. 1991. The Invention of Argentina. University of California Press, Berkeley.

Stankey, G.H. 1995. The pursuit of sustainability: joining science and public choice. The George Wright Forum 12(3): 11–18.

Sudworth, G.B. 1908. Forest Trees of the Pacific Slope. United States Government Printing Office, Washington DC. 441 p.

Wilkinson, C.F. 1992. Crossing the Next Meridian: Land, Water, and the Future of the West. Island Press, Washington DC. 376 p.

Waring, R.H. and Franklin, J.F. 1979. Evergreen coniferous forests of the Pacific Northwest. Sci. 204(29): 1380–1386.

Wright, K.W. 1960. Sitka-Spruce Weevil. Forest Pest Leaflet 47 USDA Forest Service. 6 p.

New Forests **18:** 111–112, 1999.

Epilogue

KATHLEEN KAVANAGH
College of Forestry, Oregon State University, Corvallis, OR, USA

Frequently, when examining planted forests, the tendency is to limit the scope of examination to the most simplified forest monoculture that contrasts sharply when compared to natural forests. Through this narrow focus, conclusions are drawn about ecological imbalance and lack of resiliency inherent in these systems. Planted forests occur on every continent, in all forest types and under many different objectives. This book has taken a broad outlook at the variety, nature and significance of planted forests around the globe. In recognition that planted forests, in all their forms, are rapidly becoming a major component of our forested landscape, a need was perceived to examine their role in our quest for forest sustainability. In these papers, consideration is given to the act of planting a forest and the range of species types, along with societal objectives intermixed with political, cultural and biological impacts. It is evident the nature of these forests are as complex as the biophysical forces and societal objectives that shape their development. By making a broader examination of planted forests, we are better able to realistically predict potential outcomes in our quest for sustainability.

In placing planted forests on a continuum from simplified to complex, we acknowledge they will never replace our native forests. It is this recognition of a need for balance with nature, that lends optimism for future forest management. Julian Evans observes that planted forests are not a replacement for natural forests but a complement to them. How does this fit in with the "ideologies associated with the Western industrial world" that according to William Robbins "have profoundly influenced the shaping of the landscape of the Pacific Northwest of the U.S. as it exists in the late twentieth century"? The need for balance with nature was not recognized in these "planting societies" that were focused on control and dominance. These needs, which can be compared to the pioneering mentality that shaped human expansion throughout the globe, are still in our living memory. Unlike past generations, our recognition and study of the impacts of planting forests and the natural processes abundant in both planted and native forests, further illustrate our commitment to bringing balance to our management activities. As experience is accumulated under this new paradigm, we will have an increased base of knowledge to further refine our forest management techniques.

All papers here, whether they deal with describing a type of planted forest, an objective for planting a forest or the impacts of planted forests, recognize the connectiveness to the natural system. Neils Elers Koch and J.P. Skovsgaard describe the "integration of forest functions such as wood production, biodiversity, recreation, amenity values, groundwater protection, and carbon fixation." Planting is not restricted, but seen as component in balance with other "natural-silvicultural" techniques. Balance can be maintained by using well-adapted local species or exotics with proven capability, accelerating the development of a "real" forest, and visualizing the outcome on both a local and landscape scale. In another context, David South connects human population and wood consumption rates to a need for an examination of societal needs for planted forests. One option is meeting this increasing demand by planting trees on marginal pasture lands thereby limiting the impact on natural forests. Recognition of these linkages in time and space will continue to encourage more holistic examinations of the impacts of planted forests.

In discussions of biological, social, political, and economic issues underlying planted forests there is an awareness of the need to maintain a balanced, living memory of how things were, mindfulness of the connectiveness of our decisions and actions, and the recognition of need to enter a quest for sustainability. All of these factors lend hope for future forest management decisions and for sustaining our societies.

FORESTRY SCIENCES

1. P. Baas (ed.): *New Perspectives in Wood Anatomy.* Published on the Occasion of the 50th Anniversary of the International Association of Wood Anatomists. 1982 ISBN 90-247-2526-7
2. C.F.L. Prins (ed.): *Production, Marketing and Use of Finger-Jointed Sawnwood.* Proceedings of an International Seminar Organized by the Timber Committee of the UNECE (Halmar, Norway, 1980). 1982 ISBN 90-247-2569-0
3. R.A.A. Oldeman (ed.): *Tropical Hardwood Utilization.* Practice and Prospects. 1982 ISBN 90-247-2581-X
4. P. den Ouden (in collaboration with B.K. Boom): *Manual of Cultivated Conifers.* Hardy in the Cold- and Warm-Temperate Zone. 3rd ed., 1982 ISBN Hb 90-247-2148-2; Pb 90-247-2644-1
5. J.M. Bonga and D.J. Durzan (eds.): *Tissue Culture in Forestry.* 1982 ISBN 90-247-2660-3
6. T. Satoo: *Forest Biomass.* Rev. ed. by H.A.I. Madgwick. 1982 ISBN 90-247-2710-3
7. Tran Van Nao (ed.): *Forest Fire Prevention and Control.* Proceedings of an International Seminar Organized by the Timber Committee of the UNECE (Warsaw, Poland, 1981). 1982 ISBN 90-247-3050-3
8. J.J. Douglas: *A Re-Appraisal of Forestry Development in Developing Countries.* 1983 ISBN 90-247-2830-4
9. J.C. Gordon and C.T. Wheeler (eds.): *Biological Nitrogen Fixation in Forest Ecosystems.* Foundations and Applications. 1983 ISBN 90-247-2849-5
10. M. NÇmeth: *Virus, Mycoplasma and Rickettsia Diseases of Fruit Trees.* Rev. (English) ed., 1986 ISBN 90-247-2868-1
11. M.L. Duryea and T.D. Landis (eds.): *Forest Nursery Manual.* Production of Bareroot Seedlings. 1984; 2nd printing 1987 ISBN Hb 90-247-2913-0; Pb 90-247-2914-9
12. F.C. Hummel: *Forest Policy.* A Contribution to Resource Development. 1984 ISBN 90-247-2883-5
13. P.D. Manion (ed.): *Scleroderris Canker of Conifers.* Proceedings of an International Symposium on Scleroderris Canker of Conifers (Syracuse, USA, 1983). 1984 ISBN 90-247-2912-2
14. M.L. Duryea and G.N. Brown (eds.): *Seedling Physiology and Reforestation Success.* Proceedings of the Physiology Working Group, Technical Session, Society of American Foresters National Convention (Portland, Oregon, USA, 1983). 1984 ISBN 90-247-2949-1
15. K.A.G. Staaf and N.A. Wiksten (eds.): *Tree Harvesting Techniques.* 1984 ISBN 90-247-2994-7
16. J.D. Boyd: *Biophysical Control of Microfibril Orientation in Plant Cell Walls.* Aquatic and Terrestrial Plants Including Trees. 1985 ISBN 90-247-3101-1
17. W.P.K. Findlay (ed.): *Preservation of Timber in the Tropics.* 1985 ISBN 90-247-3112-7
18. I. Samset: *Winch and Cable Systems.* 1985 ISBN 90-247-3205-0
19. R.A. Leary: *Interaction Theory in Forest Ecology and Management.* 1985 ISBN 90-247-3220-4
20. S.P. Gessel (ed.): *Forest Site and Productivity.* 1986 ISBN 90-247-3284-0
21. T.C. Hennessey, P.M. Dougherty, S.V. Kossuth and J.D. Johnson (eds.): *Stress Physiology and Forest Productivity.* Proceedings of the Physiology Working Group, Technical Session, Society of American Foresters National Convention (Fort Collins, Colorado, USA, 1985). 1986 ISBN 90-247-3359-6

FORESTRY SCIENCES

22. K.R. Shepherd: *Plantation Silviculture.* 1986 ISBN 90-247-3379-0
23. S. Sohlberg and V.E. Sokolov (eds.): *Practical Application of Remote Sensing in Forestry.* Proceedings of a Seminar on the Practical Application of Remote Sensing in Forestry (Jnkping, Sweden, 1985). 1986 ISBN 90-247-3392-8
24. J.M. Bonga and D.J. Durzan (eds.): *Cell and Tissue Culure in Forestry.* Volume 1: General Principles and Biotechnology. 1987 ISBN 90-247-3430-4
25. J.M. Bonga and D.J. Durzan (eds.): *Cell and Tissue Culure in Forestry.* Volume 2: Specific Principles and Methods: Growth and Development. 1987 ISBN 90-247-3431-2
26. J.M. Bonga and D.J. Durzan (eds.): *Cell and Tissue Culure in Forestry.* Volume 3: Case Histories: Gymnosperms, Angiosperms and Palms. 1987 ISBN 90-247-3432-0
    Set ISBN (Volumes 24–26) 90-247-3433-9
27. E.G. Richards (ed.): *Forestry and the Forest Industries: Past and Future.* Major Developments in the Forest and Forest Industries Sector Since 1947 in Europe, the USSR and North America. In Commemoration of the 40th Anniversary of the Timber Committee of the UNECE. 1987 ISBN 90-247-3592-0
28. S.V. Kossuth and S.D. Ross (eds.): *Hormonal Control of Tree Growth.* Proceedings of the Physiology Working Group, Technical Session, Society of American Foresters National Convention (Birmingham, Alabama, USA, 1986). 1987 ISBN 90-247-3621-8
29. U. Sundberg and C.R. Silversides: *Operational Efficiency in Forestry.* Volume 1: Analysis. 1988 ISBN 90-247-3683-8
30. M.R. Ahuja (ed.): *Somatic Cell Genetics of Woody Plants.* Proceedings of the IUFRO Working Party S2.04-07 Somatic Cell Genetics (Grosshansdorf, Germany, 1987). 1988 ISBN 90-247-3728-1
31. P.K.R. Nair (ed.): *Agroforestry Systems in the Tropics.* 1989 ISBN 90-247-3790-7
32. C.R. Silversides and U. Sundberg: *Operational Efficiency in Forestry.* Volume 2: Practice. 1989 ISBN 0-7923-0063-7
    Set ISBN (Volumes 29 and 32) 90-247-3684-6
33. T.L. White and G.R. Hodge (eds.): *Predicting Breeding Values with Applications in Forest Tree Improvement.* 1989 ISBN 0-7923-0460-8
34. H.J. Welch: *The Conifer Manual.* Volume 1. 1991 ISBN 0-7923-0616-3
35. P.K.R. Nair, H.L. Gholz, M.L. Duryea (eds.): *Agroforestry Education and Training.* Present and Future. 1990 ISBN 0-7923-0864-6
36. M.L. Duryea and P.M. Dougherty (eds.): *Forest Regeneration Manual.* 1991 ISBN 0-7923-0960-X
37. J.J.A. Janssen: *Mechanical Properties of Bamboo.* 1991 ISBN 0-7923-1260-0
38. J.M. Bonga and P. Von Aderkas: *In Vitro Culture of Trees.* 1992 ISBN 0-7923-1540-5
39. L. Fins, S.T. Friedman and J.V. Brotschol (eds.): *Handbook of Quantitative Forest Genetics.* 1992 ISBN 0-7923-1568-5
40. M.J. Kelty, B.C. Larson and C.D. Oliver (eds.): *The Ecology and Silviculture of Mixed-Species Forests.* A Festschrift for David M. Smith. 1992 ISBN 0-7923-1643-6
41. M.R. Ahuja (ed.): *Micropropagation of Woody Plants.* 1992 ISBN 0-7923-1807-2
42. W.T. Adams, S.H. Strauss, D.L. Copes and A.R. Griffin (eds.): *Population Genetics of Forest Trees.* Proceedings of an International Symposium (Corvallis, Oregon, USA, 1990). 1992 ISBN 0-7923-1857-9

FORESTRY SCIENCES

43. R.T. Prinsley (ed.): *The Role of Trees in Sustainable Agriculture.* 1993 ISBN 0-7923-2030-1
44. S.M. Jain, P.K. Gupta and R.J. Newton (eds.): *Somatic Embryogenesis in Woody Plants,* Volume 3: Gymnosperms. 1995 ISBN 0-7923-2938-4
45. S.M. Jain, P.K. Gupta and R.J. Newton (eds.): *Somatic Embryogenesis in Woody Plants,* Volume 1: History, Molecular and Biochemical Aspects, and Applications. 1995 ISBN 0-7923-3035-8
46. S.M. Jain, P.K Gupta and R.J. Newton (eds.): *Somatic Embryogenesis in Woody Plants,* Volume 2: Angiosperms. 1995 ISBN 0-7923-3070-6
    Set ISBN (Volumes 44–46) 0-7923-2939-2
47. F.L. Sinclair (ed.): *Agroforestry: Science, Policy and Practice.* Selected Papers from the Agroforestry Sessions of the IUFRO 20th World Congress (Tampere, Finland, 6–12 August 1995). 1995 ISBN 0-7923-3696-8
48. J.H. Goldammer and V.V. Furyaev (eds.): *Fire in Ecosystems of Boreal Eurasia.* 1996 ISBN 0-7923-4137-6
49. M.R. Ahuja, W. Boerjan and D.B. Neale (eds.): *Somatic Cell Genetics and Molecular Genetics of Trees.* 1996 ISBN 0-7923-4179-1
50. H.L. Gholz, K. Nakane and H. Shimoda (eds.): *The Use of Remote Sensing in the Modeling of Forest Productivity.* 1996 ISBN 0-7923-4278-X
51. P. Bachmann, M. Köhl and R. Päivinen (eds.): *Assessment of Biodiversity for Improved Forest Planning.* Proceedings of the Conference on Assessment of Biodiversity for Improved Planning (Monte Verità, Switzerland, 1996). 1998 ISBN 0-7923-4872-9
52. G.M.J. Mohren, K. Kramer and S. Sabaté (eds.): *Impacts of Global Change on Tree Physiology and Forest Ecosystems.* Proceedings of the International Conference on Impacts of Global Change on Tree Physiology and Forest Ecosystems (Wageningen, The Netherlands, 26–29 November 1996). 1997 ISBN 0-7923-4921-0
53. P.K.R. Nair and C.R. Latt (eds.): *Directions in Tropical Agroforestry Research.* Selected Papers from the Symposium on Tropical Agroforestry organized in connection with the annual meeting of USA (Indianapolis, U.S.A., 5 November 1996). 1998 ISBN 0-7923-5035-9
54. K. Sassa (ed.): *Environmental Forest Science.* Proceedings of the IUFRO Division 8 Conference Environmental Forest Science. (Kyoto, Japan, 19–23 October 1998). 1998 ISBN 0-7923-5280-7
55. S.M. Jain, P.K Gupta and R.J. Newton (eds.): *Somatic Embryogenesis in Woody Plants,* Volume 4. 1999 ISBN 0-7923-5340-4
56. J.R. Boyle, J.K. Vinjum, K. Kavanagh, E.J. Jensen (eds.): *Planted Forests: Contribution to the Quest for Sustainable Societies.* 1999 ISBN 0-7923-5468-0
57. K. von Gadow and G. Hui: *Modelling Forest Development.* 1999 ISBN 0-7923-5488-5
58. J. Abildtrup, F. Helles, P. Holten-Andersen, J.F. Larsen and B.J. Thorsen (eds.): *Modern Time Series Analysis in Forest Products Markets.* 1999 ISBN 0-7923-5524-5
59. S.M. Jain, P.K. Gupta and R.J. Newton (eds.): *Somatic Embrygenesis in Woody Plants,* Volume 5. 1999 ISBN 0-7923-5553-9
60. D. Auclair and C. Dupraz (eds.): *Agroforestry for Sustainable Land-Use.* Fundamental Research and Modelling with Emphasis on Temperate and Mediterranean Applications. 1999 ISBN 0-7923-5799-X

FORESTRY SCIENCES

61. F. Helles, P. Holten-Andersen and L. Wichmann (eds.): *Multiple Use of Forests and Other Natural Resources*. Aspects of Theory and Application. 1999 ISBN 0-7923-5859-7

KLUWER ACADEMIC PUBLISHERS – DORDRECHT / BOSTON / LONDON